Process in Geomorphology

Process in Geomorphology

Edited by Clifford Embleton

Reader in Geography, King's College, University of London

and John Thornes

Reader in Geography, London School of Economics

A Halsted Press Book

JOHN WILEY & SONS
New York

© Edward Arnold (Publishers) Ltd 1979

Published in the U.S.A.
by the Halsted Press, a Division of
John Wiley & Sons, Inc., New York

Library of Congress Cataloging in Publication Data

Main entry under title:

Process in geomorphology.

   "A Halsted Press book."
   Bibliography: p.
   Includes index.
   1. Geomorphology. I. Embleton, Clifford.
II. Thornes, John B.
GB401.5.P747   551.4   79-18747

ISBN 0–470–26807–7
ISBN 0–470–26808–5 pbk.

Printed in Great Britain

Contents

Contributors

Clifford Embleton, Reader in Geography, King's College, University of London
John Thornes, Reader in Geography, London School of Economics
Denys Brunsden, Reader in Geography, King's College, University of London
Andrew Warren, Senior Lecturer in Geography, University College London
Malcolm Clark, Analyst Programmer, Imperial College, University of London
Brian Whalley, Lecturer in Geography, The Queen's University of Belfast

Acknowledgement

The publisher and editors would like to thank the following for permission to quote or modify copyright material (figure numbers refer to our publication, full citations can be found by consulting captions and references).

Allen & Unwin (Publishers) Ltd for figs 3–6, 7–15 and 7–24; American Association for the Advancement of Science for fig. 3–18; American Geophysical Union for figs 3–9, 11–4B, 11–8 and 11–9; American Society of Agricultural Engineers for fig. 7–31; American Society of Civil Engineers for figs 3–3, 3–7, 5–9, 7–16, 7–19, 7–26 and 11–7; Association Internationale des Sciences Hydrologiques for fig. 3–14; P. H. Banham for fig. 8–12E; Blackwell Scientific Publications for figs 5–10, 5–15, 5–23, 7–25 and 7–27; B. J. Black for fig. 7.27; G. S. Boulton for figs 8–7B and 8–9D & E; J. S. Bridge for fig. 7–5; Cambridge University Press for figs 4–13, 5–1, 5–13, 5–14 and 5–19; S. C. Colbeck for fig. 3–10A & E; Colorado State University for fig. 7–11; R. U. Cooke for fig. 4–15; Crane, Russak & Co. Ltd for fig. 5–18; Czechoslovak Academy of Sciences for figs 5–11 and 6–11; A. M. Dowden Inc. for figs 12–3 and 12–7; Elsevier Publications (N.Y.) for figs 3–6 and 4–11; W. H. Freeman & Co. Ltd for fig. 4–6E; Gebrueder Borntraeger Verlags Buchhandlung for fig. 7–28; Geological Society of America for figs 4–5B, 4–6D, 6–9, 10–3, 10–4, 10–7 and 10–18; Geological Society of London for figs 8–10 and 8–11J–K; The Geologists' Association for fig. 5–21; Generalstabens Litografiska Anstalt for fig. 9–8; R. Greeley for fig. 3–18; Mrs H. Haefeli for fig. 5–17; Ellis Horwood Ltd for fig. 3–19C; The Institute of British Geographers for figs 4–16, 4–17, 7–4, 7–30, 8–2 and 8–8; The Institution of Civil Engineers for figs 5–6, 5–13 and 5–22B; International Association of Hydrological Sciences for figs 9–5 and 9–6; International Glaciological Society for figs 3–10A & B, 3–12A–C, 3–13A–C and 9–1; Kansas Geological Survey for fig. 10–5; M. J. Kirkby for figs 5–26C, 12–3 and 12–7; J. H. Leeman Ltd for figs 8–7B, 8–9D & E and 8–12E; The London School of Economics for fig. 6–3; McGraw-Hill Book Co. Ltd for figs 3–1, 4–6F, 4–7 and 6–2; Macmillan (Journals) Ltd for fig. 4–15; Macmillan Publishers Ltd for fig. 4–3B, Masson Editeur for fig. 4–3B; W. H. Mathews for fig. 3–12C; National Research Council of Canada for fig. 5–16C; Newnes-Butterworths for fig. 7–9; North-Holland Publishing Co. for figs 7–2 and 7–7; Oliver & Boyd for figs 4–3A & B and 4–8; Pergamon Press Ltd for fig. 7–13; M. F. Perutz for fig. 3–12A; Plenum Publishing Corp. for figs 4–3A & B and 4–8; C. F. Raymond for fig. 3–13B; L. Reynaud for fig. 3–13C; Royal Geographical Society for fig. 8–6; The Royal Society for fig. 8–4; R. P. Sharp for fig. 3–12B; Societe Geologique de Belgique for fig. 6–10; Society of Mining Engineers of AIME for fig. 4–6A & B; Soil Conservation Society of America for figs 7–32 and 7–33; Soil Science Society of America for figs 4–6C and 7–6B; A. W. Skempton for fig. 5–17; State University of New York at Binghamton for figs 5–18 and 8–3; D. Tabor for fig. 8–4; J. B. Thornes for fig. 7–5; United States Geological Survey for figs 3–12D, 5–22A, 7–8, 7–14 and 7–23; University of Chicago Press for figs 5–26B and 10–16; University of Toronto Press for figs 4–14 and 5–16C; John Wiley & Sons Ltd for figs 4–10, 7–12, 7–29 and 12–2; Wiley Interscience for fig. 1–3; Williams & Wilkins Co. for figs 4–6C and 7–6A and Yorkshire Geological Society for fig. 5–5.

List of plates

Notation in mathematical formulae

Except where otherwise defined, the following general conventions are used throughout the book. The list is far from exhaustive; to list every quantity or term, including subscripts, would greatly increase the length of the list and would be self-defeating. As far as possible, we have used traditional forms of notation, such as τ for shear stress and μ for both friction and viscosity, even where this results in several parameters having the same symbol. Thus there are inevitably overlaps in usage, so that terms must be defined on each occasion that they are introduced in the text. This list merely attempts to bring some uniformity to the use of the commoner symbols.

C	Centigrade
J	joule $(\text{kg m}^2\,\text{s}^{-2})$
K	Kelvin
M	mega –
N	newton $(\text{kg m s}^{-2}=\text{J m}^{-1})$
W	watt $(\text{kg m}^2\,\text{s}^{-3}=\text{J s}^{-1})$

cal	calorie
cm	centimetre
g	gram
ha	hectare
hr	hour
kg	kilogram
km	kilometre
l	litre
m	metre
mg	milligram
ml	millilitre
mm	millimetre
s	second
t	tonne
yr	year

kN	kilonewton
μm	micrometre
Hz	hertz (cycle per second)
Pa	pascal
Pl	poiseuille

A	cross-sectional area; basin area
C	wave celerity
D	diameter; particle size; coefficient of thermal conductivity; Dean's parameter
Dd	drainage density
E	modulus of elasticity
F	force; Froude number; safety factor
G	gravitational constant
H	heat of fusion
I	rainfall intensity; immersed weight
K	hydraulic conductivity; von Karman constant
KE	kinetic energy
L	slope length; unit length
M	mass
P	pressure; energy flux
PE	potential energy
Q	discharge; aeolian sand transport
R	Reynolds number
S	shear strength
T	temperature; wave period
V	velocity; volume
W	weight; power
Z	gravitational potential energy

a	acceleration; area
c	cohesion; volume concentration
d	distance, depth, thickness
f	infiltration rate; fall time of a sediment particle; Darcy-Weisbach factor
g	acceleration due to gravity
h	vertical interval, height, depth
i	hydraulic gradient
k	coefficient; permeability
l	length
m	mass
p	pressure (water, ice, etc.); probability
q	discharge per unit area
r	radius; roughness
s	slope
t	time
u	pore-water pressure
v	velocity, volume
w	width
x, y, z	coordinate axes

Δ	(delta)	change
Θ	(theta)	sediment transport

α	(alpha)	slope, angle
β	(beta)	beach slope
γ	(gamma)	bulk density, unit weight

ε	(epsilon)	strain; diffusion
θ	(theta)	angle
λ	(lambda)	coefficient of expansion; wave-length
μ	(mu)	friction; viscosity
μ_*	(mu)	shear velocity
π	(pi)	coefficient $(3.14159\ldots)$
ρ	(rho)	density (fluids)
σ	(sigma)	normal stress; surface tension
τ	(tau)	shear stress; shear strength; tractive force
υ	(upsilon)	kinematic viscosity
ϕ	(phi)	angle of friction; potential energy
ψ	(psi)	tension, suction
ω	(omega)	sediment fall velocity

Preface

The idea of this book grew out of a course of lectures given to second and third year under-graduates in King's College and the London School of Economics as one of the 'options' in geomorphology. In spite of the appearance of several texts in the last ten years or so dealing with particular systematic aspects such as fluvial geomorphology or glacial geomorphology, there has been a lack of a detailed and comprehensive text on 'process geomorphology'. The last twenty years have seen a rapid expansion of interest in geomorphological processes. Prior to the 1950s, studies of the operation and nature of particular processes were relatively few. Many of the most important investigations had been undertaken by engineers (for example, studying fluvial processes in connection with river control, the weathering of building materials or the stability of slopes), and there was little understanding of many fields of process geomorphology such as the mechanics of ice movement and glacial erosion, the formation of desert dunes or hillslope processes; still less was the interaction of these processes in particular environments or their rates of action appreciated. The growth rate of work in process geomorphology can be judged by the fact that, of the 1100 references cited in this book, 68 per cent relate to publications since 1960.

This book is concerned solely with exogenic processes, broadly those modifying the Earth's surface from the outside and powered principally by solar or gravitational energy. Although geothermal energy is relevant to a few exogenic processes, such as glacier sliding and processes in permafrost areas, processes depending on the interior energy sources of the Earth, including the whole range of tectonic activity, will not be considered. The book begins with some general considerations of the nature of energy, forces and resistances, the properties of materials, and the nature of fluid flow. It then deals with groups of processes systematically —weathering, mass movements, fluvial, glacial, nival, aeolian and marine—before a final chapter reviews their position in the environment and their interrelationships.

Many people apart from the authors of the book have helped in its production, and we are grateful to all of them. Most of the line drawings have been prepared by Roma Beaumont, Alison Hine and Gordon Reynell in the King's College drawing office, and among those who have helped in the typing we would like to thank Victoria Blank, Bridget O'Donnell, Anne Rogers and Hilary Salter.

Clifford Embleton
John Thornes

18 December 1978

1
Introduction

J. B. Thornes

In geomorphology the word process is a noun used to define the dynamic actions or events in geomorphological systems which involve the applications of forces over gradients. Such actions are caused by agents such as the wind and falling rain, waves and tides, river- and soil-water solutions. Where the forces exceed the resistances in these natural systems, change occurs by deformation of a body, by change in position or by change in chemical structure. Change in the form of the Earth's surface indicates the operation of such processes, but lack of observable change need not imply that no processes are operating. This may be because the rate of operation of the process is small, because the ratio of force to resistance is small, resulting perhaps in the dissipation of energy in friction, or because the forces and resistances are fairly balanced.

In geomorphology a major task is to understand the relationships between form and process. They may be expressed in a cause-and-effect chain or, for example, by the rates of change of operation of one with respect to rates of change of the other. Simple representations of these relationships are called models. Process-response models, where the word 'response' means form or structure, are often used to describe and explain the relationships. The models may be reached inductively, that is, by the observation and combination of particular instances, or by deduction, in which one proceeds from known or assumed relationships in the physical world to argue about the consequences of these relationships. Sometimes one proceeds by experimentation which may involve either or both of these approaches, as outlined in the second section of this chapter.

Processes in relation to space and time

Geomorphologists consider processes at various levels of spatial resolution. At the finest level one may be concerned, for example, with the filling of pore spaces in the soil by moving water. An example of a process at a very coarse scale is thermal convection below the Earth's crust. In this way processes are understood to be spatially nested; infiltration occurs within the soil, the migration of the weathering front within a hillside and so on. Geomorphologists have chosen, at least in recent years, to consider levels of resolution mainly of the order of a few hundred square kilometres or less, especially when investigating processes. Because of variations in the substratum upon which the processes operate and because the type and intensity of the processes vary from place to place, the resulting landforms are highly variable. What we view as the dominant process depends on the spatial scale at which we observe the surface. Looking at a few square metres of rock-cut platform, we might perceive the most important process to be the feeding habits of marine organisms. On the broader scale of about 1 km², the pattern and dynamics of wave attack may be mainly responsible for form. At the largest scale, covering several thousand square kilometres, regional isostatic adjustment might be most important.

Processes also occur at varying rates. A force may be applied rapidly or slowly; its magnitude may be large or small. If we observe a particular form or assemblage of forms over a short period, there may be little change either because the forces are small or because they have not operated for a sufficient length of time. If the nature or rate of operation of the processes change, there may be a corresponding change in the landforms. Sometimes the response is instantaneous, as when a large flood passes through a channel. At other times the response may be quite slow or there may be 'dead time' when nothing happens to landforms to reveal the change in process. The time taken for the system to respond to externally imposed changes is called its reaction time. Reaction time in geomorphological systems may be intimately tied up with spatial resolution as well as the time-scale involved. If an effect has to 'diffuse' through the landscape, as a knickpoint moves in a river, then the distance over which the diffusion takes place and the rate of diffusion are important controls on the reaction time.

Just as forms are influenced by process, so processes are influenced by form. The terrain over which an ice sheet moves will help to determine the hydraulic conditions at the bed and this in turn will affect the nature and rate of movement. Steepening of a river-channel gradient by deposition may increase the velocity and decrease the depth, which then affect the amount of scour. Such a series of events, when one change begets another which itself affects the original change, is called feedback. When the result is to reinforce the initial change the feedback is said to be positive. This occurs, for example, in the growth of a snow patch. The patch may, through nival processes, cause enlargement of the hollow which then catches more snow. When the sequence leads to damping of the initial deviation, the feedback is said to be negative. Thus, for example, in a gravel stream during the passage of a flood wave, the channel tends to become wider and shallower until the depth falls so much that transport is inhibited by increased friction and narrowing follows. Negative feedback tends to lead to a steady condition of form properties. Systems experiencing positive feedback or responding rapidly to some externally imposed change in the nature or rate of operation of the processes are said to show transient or dynamically unsteady behaviour. The investigation of process-form relationships by observation of form may be most profitable at these times.

Approaches to the investigation of process

Geomorphologists have been concerned with investigations of process in three frameworks. First, they have sought to account for an evolution of forms in terms of an evolution of processes. The main intellectual thrust of this activity has been in determining the sequence of forms and deposits. The assumption is that the processes are understood. A second group has focused on the spatial and temporal variations of the processes and particularly on problems of the magnitude and frequency of events. The third group has considered mainly the relationships between form and process. Throughout, however, it has been generally recog-

Plate I (*upper*) A sharp wave front moving through an ephemeral channel after an intense storm. The white front results from the escape of air under hydrostatic pressure. (J.B.T.)

(*middle*) A laboratory model of the same phenomenon showing the wetting front developing in the subsurface materials in relation to the opportunity time for infiltration. (E. Cory)

(*lower*) Microfilm output of a computer simulation of the discharge in a large ephemeral channel under identical conditions. The extreme left-hand peak represents the upstream end of the channel. The lower peaks show the same simulated wave front farther down the channel. Diminution of flow results from infiltration into the bed. (G. Butcher, J. B. Thornes)

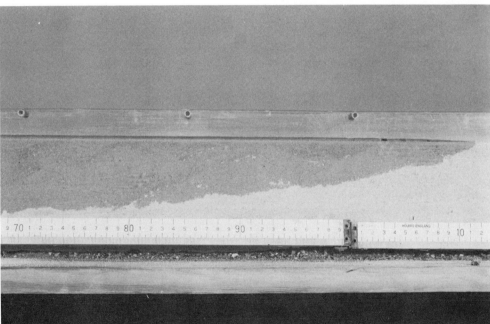

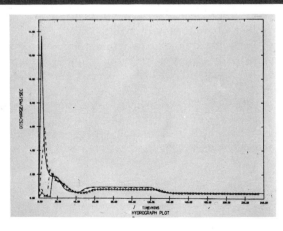

nized that the problems of investigation of processes in the field, in the laboratory and in the office are essentially different (Plate I). Certainly, all these modes of working are necessary if advances are to be made within the frameworks outlined above.

Fieldwork

Fieldwork is crucial to studies of process because it confronts the investigator with associations and events of geomorphological significance which are outside his range of experience and hence prompt his curiosity. It was this type of confrontation together with a steadiness of thought and accuracy of observation that made Gilbert one of the greatest pioneers in process geomorphology. The same confrontation led Leopold to his remarkable investigations of fluvial processes and forms (Leopold, 1978). Although fieldwork can be an informal, some-times almost incidental, activity, it is important because it provides the data for the examina-tion of empirical relationships, for the parameterization of theoretical models and for the verification of existing theoretical concepts. These are, or should be, much more formal activi-ties based on carefully developed field programmes, in which objectives and procedures are precisely defined.

Most commonly, processes have been inferred from the direct observation of form, or change of form in the field. Early workers such as Agassiz and Gilbert laid the foundations of modern process studies by direct field observation combined with seemingly effortless in-duction of the large-scale processes operating over wide areas. Throughout the first half of the twentieth century inferences were drawn from field evidence about the nature and rate of operation of processes. As examples, consider the inferences drawn by Wayland (1947) as to the processes responsible for tropical weathering from the extensive plains of low lati-tudes or King's (1962) inferences about the occurrence and magnitude of sheet wash from his observations of pediplains. Such inferences continue to form an essential part of geomor-phological thinking because they lead to the development of hypotheses that can be tested in a more rigorous framework. Recent examples include the comparison of river-channel forms before and after floods (e.g. Thornes, 1976; Anderson and Calver, 1977), and upstream and downstream from reservoirs (Gregory and Park, 1974). A major difficulty in this pro-cedure arises from the need to assume a unique relationship between form and process. This has rarely been demonstrated and different processes may give rise to forms that look very much alike. Moreover, the same process, through its variability in time and space, may result in widely differing forms. Finally we have to admit that the form–process relationship as we see it in the field is heavily conditioned by structure, lithology and, in most places, by human activity. Some of these effects can of course be minimized by careful and well designed experiments.

A further major difficulty which we face when inferring process from form is the multi-plicity of processes that occur at one point in time. There may be great difficulty in disentang-ling the effects of different processes. On most hillsides, weathering, creep and erosion by running water occur together. One also has to accept that fossil evidence indicates substantial variations of climate in the past 2 million years. Unless one assumes a very rapid response to such changes, one has to admit to a mixture of forms, one set superimposed on the other, resulting from a mixture of transient processes. A cursory observation of almost any landscape confirms that this is the case.

It seems to follow that the field offers a primer, a challenge to our explanatory powers, a realistic tableau of what we have to explain, rather than the explanation itself. It does, however, offer more. In recent years geomorphologists have realized the need to validate the models of process that have been developed and to attempt to verify them. Validation involves two operations: first, the appreciation of the likely domain over which process inputs operate in terms of magnitude and frequency; secondly the determination of constants (parameters) in physical relationships. Examples of the latter include Manning's n or the exponents in velocity equations for overland flow. Verification involves checking the expected outcomes of process models against the field data. Do the models propounded on the basis of flume experiments hold over a satisfactory range in the natural environment? Do the predicted zones of scour and deposition along a model beach actually occur in the life-sized version? These questions require much more rigorous designs of field observation than the 'priming' activities mentioned earlier and in recent years the main effort in field studies has been in this direction. Among the most successful examples of fieldwork have been observations on the mechanics of glacier movements (e.g. Meier, 1960); the evaluation of hydrological processes in small catchments, such as the Waggon Wheel Gap experiments (Bates and Henry, 1928) and studies of process rates in the coastal zone (McClean, 1967).

The major difficulties here are the inherent sampling problems and the cost of providing and maintaining expensive instrumentation. It is a paradox that the very variability of the field environment, which one seeks to incorporate into theories, makes the assessment of 'real' conditions a virtual impossibility. The other major problem lies in the severe technical constraints linked with the constant need not to interfere with the process under observation. Soil creep, bedload transport and soil-water movement illustrate these problems well. The coming of the neutron probe and the consequent breakthrough in the measurement of soil moisture exemplifies the severity of these technical limitations.

Laboratory investigations

The field provides data on the spatial and temporal ranges of inputs and parameters for models. It enables interrelationships among complex processes to be observed and it is the standard against which our understanding of process- and form-relationships must finally be judged.

In the laboratory the emphasis is on the control of events in order to observe phenomena which, though present in the real world, cannot be examined because of their interaction with other land-forming processes. Moreover, the laboratory offers the opportunity to scale down space and speed up time, two of the major stumbling blocks in field experimentation. The third major role that laboratory work plays is as a first test of the adequacy of a theoretical model or a particular set of relationships within it. Finally, it offers an opportunity for simple empirical experiments to be carried out in anticipation that new and hitherto unexpected associations may be observed.

One of the areas of investigation in which laboratory studies have been most widely and successfully applied is that of hydraulic processes, and the role of the laboratory flume has been pre-eminent. Other investigations that have made use of controlled laboratory conditions include the early work of Griggs (1936b) on arid-zone weathering, experiments by Lewis and Miller on model glaciers (1955) and by van Burkalow (1945) on mass movement. Perhaps one of the most exciting and largest-scale experiments in recent years has been Schumm's attempt to produce an entire drainage basin system under cover (Schumm, 1973). Overall, however, geomorphologists appear to have been somewhat reluctant to carry process investigations into the laboratory.

Two types of study have been carried out. In the first, the mass, length and time properties of the model are maintained identical to those in the field but the inputs may be controlled for the purposes of the particular experiment. Under these conditions observations may be possible in the laboratory which are not feasible in the field. Well known examples include the investigation of erosional phenomena by sprinkler devices (e.g. Mosley, 1973) or the experiments to study the effects of freeze–thaw on rock particles (e.g. Potts, 1970). In the other type of laboratory investigation the models are scaled-down versions of the real-world processes.

The latter type are based on the concept of similarity. The basic proposition is that the fundamental units of the laboratory model bear a constant ratio to those of the real-world prototype. For length (l), mass (m) and time (t), these relationships may be expressed by:

$$l' = \lambda_l l''$$
$$m' = \lambda_m m''$$
$$t' = \lambda_t t''$$

where l' and l'' indicate length in the model and the prototype respectively; similarly for m', m'' and t', t''. Systems related to each other by proportionality of length are said to be geometrically similar. Those which are proportionately related in both length and time are kinematically similar, while systems related to each other by all three characteristics are dynamically similar. An important characteristic of dynamic similarity is that it implies identity of all the dimensionless laws governing the model and the prototype phenomena. In a truly dynamically similar model, therefore, one can study any property of the phenomenon under investigation.

In reality, major difficulties exist because in most hydraulic models the prototype fluid (water) is used and the value of gravity in the model is the same as that of the prototype because both are on Earth. Given that this is the case, the model builder has to ensure the dynamic similarity of at least those properties that are going to be observed. For example, if the width–depth ratio of a river channel is large, the mechanical properties in the central regions of the flow are not dependent on the ratio and therefore there is no need to achieve identity in this ratio.

The effort required to meet these conditions often leads to considerable technical complexity in laboratory models. Ramberg's (1964) attempt to simulate folding in the moraines of piedmont glaciers is an example of this. Here it is argued that the driving agency of an active glacier is almost entirely the difference in potential energy in different points in the ice body because inertial terms in the fluid dynamic equations are insignificant. The glacier model is whirled round in a centrifuge because the potential difference between two points is directly proportional to the centripetal acceleration. The arrangement is shown in Figure 1.1. Compared with models at rest in the field of gravity, strong and highly viscous viscoelastic materials can be used in the centrifuge. These do not flow under gravity and so the initial stage of the model can be constructed without sagging such as would occur in imitation substances soft enough to permit maturing of models at rest in the normal field of gravity. After the structures have developed, the results can be studied in detail without further appreciable deformation.

Office work

Investigations of processes by 'armchair' geomorphologists take three forms: the analysis of field data in order to investigate empirical relationships, the development of theories based on fundamental physical principles and the development of laboratory-type experiments on digital or analogue computers. All three levels have made highly significant contributions to the understanding of geomorphological processes.

In looking at *empirical* relationships emphasis had been laid on the relationships between 'factors' or agents and the response variables. Characteristically the work proceeded through multiple regression and correlation analysis. Although the methods have recently become

A

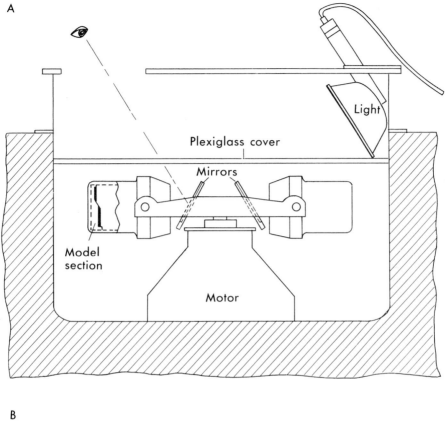

B

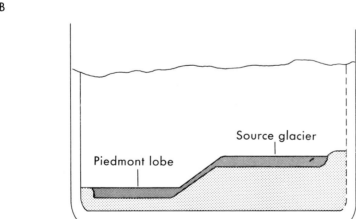

Figure 1.1 Diagram to show Ramberg's (1964) experimental equipment for the study of folded moraines in piedmont glaciers. **A:** Section through centrifuge bowl showing position of glacier-model section. **B:** Larger view of glacier-model section (length 10 cm).

more sophisticated, for example by incorporating temporal and spatial lags in the estimated relationships (Thornes and Brunsden, 1977), the procedure remains essentially inductive. Although the work was initiated by Strahler (1954), it is best exemplified in Melton's (1957) study of the relationships among elements of climate, surface properties and geomorphology. Melton later (1958) showed how the building up of correlation matrices between variables could give important indications as to the relationships among process variables particularly with respect to the direction of causation. Multiple regression techniques have also been widely used in fluvial process studies. For example, Walling (1971) analysed a series of 1200 suspended-sediment samples obtained during storm runoff in the Rosebarn catchment in Devon, England. From this he created a four-variable regression model to account for variations in the logarithm of concentration. This in turn led to conclusions about process. For example, he observed that the time of taking the sample during the storm hydrograph and the storm-flow discharge are most important in accounting for the variations in levels of concentration.

Notwithstanding the question of the availability of data, the approach does have some limitations. One of these is the rational choice of variables for inclusion in the regression model in the first place, which tends to suggest that the essential characteristics of the process are already known. Another is that high levels of correlation, taken alone, tell us little about

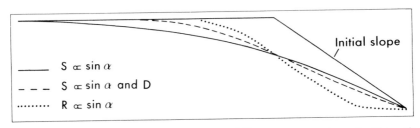

Figure 1.2 Characteristic forms produced under different process laws in Young's (1963) deductive models. S=rate of transport; D=distance from slope crest; R=rate of direct removal; α=slope angle.

the mechanics involved, and feedback and hysteresis effects often obscure fundamental interactions. Only where there is a complete lack of knowledge of a situation is the 'whirlpool' type of approach of the mid-1960s, which sucks in all the data and spews out all its combinations, remotely justified. In its defence, it has to be said that the statistical manipulation of data prevalent in the last decade drew attention to the real lack of data and the haphazard way in which such data were collected. In this respect the approach did the subject a great service.

In *theoretical* work, the basic procedure is to deduce from axioms or laws the behaviour of processes or the effects of these processes on landform change. Developments in this area over the last decade are reviewed by Thornes (1978). Although the application of theory to landform evaluation has been slow, the techniques have been well developed in the associated sciences of hydraulics and, to a lesser degree, sedimentation.

An excellent example is Young's (1963) study of slope evolution. In this, Young sought to determine the slope profiles that would characterize combinations of process laws with predetermined initial and boundary conditions. Young developed a set of laws in the form of simple equations that were assumed to be analogous to the processes involved in terms of their effects on a predetermined profile. Figure 1.2 shows how an initially straight slope responded to three different process laws. In case one, transport was proportional to the sine of the slope angle; in two, proportional to sine slope angle and distance from divide, and in three, the rate of removal of material was proportional to sine slope angle. A slightly different type of investigation deals with the deduction of the behaviour of a process or agent from laws or axioms. In 1938 Horton developed a set of equations for the discharge of a spatially varied unsteady surface flow due to a uniform rate of rainfall excess or water supply rate.

A major element of theoretical work, and one that causes some difficulty, is the transformation of a verbal statement into a symbolic statement, usually mathematical. This is because

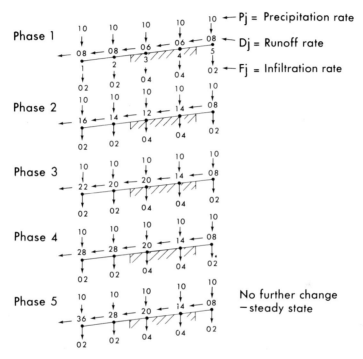

Figure 1.3 Iteration of a simple continuity model for overland flow showing six slope segments and six time phases. Shaded segments represent zones of higher infiltration rate. Based on Ahnert's (1977) model.

the mathematical formulation eases the process of deduction and minimizes the likelihood of ambiguities. This transformation often takes the form of a set of differential equations which then have to be solved, bearing in mind the initial and boundary conditions. As the situations become more complex it is increasingly difficult to represent the verbal statements by equations and even more difficult to obtain solutions. To overcome these problems the processes under investigation may be simulated on a digital computer. This operation has the added advantage that elements of uncertainty in the inputs or parameters can be fairly easily accommodated.

A common feature of computer simulations is the need to divide space and time into small increments. These are operated on successively in a process known as iteration. The process laws are applied at one time-interval to all spaces and then, as the time-unit changes, material is transferred between spaces. Ahnert (1976) has a complex computer program that simulates the effects of 'processes' on a three-dimensional hypothetical land surface. A sub-routine in this program computes surface runoff in an iterative fashion and illustrates the procedure quite well. The sub-routine is described in Ahnert (1977). The runoff present at each point is determined by a mass-balance equation:

$$D_{j,t} = P_{j,t} - F_{j,t} + D_{j+1, t-1} \tag{1.1}$$

In this equation, j is the slope section; so $j=1$ is the bottom section and $j=2$, $j=3$, etc., are sections successively farther upslope. The current time unit is t, $D_{j,t}$ is the runoff discharge from the jth section at time t, $P_{j,k}$ and $F_{j,k}$ are the precipitation and infiltration respectively. Figure 1.3 shows the runoff development over six time-phases with a uniform precipitation (10) but a spatially varying infiltration rate (2 at some points, 4 at others). Comparison of phases 4 and 5 shows that steady state is reached when the precipitation has uniformly continued for as many time-units as there are sections on the slope. If the rain falls over fewer units than there are space-units, steady overland flow does not occur. Time- and space-scales are related through the average flow velocity.

Concluding remarks

Although the various approaches have been separated, most successful geomorphological investigations of processes and their effects involve elements of several different approaches. In *macro-scale models* the deductive element is most important because there is little hope of observing the landscape over a large enough space or over a long enough time to define gross developments empirically or to verify, for that matter, the predictions of theories. Our only hope is that the process laws on which such theories are developed are correct and appropriately parameterized. Along the shore, under a glacier and in the drainage basin, at the *meso-scale level*, the assemblages of different processes, operating at widely varying rates in a specific spatial context, may be investigated. Here the vagaries of observation and the inherent variety of the controlling agents mean that statistical relationships and empirical work assume a greater significance. The time-scales are also shorter and the 'average' condition has to be replaced by a distribution of temporal and spatial magnitudes and frequencies based on events. Finally, at the *micro-scale level* emphasis is on the revelation of new processes, the careful observation of mechanisms and the calibration of known relationships in the field and laboratory.

2

Energy, forces, resistances and responses

C. Embleton and W. B. Whalley

In the last few years the Earth sciences have seen a move towards greater quantification and increased aspiration towards status as a science. A reflection of this is our concern with geomorphological processes and consideration of the dynamics of the landscape. Putting numbers into geomorphological explanations is an integral part of such an approach but it involves more than this—a need to look at problems in terms of a mechanism. We might ask questions such as 'how much material is this river transporting in solution and on its bed?' or 'what is the rate of creep of this hillslope?' These are perfectly valid questions but the answers, though quantitative, say nothing about *how* the process works. Similarly, quantitative description of a landform, a drumlin for instance, will not tell us about the *mechanism* of its formation. To understand the evolution of parts of a landscape we need to explain it, or least offer one possible explanation of how it has evolved. It is possible to work in the way a physicist might tackle the problem—by investigating a phenomenon as something explicable by the laws of physics and chemistry. In other words, we can attempt an explanation by offering a deterministic model of the process. It is not possible here to discuss probabilistic and deterministic models of explanation but it can be argued that they are complementary views and not exclusive; a mechanism will be deterministic and obey the laws of physics yet the long-term result may only be explicable as a stochastic process. Thus, this chapter will be concerned with the application of physical principles to geomorphology.

As an example of this viewpoint, consider the expression known as the Manning equation which is concerned with some stream-channel variables:

$$v = \frac{r^{2/3} s^{1/2}}{n} \quad \text{(for metric units)}$$

$$(2.1)$$

where v is the mean channel velocity, r is the hydraulic mean radius (the area of the cross-section divided by its 'wetted' perimeter), s is the slope of the water surface and n is a constant known as Manning's 'n', its value depending upon the channel-bed roughness. This is an empirical equation, evolved from analysis of stream-flow records. It means that, after measuring v, r and s, a characterizing value of the stream bed (i.e. n) can be found. Alternatively, v can be found after measuring r and s, a suitable value of n being obtained from tabulated values. Again, it tells us nothing *in itself* of the processes involved in stream behaviour other than that these variables are numerically linked.

An alternative approach is by the physics of the problem. A stream moving over its bed can be observed to move particles if the velocity is great enough; in other words by virtue of its inherent energy, it supplies a tractive force. This force, denoted by τ, is equal to the component of water mass given by the product of γ the density, r the hydraulic mean radius and s the surface slope as before:

$$\tau = \gamma r s \qquad\qquad (2.2)$$

Though it is perhaps difficult to appreciate, an equation such as this *in itself* offers an explanation of the mechanism of river behaviour. What we see when a river has sufficient energy to move particles on the bed is a manifestation of this equation. This is the most important consequence of such an approach to geomorphology; certain expressions can be derived that explain both what happens in physical terms and also describe numerically what we can see. The same applies to chemical processes; for instance, H_2O is not only a formula for water but it *explains* some of its properties both physical and chemical, provided that one can understand the language of the formula.

It is possible to have a critical value of τ at which sediment motion occurs and relate this to the size of particle moved. In this way we can still obtain numbers that describe the action of the process, but the expression also says something about the physics of the problem, the force that moves particles. It is possible to derive other expressions of greater complexity by considering the mechanics of the process. The isolated example here is to show the need, not just for numbers alone, but for a physical interpretation of the problem. Indeed, a theoretical system can be deductively produced which models certain aspects of geomorphology. A good illustration of this is Kirkby's (1977b) modelling of slope-wash processes. Given such a model it is necessary to test it, in whole or part, for validity; the model will enable predictions to be made, and appropriate field and laboratory work need to be carried out. This deductive and testing approach is one that demands a knowledge of the physical principles and as such is potentially a very powerful tool. It is with physical principles that the present chapter is concerned.

Terms such as force, stress and work have everyday meanings; in mechanics they are not usually different but more precisely defined. 'Everyday' includes geomorphology. How much energy is in a 'high-energy' beach, for example? If we talk about a 'powerful' stream, this conveys only a rough comparative meaning though the power of a stream can be precisely defined and the concept can thus have a quantitative meaning. Later parts of this chapter will look in some more detail at terminology; first, however, the basic sources of energy available for geomorphological processes will be examined.

Sources of global energy

All known forms of energy on the earth are ultimately related to certain basic sources. These comprise solar radiation, atomic energy, chemical energy, gravity and the energy of the Earth's rotation. Derived from these basic sources are the immense store of heat energy inside the Earth and the Earth's electromagnetic field. Some forms of energy, such as geomagnetism, are of only indirect interest to the geomorphologist; others, such as gravity and solar radiation, are fundamental; and it is on these that this section will concentrate. Table 2.1 gives some comparative estimates of the energy levels available; the units of measurement (J, joules) are defined on p. 23.

Table 2.1   Orders of magnitude of some basic energy sources at the Earth's surface (after Bott, 1971 and others)

	Joules per year
Solar energy	10^{25}
Geothermal energy	10^{21}
Volcanic activity	10^{18}
Elastic wave energy released by earthquakes	10^{18}

Solar energy

Its dominating position in the hierarchy of Table 2.1 is evident. More precisely, the solar radiation received amounts to about 7000 times the heat flow from the Earth's interior. Even so, the Earth intercepts only about 0.5×10^{-9} of the total energy emitted from the sun. At the top of the atmosphere, a surface normal to the solar beam receives on average 0.83×10^5 J m² s⁻¹. This is the so-called solar constant, though it varies seasonally by about 7 per cent because of the eccentricity of the Earth's orbit, and smaller variations of 1–2 per cent over longer periods corresponding to sunspot cycles have been detected. Of the total radiation entering the upper atmosphere, about 23 per cent is reflected back into space by clouds, 6 per cent is reflected by dust particles, air and water vapour, and about 7 per cent by the Earth's surface, depending on its albedo. A further 14 per cent is absorbed by the air and 3 per cent by clouds, leaving only 47 per cent to be absorbed by the Earth. Only a minute part penetrates more than a few metres.

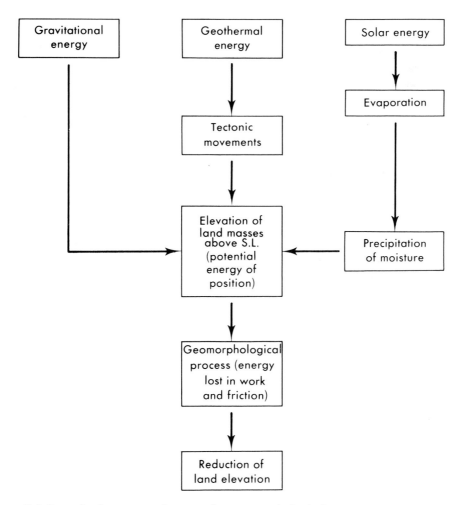

Figure 2.1 Some basic sources of energy for geomorphological processes.

Solar radiation is the principal source of energy for all climatic and hydrological processes. It is the basic control on atmospheric and surface temperatures, and on the redistribution and movement of moisture. It is the driving force behind the hydrological cycle since it provides the lift, against gravity, of water vapour by evaporation, to form precipitation (Figure 2.1). Variations in the quantity and quality of solar radiation are likely to be the fundamental cause of climatic change, whose repercussions include the periodic ice ages which the Earth has experienced throughout its geological history.

Table 2.2   Stores and transfers of water on or near the surface of the Earth

Stores:	Oceans	$1\,322\,000 \times 10^3$ km³
	Glacier ice	29 200
	Lakes, rivers	230
	Atmosphere	13
	Groundwater	8 470
Transfers:	Evaporation from oceans	361×10^3 km³ year^{-1}
	Evaporation from land areas	62
	Precipitation	423
	Runoff	37

The magnitude of the water transfers brought about by solar energy is immense. Table 2.2 provides some estimates of the stores and transfers involved in the global hydrological cycle. From the point of view of geomorphology, surface-water transfer, including discharge of glacier ice, is the most significant. The bulk of the precipitation feeding river and glacier systems falls on elevated land, and during its return to the oceans through these sysems is provided with gravitational energy enabling channel erosion and sediment transport to take place (Figure 2.1). As the mean elevation of the world's land areas is about 800 m above sea level, 37 000 km³ of water runoff per year can theoretically provide total energy of 3×10^{20} J year^{-1} ($=10^{15}$W).

Gravitational energy

The term 'gravity' as commonly understood represents a combination of forces: (i) the attractive force, G, between the earth and other bodies, acting vertically by definition and at right angles to mean sea level; (ii) the centrifugal force set up by the Earth's rotation, amounting to about 0·4 per cent of G; (iii) the attractive effect of other heavenly bodies, notably the sun and the moon. The third of these forces is so small (about 0·00001 per cent of G) that for many purposes it is neglected; also, unlike the first two, it varies temporally. The principles of gravitational attraction were, as is well known, first elucidated by Newton, who showed that the attraction varies with mass and with the square of the distance. In the case of two bodies, mass m and M, whose centres are separated by distance d which is large in comparison with their own dimensions, the force, F, is proportional to mM/d^2. At the point occupied by the centre of m,

$$F=GMd^{1/2}=G\gamma vd^{1/2} \tag{2.3}$$

where γ is the density of M, v is its volume, and G is the gravitational constant. In the case of the Earth, $\gamma=5\cdot517$, M is about 5977×10^{24} kg, and $G=(6\cdot673\pm0\cdot003) \times 10^{-11}$ N m² kg^{-2} (Bomford, 1971).

Observed gravity at the Earth's surface varies with latitude (because of the Earth's generally spheroidal shape) and height above mean sea level (i.e. with distance from the 'centre' of the Earth). Average sea-level values at the equator are about 978 gals (1 gal$=10^{-2}$ m s^{-2}) compared with 983 gals at the poles, a variation of about 0·5 per cent. The decrease of gravity with height is about 1 gal for 3000 m, or about 0·3 per cent between sea level and the highest mountain peaks on earth. Such small variations have not been shown to have any geomorphological significance, though larger variations of G through geological history are both possible and important in their implications. Theories of an expanding Earth would involve a decrease in G. The gravitational attraction of the moon and the sun, though small in comparison with Earth gravity, does have an important effect on sea level, creating tides. The tidal range over the Earth varies from zero to about 15 m, so that wave energy on coasts is spread over a greater vertical and horizontal range than would otherwise be the case. There are also Earth tides—a bulge of about 0·3 m is raised in the crust on the side of the Earth facing the moon—but these have no known geomorphological effect apart from the possibility that they may trigger earthquakes in unstable areas.

Because of the existence of gravity, every object on the Earth's surface possesses potential energy proportional to its mass (m) and its position (Figure 2.1). Sometimes termed 'elevation energy' or 'energy of position', it is the product mgh, where h is the vertical interval between the object and the lowest point to which it can move (see p. 17). For geomorphological purposes, h is equivalent to height above sea level, ignoring the few land areas that descend locally below this datum. The enormous potential energy of the water annually precipitated on high land areas has already been mentioned. The potential energy of a stream is converted by flow into kinetic energy (see p. 22) and this in turn is mostly dissipated in the heat of friction due to turbulence and friction with the stream bed and sides.

Gravity also provides the motive power for glaciers. Andrews (1972) shows that total theoretical glacier power ranges from about 0·2 to 1·5 W for velocities between 5 and 100 m year^{-1}. Although streams and rivers flow much faster than glaciers, this is more than counterbalanced by the greater shear stresses at the bases of glaciers.

Geothermal energy

Just over a century ago, Lord Kelvin made his famous deduction on the age of the Earth (more precisely, the length of time since the Earth consolidated from a molten state), based on the assumption that it was steadily cooling without further significant addition of heat. His answer, between 20 and 400 million years (in 1897 (Kelvin, 1899), he narrowed the range to 20–40 million years) initiated great controversy and highlighted, in its great under-estimation, the magnitude of the factor omitted, namely, internal heat generated by radio-activity. There is no evidence that the Earth is steadily cooling; rather, an approximate state of equilibrium is thought to exist, heat escaping at the surface at about the rate at which it is being produced in the interior. In fact, there is a sufficient supply of geothermal heat to provide the energy for volcanic activity, earthquakes, mountain building and to maintain the Earth's electromagnetic field. The heat of radioactive decay is generated mostly by the isotopes ^{238}U, ^{235}U, ^{232}Th and ^{40}K. These are the most important isotopes by virtue of their relative abundance, their heat-producing capacity, and the fact that their half-lives (from 0·7 to 13·9 years $\times 10^9$) are of a similar order of magnitude to the age of the Earth. Because of their strong concentration in granitic and intermediate-type igneous rocks, continental crust of average thickness could itself account for as much as two-thirds of the average observed heat-flow at the surface, and the remainder could be generated in the upper mantle. In oceanic areas, however, the source must be in the mantle.

Apart from radioactivity, other possible sources of geothermal heat include the heat produced by accretion, due to collision of other bodies, during the Earth's formation, and the conversion of part of the Earth's rotational energy to heat. The Earth's rotational energy is being steadily dissipated by tidal interaction with the moon and, less significantly, with the sun. In earliest Earth history, the rotational period may have been as little as 2–4 hours. A slowing down from a 3-hour to a 24-hour period would release energy of 1.5×10^{31} J, equivalent to $2500 \, \mathrm{J \, g^{-1}}$ for the whole Earth. Assuming that 90 per cent of this energy were dissipated by ocean tides, the remainder could produce a rise of Earth temperature of the order of 200°C.

Actual temperatures in the Earth below the level reached by deep drilling (e.g. the Mohole) are a matter of estimation and speculation. Rates of geothermal heat loss obtained from depths great enough to overcome any residual cold from the last glaciation have been measured from over 3000 localities but unfortunately these are not well distributed over the Earth's surface (Lee and Uyeda, 1965). Lee (1963) computes the world mean heat flow at $62.79 \pm 6.3 \, \mathrm{J \, m^2 \, s^{-1}}$; the total heat energy escaping from the interior to the surface by conduction amounts to 10^{21} J year^{-1}. Volcanic activity releases about 0.2 per cent of this. Considerable spatial variations in heat flow are known. Parts of ocean ridges appear as pronounced 'hot spots', yielding heat-flowrates of over $300 \, \mathrm{J \, m^2 \, s^{-1}}$. Other areas with higher-than-average heat flow include the Tertiary volcanic areas and young orogenic belts. The Pre-Cambrian shields show lower-than-average rates.

The fact that the Earth's interior is hot and mobile is, fundamentally, the cause of tectonic activity at the surface. Geological evidence shows that vertical crustal movements, opposed to gravity, can in some places total 3000 m or more in the last 20–30 million years, while horizontal displacements of tens of kilometres in the same time have also occurred. Even larger horizontal movements are involved in continental drift. Bott (1971) estimates that the elastic wave energy released by earthquakes alone is of the order of 10^{18} J year^{-1}. Uplift through 1 m of a rock mass, density 2.8, of dimensions $10 \times 10 \times 1$ km, would involve expenditure of energy amounting to 2.7×10^{15} J.

Geothermal energy as the origin of endogenic forces is therefore fundamental in considering the Earth's surface. It provides the vertical elevation of land masses which confers the potential energy of position (Figure 2.1) on any object. There is, however, one other significant aspect of geothermal energy. The dissipation of geothermal heat is normally unimpeded at the Earth's surface, in the case of both oceanic and land areas, but one exception to this is the land areas that are covered by temperate glacial ice. Temperate glaciers (see Chapter 8) are approximately at pressure melting point throughout their thickness; their vertical temperature profile therefore shows a slight increase of temperature from the base of the ice to the surface. This reversed temperature gradient prevents the escape of geothermal heat, which therefore has to be used in melting the base of the ice. A thin layer of basal meltwater is created, assisting such glaciers to slide over the bedrock, and there are important implications for processes of erosion, transport and deposition at the ice–rock junction. Moreover, if the basal meltwater cannot escape as fast as it is being produced, unstable conditions will result leading to accelerated movement or surging (p. 64). Not all ice sheets or glaciers are temperate; at high altitudes and in polar regions, ice may be frozen to bedrock and its temperatures then decrease from the base upwards, permitting geothermal heat to escape, but it is now known that, even in Antarctica, considerable areas of the polar ice sheet are temperate in their lower layers and basal meltwater is present (Figure 8.1).

Geomagnetism

Fluid motions in the Earth's core are likely to be the source of the Earth's magnetic field: electromagnetic energy results from the conversion of geothermal energy. There have been periodic reversals of polarity, whose cause is unknown. Although of no direct importance to geomorphological processes, paleomagnetic studies have been of great value in studying continental displacements and in dating certain rocks.

Changes in energy levels

There are important differences between the three basic energy sources (p. 12) in terms of their variability through time. Although we speak of the 'solar constant', there is a strong probability that the input of solar energy into the Earth system fluctuates considerably over time, and that this is a major cause of climatic change. The time-scale of climatic change includes minor fluctuations occurring over $10-10^2$ years, measurable in historic time, medium-scale changes over 10^3 years (e.g. oscillations within Late- and Post-glacial time), glacial-interglacial changes in the Quaternary of the order of 10^4-10^5 years, and long-term changes of the order of 10^6 years or more, evident in the geological record. Unlike solar radiation, the other two basic energy sources, geothermal heat and gravity, may not have varied significantly in recent geological time; in terms of geomorphological processes acting over the last million years, or even over the whole Cenozoic period in which practically all the world's present landforms evolved, both gravity and mean world heat flow can be taken as constant. Geothermal energy, however, is unequally distributed at the Earth's surface, so that certain regions are tectonically or volcanically active, others relatively inactive at any given moment. Patterns of activity have thus varied over time, so that, for instance regions undergoing rapid uplift in one epoch may subsequently become relatively stable, rates of denudation then overtaking rates of earth movement, resulting in reduction of relief amplitude.

Figure 2.2 shows the main links between the basic energy sources and geomorphological processes. Changes in the potential energy of elevation can result from both land- and sea-level changes, which in turn are controlled by both tectonic and climatic events. The latter, however, also act directly on geomorphological processes by affecting the type of activity (e.g. fluvial, aeolian, glacial) and by affecting rates of action (e.g. varying river or glacier discharge, or rates of weathering). At any given time, some regions may be characterized by high rates of geomorphological activity (*high-energy environments*), others by relatively slow rates of landform change (*low-energy environments*). An example of the former is the Karakoram Himalaya (Hewitt, 1972) where the high amplitude of relief, and therefore prevalence of steep slopes, is related to recent and continuing tectonic uplift (geothermal energy) and to rapid rates of valley cutting (abundance of precipitation in the monsoon season, and of snow and glacier melt, linked to climatic régime). At the other extreme, parts of the Sahara desert characterized by negligible relief and intense aridity represent low-energy environments, where rates of landform change are at present comparatively slow. The availability of energy for geomorphological activity will change through time. As landscapes are lowered by denudation, the potential energy of runoff, glacier flow, mass wasting and so on is reduced; rates of denudation decrease, and there may also be a climatic feedback in that reduction of altitude may well reduce precipitation (as Davis suggested in 1899). Working in the opposite sense are new inputs of energy, such as tectonic uplift, fall of sea level or increase in precipitation owing to climatic change. The nature and effects of such changes in energy levels will now be discussed in some further detail.

Land- and sea-level change

The basic causes, tectonic and climatic, have long been identified. One effect of tectonics is relatively simple: uplift or subsidence of the crust. Table 2.3 gives some examples of contemporary or recent tectonic uplift in non-glaciated areas (to avoid examples of the complication

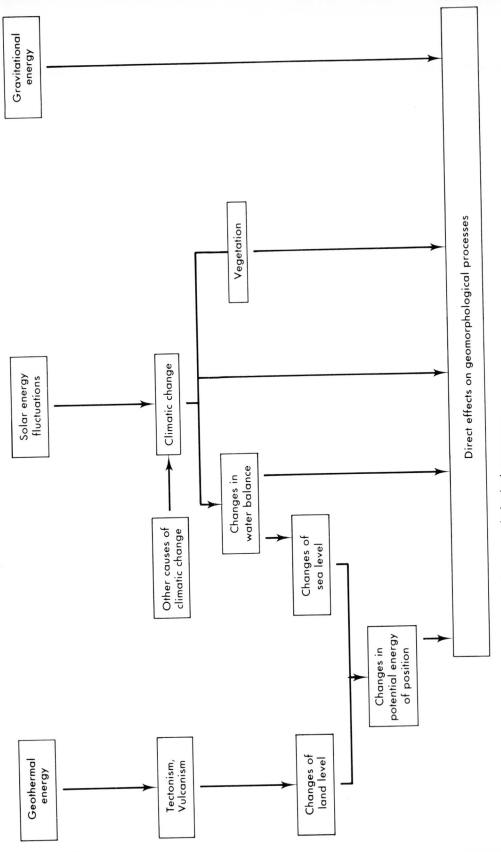

Figure 2.2 Links between energy sources and geomorphological processes.

of glacio-isostatic uplift). The data vary greatly in reliability, and extrapolation is dangerous since rates are unlikely to be constant over significant intervals of time, but it is worth pointing out that, at the rate of 50 mm year^{-1} and ignoring denudation, the present average altitude of the Himalayan peaks (say 8000 m) could be accounted for in only 160 000 years, or the duration of the last glacial and interglacial period. Schumm (1963) presents some evidence to suggest that average maximum rates of denudation in major mountain ranges may be of the order of 1 mm year^{-1}, and concludes that present rates of orogeny may be operating several times faster than denudation in tectonically active areas. Tectonic activity can also affect sea level. Alterations in the shapes of ocean basins brought about by plate movements can affect world sea level; so also can the growth of major submarine mountain systems such as the Mid-Atlantic Ridge.

Isostatic changes of land and sea level can result from denudation of land-masses, which are thereby unloaded, and from the deposition of the resulting sediments on the sea floor. Deposition of sediment in the ocean basins has an immediate eustatic effect (rise of sea level) before isostatic compensation commences. The latter then causes the land to rise, though adjustment is probably intermittent rather than continuous (Pugh, 1955). In a region such as the Himalaya undergoing strong erosion, isostatic uplift owing to rock removal by denudation will be taking place at the same time as tectonic uplift, and both processes contribute

Table 2.3 Examples of rates of uplift in non-glacial areas

Area	mm year^{-1}
Hong Kong	1–5
California	4–13
Krivoi Rog, USSR	11
Indo-Gangetic plain	*c.* 11
SE Alaska, 1922–60	40
Himalaya	*c.* 50

to the figure quoted in Table 2.3. The other main causes of land- and sea-level change are climatically induced and related to the amount of water stored on land areas at any given time in the form of glacial ice. There is first of all the glacio-eustatic effect, the transfer of water between the oceans and glacial ice. Table 2.2 shows that, at present, about 2 per cent of the earth's water store is locked up as glacial ice, of which over 98 per cent is located in Greenland and Antarctica. If all were to melt, Fairbridge (1961) estimates that a net rise of sea level of about 50 m would result. In glacial periods of the Pleistocene, the maximum fall of sea level is thought to have been about 145 m according to Curray (1965). Precise figures cannot be given because (i) we do not know Pleistocene ice volumes accurately and (ii) the amount of isostatic compensation owing to the loading or unloading of the oceanic floors by sea water is difficult to assess. A second result of world glaciation is the glacio-isostatic effect, the local and temporary depression of the Earth's crust caused by ice loading, and the rebound associated with deglaciation (Table 2.4). A considerable time-lag is involved, so that areas deglaciated 10 000 years ago are still rising. The figures in Table 2.4 refer to uplift of the land relative to sea level; as the latter is presently rising by 1–3 mm year^{-1}, the net uplift is correspondingly less.

The effects of changes of land- and sea-level on geomorphological processes are far from simple or predictable, and there is no space to examine the matter fully at this point. It is, of course, clear that fall of sea level relative to the land will increase the potential energy

Table 2.4 Examples of rates of uplift in glacio-
isostatically depressed areas

Area	mm year[-1]
Central Scotland	? 3–4
Gulf of Bothnia	9
SE Hudson Bay	13
Ellesmere Island	7
Northern shores of Great Lakes	5

of elevation. It is equally clear that some processes will be wholly unaffected—aeolian or sub-surface processes, for instance. The effects on fluvial systems are much more complex and not well understood. Sea level is the ultimate base level exercising a control on all through-flowing river systems. A rise in sea level, or land subsidence, will drown the lower part of the river system, and, if sediment supply is sufficient, aggradation of the drowned portion will commence as a delta is built out at the head; fluvial deposits will then accumulate above the delta until a new profile of equilibrium is established (Figure 2.3A). Reservoir studies (Leopold, Wolman and Miller, 1964) have shown that the effect on the river system will be limited to the distance upstream that this new profile extends: above this point (P), the remainder of the system will be unaffected. Essentially, the energy available to the stream will be reduced below this point as the gradient is lessened. Lowering of sea level (or land uplift) might be thought simply to have the opposite effect, but this is not necessarily the case. Much depends on the gradients of the surface over which the river is extended to reach

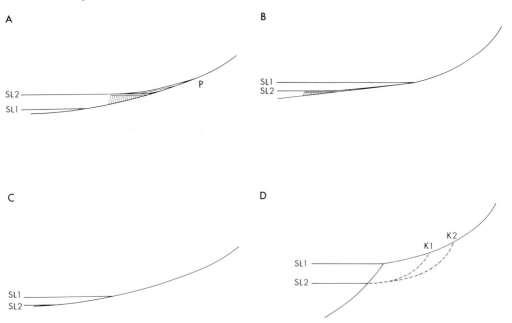

Figure 2.3 Effect of base-level change on river long profiles. **A**: rising base-level and aggradation; **B**: falling base-level and aggradation; **C**: falling base-level: no change in stream energy; **D**: falling base-level: increase in stream energy.

a more distant coastline. If the gradient is less than that of the previous lowermost river segment (Figure 2.3B), aggradation because of reduction of energy may result; Figure 2.3C shows a hypothetical case where no change in energy conditions occurs, while only in the case where steepening of the profile results (Figure 2.3D) will an increase in energy be available to the stream. In this case, the steeper gradient may be propagated headwards up the stream system as a knickpoint (K1, K2), with accompanying terrace formation. Valley-side slopes will be steepened, increasing the relief energy of the drainage basin.

Climatic change

In the last 1–2 million years alone, the Earth has experienced great changes in mean world temperatures, in the spatial distribution of precipitation and in the patterns of climatic zones. Glacial and interglacial periods have alternated in the middle and higher latitudes of humid areas; pluvial and dry phases alternated in lower latitudes; climatic zones expanded or contracted (e.g. Büdel, 1957) during the Pleistocene. These and the smaller-scale oscillations referred to on p. 17 had a profound effect on geomorphological processes, though the precise nature and extent of some of the effects are by no means agreed (Stoddart, 1969; Vita-Finzi, 1973). Undoubtedly, climatic change affected vegetational patterns (Fig. 2.2) and thereby indirectly influenced some processes. A vegetated surface will offer greater friction to runoff than a bare surface, for instance, reducing the stream energy available for erosion and transport. Plant roots can help in slope stabilization and retard the rate of debris supply to stream channels. The relationships between climate and weathering are equally important (Chapter 4), and climatic régime can act as a major control on mass movement (Chapter 5), for example, the frequency of rockfalls in areas susceptible to alternate freezing and thawing conditions. The influence of climate on runoff levels and régimes is fundamental. The interrelationships of precipitation, evapo-transpiration and infiltration determine the amount of water available to enter the river system, the nature of any seasonal fluctuation in supply, the frequency and magnitude of flood events and so on, while the sediment yield of any drainage basin is also largely determined by climatic parameters.

The possible relationships between climate and process are so numerous that no useful purpose would be served in extending the list. However, a brief example of the effects of changing climate on the energy available in a river system will be given to show the complexity of the responses induced. The Thames valley possesses a fine suite of terraces, a floodplain and buried channels dating from the Quaternary. Relationships with glacial and periglacial deposits, archaeological, palynological and geomorphological evidence have yielded a chronological framework for the region into which many of the terraces and the floodplain can be fitted. The Thames received direct accessions of meltwater and outwash in certain glacial stages. At such times, vegetation cover was restricted, mass wasting was active in periglacial conditions, and the river channels were copiously supplied with debris. Discharge rose during deglaciation and transport of the debris through the system increased, leading to extensive deposition in the lower reaches. In the following interglacial, both load and volume decreased, and adjustments of channel cross-section and gradient resulted. The climatic changes were accompanied by, and partly gave rise to, changes of relative land and sea level. The terraces and valley fills that are now left were the combined result of all these changes and their interpretation is not only complex but also controversial. Certain combinations of conditions (falling base-level, increasing discharge, reduction of load) represent a greater input of energy into the system; others conversely represent a reduced input.

Turning to broader considerations, it is agreed that climate, as well as relief, plays a part in determining regional rates of denudation. In regions of similar relief and lithology, rates of denudation will vary, for instance, according to availability of water, vegetation cover, and presence or absence of glacial ice. There are differences of opinion, however, over the degree of correspondence between climatic type and denudation rates, partly because of the insufficiency of reliable data on the latter. Table 2.5 presents some figures suggested for four different regions, and the lack of agreement is striking. Most workers, however, appear to regard the mountains of the humid tropics or sub-tropics as the most active environments geomorphologically. It is interesting to note that most workers also find that, in tectonically active regions, rates of uplift can exceed, sometimes by an order of magnitude, the rates of denudation common over most of the Earth's surface.

Table 2.5   Comparison of some suggested rates of denudation

	mm year^{-1}				
	Menard (1961)	Judson and Ritter (1964)	Strakhov (1967)	Fournier (1960)	Corbel (1964)
Himalaya	100		>17	>140	10–15
Mississippi	4·6	5	7–17	7–70	2–3
Appalachia	1·1	4·5	0·7–3·6	<8	10–15
Pacific slopes, California		9	3·6–7·2	70–140	up to 15

In this way the Earth maintains its relief; if it were otherwise, as Bloom (1969) notes, we would end up living on or in a hydrosphere.

To summarize, changes in land and sea level, and changes of climate, all of which can be referred to variations in geothermal and solar energy, cause pronounced variations, both spatial and temporal, in the energy available for geomorphological activity. The changes are interrelated in several ways: climatic change can itself be a cause of sea-level change; and as landscapes are lowered by denudation, so climatic change will occur, possibly by reduction of precipitation. At the present day, it is possible to distinguish between high-energy environments where either or both tectonic uplift and rates of denudation are high, and low-energy environments on the other hand, but it is not yet possible to present a complete or agreed regional differentiation of the world on this basis, basically because of lack of data.

Energy and forces

Having reviewed the sources of energy on the Earth and changing energy levels over time, it is now necessary to look more closely at the physical principles and concepts involved in the operation of geomorphological processes.

Energy and the conservation laws

Both kinetic and potential energy come within the province of the geomorphologist concerned with examination of processes. If an object can do *work* then it possesses energy. Thus water in a stream can do work by moving boulders on the bed, *kinetic energy* is used and the boulders present a certain resistance to the waterflow. A moving boulder also possesses kinetic energy, defined as the product of its mass (m) and half its velocity (v) squared:

$$KE = \tfrac{1}{2}mv^2 \tag{2.4}$$

Work is also done when a body is raised above some datum and this *potential energy* (as described on p. 15) is released if that body falls or rolls downhill:

$$PE = mgh \tag{2.5}$$

The unit of energy (and work) is the joule (J) (see p. ix). Note that G, the gravitational constant, is quite distinct from g, the acceleration due to gravity.

Friction in a system must be overcome; for instance, one block sliding on another produces a resistance and a proportion of the available energy is converted into another form (heat) to overcome it. There are two important Laws of Conservation. The first is the Principle of Conservation of Energy: 'the total energy in any given system is constant'. Secondly, there is the Principle of Conservation of Mass; 'the total mass of a given system of objects is constant'. The latter is true even if collisions take place between the objects.

Force and Newton's Laws of Motion

As forces and resistances are frequently measured in practice, so both geomorphological work and power (see below) depend on a knowledge of forces for their determination. This is why force is such an important concept. To appreciate fully what forces are, we need first to look at the three fundamental units, mass, length and time. Mass is often thought of as the weight of a body, but weight is dependent upon the gravitational attraction of the Earth which, as explained on p. 14, is not constant at every place. The mass of a body is a measure of its resistance to movement (its inertia), compared with a standard on a beam balance. Thus masses will be the same on the Earth and moon, even though gravitational attraction will be very different. Masses are measured in kilograms (kg), weights as kilograms force.

The standard of length is the metre (m) and of time the second (s), so that length travelled per unit gives a measure of *velocity* in metres second^{-1}. (Notice that though *speed* has the same units, it has no direction and only magnitude; velocity has both magnitude and direction. The former is called a scalar quantity, velocity a vector.) The rate of change of velocity with respect to time is called acceleration and has units of $m\,s^{-2}$, i.e. metres per second per second. The acceleration due to gravity is approximately $9.81\,m\,s^{-2}$ (p. 15).

We are now in a position to consider the concept of *force*. It is defined as the product of mass and acceleration; in SI units, it is $kg \times m\,s^{-2}$ or the newton (N). Thus, if a mass of 1 kg is accelerated at $1\,m\,s^{-2}$, the applied force is one newton. This definition follows from the second of Newton's three laws of motion:

1 Every body continues in its state of rest or uniform motion in a straight line unless acted upon by an impressed force (Fig. 2.4).
2 The change in momentum per unit time is proportional to the impressed force and takes place in the direction of the straight line along which the force acts.
3 To every action there is an equal and opposite reaction.

From the second law, force is proportional to momentum (i.e. mass $\times$ velocity) divided by time; if the proportionality constant is made equal to one, we arrive at the relationship: force equals mass times acceleration:

$$F = ma \tag{2.6}$$

Work, efficiency and power

Geomorphological processes operate when geomorphological work is done. Though this may seem a trite statement it has important consequences because the physical concept of work

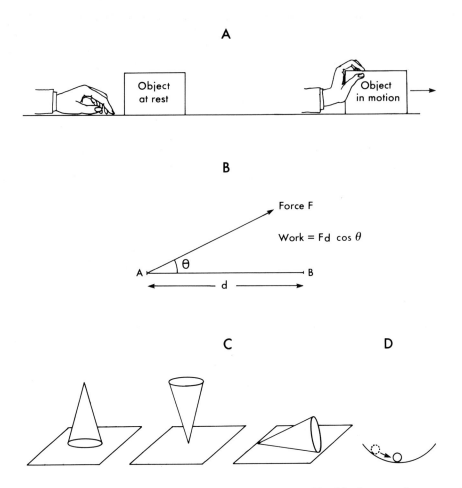

A

Object at rest

Object in motion

B

Force F

Work = Fd cos θ

A

θ

B

d

C

D

Figure 2.4A: An illustration of Newton's First Law of Motion. The block can only move when a force acts upon it and work is done only when the block moves.

B: Work is done when the object moves a distance *d* from *A* to *B* and equals $F \times d$. If the force applied (*F*) is the direction of angle θ to *AB*, the work done is *Fd* cos θ.

C: Three cases of equilibrium: the cone is stable when on its end, unstable when balanced on its apex but in neutral equilibrium when on its side—it can move by rolling, though its centre of gravity stays at the same height above the ground.

D: The ball-bearing in a saucer moves from an unstable position on the side to one of stability in the centre. This illustrates the principle that the equilibrium position corresponds to minimum potential energy.

involves us directly with masses, forces and resistances. These are frequently the things we try to measure in any assessment of geomorphological processes and will be discussed in more detail shortly. There are two other concepts that should be borne in mind when looking at the mechanics of geomorphological processes. These are *efficiency* and *power* which can be related to each other through the expression:

Efficiency=useful power output/power input (usually expressed as a percentage),

where power is defined as the rate of doing work (or work done/time taken) and measured in watts (W). It is not usual to talk of the efficiency of a geomorphological process, though it is common to talk of rates of erosion and this might be considered as a measure of the power of a process or set of processes. We will return to a discussion of this at the end of the chapter. *Work* is done when a force moves its point of application and is measured by the product of force and the distance moved by the point of application in the direction of that force:

$$\text{work done} = F \times d \tag{2.7}$$

where F is the force and d is the distance moved. If the point moves along a line acting at an angle θ to the direction of the applied force, this is modified to:

$$\text{work} = F \times d \cos \theta \text{ (Fig. 2.4)} \tag{2.8}$$

The unit of work is the newton (N) (the unit of force) times the unit of distance (the metre), i.e. a newton metre or a joule, and from this comes the definition of the watt: joule second^{-1}.

Types of force

There are various types of force though not all of these need concern us in detail here. We can distinguish five:

1 Body forces, which act throughout a body, gravity being an example.
2 Surface forces are those that one solid body can apply to another. These are important on a microscopic scale related to mechanical strength.
3 An inertial force (or effective force) is defined as $-ma$ (from $F-ma=0$). When impressed forces and the inertial forces are considered together they form a system in kinetic or dynamic equilibrium. This is d'Alembert's principle.
4 Centripetal force is that which, when directed towards a point, makes a particle describe a circular path with uniform angular velocity centred on that point.
5 Coriolis is the inertial force concerned with a change in the tangential component of a particle's velocity. Centrifugal force is the radial component in such motion.

Forces in geomorphology

When we consider a landscape in terms of processes, we are unlikely to consider it to be completely static, that is, in stable or perhaps neutral equilibrium. Rather, erosion and deposition will be occurring, though not necessarily everywhere at a given instant. Part of the complexity of geomorphology is due to temporal changes in rates (that is, the mechanical power) of processes but what actually goes on is frequently related to sedimentary processes operating in the Earth's gravity field. This is one type of force field, against which other forces operate. For example, stones falling from a cliff face will be controlled by gravity but energy will be lost by air resistance and collisions with the ground. A particle falling in a lake or the sea will act similarly except for the magnitude of the resistance. On the other hand, in the case of a fine sand particle in a dune being lifted from its neighbours, the 'active' geomorphological force is the wind acting essentially normal to the force of gravity. Other resistances to movement come from the friction and cohesion of the particle with its neighbours and its inertia.

From these sketches of processes it can be seen that forces are the important parts of geomorphological processes. Indeed, such a process may be considered as the work done by all the forces in a system. Work is done in the system as impressed forces overcome resistances of various kinds. Seen in these terms, the examination of a geomorphological process is merely the resolution and measurement of a set of interacting forces that obey certain mechanical laws. Unfortunately it is rarely as simple as a laboratory determination in dynamics! Nevertheless, if geomorphologists strive to make such measurements, a better understanding of processes will result. The next section will examine forces and mechanical properties in more detail.

Stress and strain

Any measurement must be related to a frame of reference and suitable units. For the latter, SI (Système International) units are preferred using the metre, kilogram and second as base units. Commonly, measurements are made with respect to a local horizontal (as in plane surveying) and a vertical related to the Earth's gravity. Cartesian coordinates follow from this, i.e. three mutually perpendicular axes. In some cases, polar coordinates are useful, as for example, when looking at the three-dimensional orientation of stone long axes in a sedimentary fabric, but generally speaking Cartesian coordinates are used.

Though we have been talking of forces, it is more usual to be concerned with *stress*. A stress is produced in a solid by pushing or pulling, that is, inducing deformation. The definition of stress is force per unit area, measured in newtons per square metre $(\mathrm{N\,m^{-2}})$. There are three basic types of stress. *Compressional* or *tensile stresses* are different only in the direction of application. In a cube of material, they are the forces applied to the top and bottom (or two opposite sides) to compress or stretch it per unit area of the cross-section. At any point in a substance the forces can be considered as acting on the faces of a cube with three stresses on each (Figure 2.5A). One acts at right angles and is the *normal stress*, so that it can be either compressional or tensional. On the cube, a compressional normal stress has a counterpart on the opposite face of equal magnitude. Forces acting *along* the face are resolved into two *shear stresses* at right-angles to each other. Again, opposite faces have opposing forces of equal magnitude; this can be seen in two dimensions in Figure 2.5B.

At any stressed point in a solid there are three orthogonal planes that have no shear stresses acting along them—the principal stress planes. The normal stresses that act on them are the major, intermediate and minor stress directions denoted by σ_1, σ_2 and σ_3 respectively (Figure 2.5C). It is possible to relate these stresses to each other. On any plane at an angle θ to the direction of the major principal plane σ_3 (at right-angles to the major principal stress σ_1), components τ (shear stress) and σ_n (normal stress) can be determined:

$$\tau = [(\sigma_1 - \sigma_3)/2] \sin 2\theta \tag{2.9}$$

$$\sigma_n = \sigma_3 + (\sigma_1 - \sigma_3) \cos 2\theta \tag{2.10}$$

These components will be examined again later.

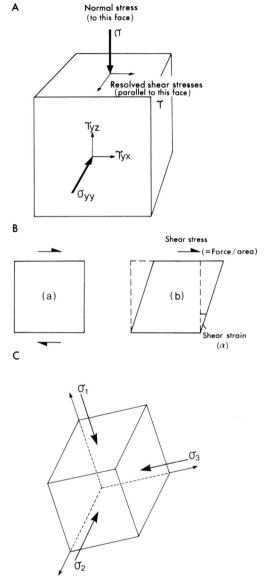

Figure 2.5A: A cube of material with a normal stress (σ) acting on the top face. The front face shows another normal stress with its subscript showing that it is acting in the *y* direction. There are two resolved shear stresses (τ) acting parallel to the face; on the front face these are acting in the *z* and *x* directions.

B: Two shear stresses acting on a block to produce a shear strain (the angular displacement α). The force applied along the top surface is equal to the force on the bottom (which could be fixed).

C: A cube, similar to that in **A** but with an arbitrary orientation showing the three stress directions σ_1, σ_2 and σ_3.

Pressure and Archimedes's principle

The third type of stress is hydrostatic stress or 'pressure'; that is, the *average* force per unit area on a body when it is immersed in a fluid (which could be air, water or oil). At any point in the fluid except at a liquid surface, the pressure can act in any direction and so is a scalar quantity. It can be shown that the pressure p at a point is given by

$$p = \rho g h \qquad (2.11)$$

where h is the height of the fluid, ρ its density and g the acceleration due to gravity. It follows that, at any horizontal level in the fluid, the pressure is the same at all points. We have used the word 'fluid' rather than liquid because air as well as water follows the same behaviour, though the influence of air pressure can frequently be disregarded in geomorphological (though not meteorological) contexts. As the density of air varies, it is necessary to standardize the pressure conditions: these are at $0°C$ and 0.76 m of a mercury column. This is standard (or normal) temperature and pressure (STP or NTP) and the air pressure of one atmosphere is $1.01325 \times 10^5 \, \mathrm{N \, m^{-2}}$. The unit known as the bar is a useful one as it is very nearly equal to one atmosphere; $1 \, \mathrm{bar} = 10^5 \, \mathrm{N \, m^{-2}}$. A new SI unit of pressure is the pascal (Pa): $1 \, \mathrm{Pa} = 1 \, \mathrm{N \, m^{-2}}$. Note that stress and pressure are different concepts, although often numerically identical.

Archimedes's principle depends on the resultant difference in upthrust between the force on the top and that on the bottom surface of a body. It can be shown that the upthrust is equal to the weight of the fluid displaced by the object. This occurs when the object is in air though it is only of importance when the fluid is a liquid, usually water. This buoyancy is important when we come to consider the forces in soils that are saturated or partially saturated with water.

Densities, and therefore exerted pressures, of a fluid depend on temperature but though this is important in, for example, meteorology and oceanography, it does not affect geomorphological processes very much, and liquids can be considered as exerting a pressure proportional to the depth. Such conditions are said to be hydrostatic. In some cases it is possible to consider a solid in the same way. This is true of ice and also of some soils, provided that the type of soil does not vary. In this case conditions are geostatic and the horizontal plane through a point is a principal plane.

Strain

After having looked briefly at stresses it is now necessary to consider deformation, the change in shape or volume of a body, produced by applied stresses. Such deformation is called *strain*. There are several types of strain but we can recognize three simple types as follows: *Longitudinal strain:* axial lengthening or shortening. The change in length (Δl) per unit length (L) is usually denoted by ε; thus for longitudinal strain:

$$\varepsilon_1 = \frac{\Delta l}{L} \qquad (2.12)$$

Volumetric strain is similarly defined:

$$\varepsilon_v = \frac{\Delta v}{V} \qquad (2.13)$$

where V is the original volume and Δv is the change in volume such as would result from hydrostatic stress.

Shear strain can also occur and can be produced by shear stresses deforming a square into a parallelogram (Figure 2.5B). If we have a cube instead of a square, shear strain will deform it to give angular deformation but with no change in volume.

Stress–strain relations; liquid, elastic and plastic behaviour

In most geomorphological conditions there are both normal stresses and shear stresses, and corresponding complex strains, though we have seen how these stresses can be related. In analysing geomorphological processes it is necessary to look closely at the stresses and resulting strains in a wide variety of materials. For instance, analysis of slope processes requires study of the stress–strain behaviour of soils, both in slow creep and rapid failure. Stresses in fluids are a different matter but the forces applied to stones on a river bed are obviously important in fluvial geomorphology. These processes will be examined further in later chapters.

It should be noted that any stress applied to a body will produce deformation. It may not be easily measurable but compression of the molecules will nevertheless occur. Thus, if a stress is applied to a body there will be a corresponding strain and so stress–strain diagrams are useful ways of looking at the behaviour of materials. Figure 2.6A shows a simple response (strain) to applied stress—in this case tensile stress (as in the stretching of a wire). In the linear portion of the graph, this behaviour is said to be elastic. In the curved section, however, an increase in stress produces a much greater amount of deformation. This is plastic behaviour and the material starts to yield at the transition. At point F failure takes place. To a first approximation, rock masses have a response of this type. It is often easier to see this type of response if we plot the rate of strain (usually denoted by ε) against stress. Here it is not possible to depict the behaviour of an elastic substance as the strain is achieved very nearly instantaneously. However, Figure 2.6B shows the curve for an ideal viscous (or Newtonian) fluid the gradient of which is the viscosity of the fluid. Thick suspensions of clays in water are liquids but they have a curved rather than linear relationship between stress and strain rate. Figure 2.6B also shows the behaviour of a plastic substance where there is instantaneous deformation when the applied stress reaches a certain level. This is the 'yield stress' of the material and is important in several geomorphological problems. A curve is also shown which illustrates the behaviour of glacier ice according to applied shear stress, a point discussed later.

These relationships are essentially idealized. They are usually correct for small stresses or strains and are correct for a range of materials from steels to concrete and treacle to water, though in geomorphological materials more complex relationships are often found. It is not possible here to investigate such relationships in detail; indeed practical difficulties often dictate a simplified explanation. For example, the hypothesis of drumlin formation according to Smalley and Unwin (1968) uses an elementary stress–strain relationship to explain not only the way in which drumlins can be initiated but their spatial pattern, a good example of the approach advocated here.

Creep and fracture

One final concept that is frequently encountered in the geomorphological literature must be mentioned, namely the process of *creep*. Figure 2.6C shows the strain–time relationship of a material with the different types of creep recognized by engineers. Creep is any time-dependent deformation under constant stress. Just how a particular material behaves (for

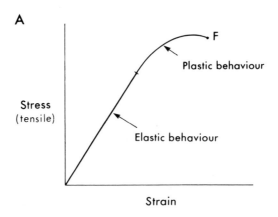

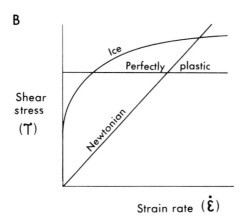

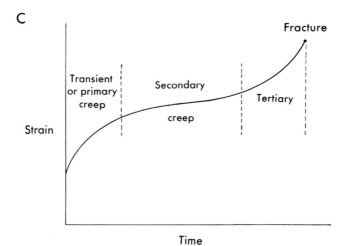

instance, the form of the curve depicted in Figure 2.6C) will depend on the plastic properties of the material and these are essentially related to its molecular configuration.

The study of the atomic and molecular nature of matter is too vast a topic for this chapter (for a simplified discussion, see Whalley, 1976); suffice it to say that it *is* important for the geomorphologist who wishes to study processes to have a basic knowledge of it. For certain branches it is a necessity. For instance, weathering processes, physical or chemical, depend on the ways in which chemical bonds are broken or cracks propagate. The resistance offered by various rock-types depends ultimately on the way in which the molecules of the constituent minerals are bonded together and how much energy is required to separate grains or break down chemical components. Here we meet the concept of surface energy; the energy per unit area of an exposed surface. If a new surface is to be revealed, that is, if a rock is to be broken down, then work must be done in putting in energy to create a new surface. Such energy is used to overcome interactions between atoms on each side of the rupture. This is true whether it is a question of rock breaking by freeze–thaw, salt crystallization, chemical weathering or a shear plane on a clay slope which is failing.

Also important in the study of rock faces is an understanding of fracture and crack propagation. *Fracture* is simply the division of a body into two or more parts and can be of two types: brittle or ductile. Brittle fracture occurs by rapid propagation of a crack with little or no plastic deformation of the material. The failure may occur either through crystals in cleavage planes or at crystal boundaries. Most rocks tend to be crystalline rather than amorphous substances. A ductile type of failure occurs with clays: a material is said to be ductile when it can sustain permanent deformation without losing its ability to resist a load. Transitions exist between the two types of failure but ductility in rocks is more the concern of the geophysicist and tectonic geologist than the geomorphologist. Some aspects of global geomorphology and geophysics such as crustal flexure and faulting are, however, very much related to properties of rocks.

Figure 2.6A has shown generally how failure may occur, and this would be true of most rocks. However, this is a very complex subject and we cannot do it justice here. Suffice it to say that processes at the crack tip are most important in the failure process and are related to the speed of propagation. There seems to be little doubt that processes such as frost and salt-crystal shatter and the critical size of certain existing cracks are related to the extension of existing cracks in rocks but as yet there has been little linking of the geomorphological and the physical processes. In effect, relatively little is known about the *geomorphological processes* concerned though something is known about the environmental controls required.

Heat, available energy, thermodynamics and entropy

Heat is a form of energy as noted on p. 12. Temperature is a measure of the hotness or coldness of a body and is strictly measured in kelvins (K) which have the same interval as degrees Centigrade (°C), but whose zero point is -273.16°C. Heat is produced as a waste

Figure 2.6A: A diagram of tensile stress applied to a wire plotted against the strain from that stress, showing elastic and plastic behaviour and its failure at point F.

B: A plot of shear stress (τ) against the rate of strain ($\dot{\varepsilon}$) for three different materials: a Newtonian fluid (such as water), a perfectly plastic body and the 'intermediate' behaviour of ice.

C: Different types of creep shown in a generalized diagram. The time to fracture may be many thousands of hours for a metal and even longer for rocks but much shorter for clays. It is a temperature-dependent property.

product of many processes as friction is overcome, and sometimes this has important geomorphological consequences. Glacier sliding, for instance, as well as processes of glacial erosion and deposition, are affected by frictional heat and the thermal properties of ice (Chapter 8).

Heat energy has another important role to play in the field of weathering. The solubility of salts in water varies according to temperature, and the crystallization of certain salts from groundwater often has disastrous effects on concrete or brick-work in hot climates (e.g. Goudie, 1977). In general, however, chemical reactions are temperature-dependent and the hotter the conditions the faster will be the reaction. A rule-of-thumb is that the rate will double for a 10 K rise in temperature. This does not mean that there will necessarily be a twofold increase in the quantity of a product, clay from granite for instance, as other environmental conditions (e.g. surface area of particles) will play a part.

Energy relationships in chemical processes are important and are slowly coming to be recognized as such in the geomorphological literature, for instance, in the recent work of Curtis (1976a, 1976b). Chapter 4 examines weathering processes in detail. We have seen how to combine the fundamental units of mass, length and time to give momentum, force and energy, but to consider the form of energy known as heat we need to look not only at temperature but an associated variable, entropy. One can be used to define the other. Entropy, an often puzzling quantity, is usually associated with the energy in a system which is not available to do work. Its precise definition need not concern us but note that for a (reversible) change in a system, the change in entropy (ΔN) is equal to the energy absorbed (ΔY) divided by the thermodynamic temperature (measured in kelvins):

$$\Delta N = \frac{\Delta Y}{T} \tag{2.14}$$

Entropy is sometimes viewed as a measure of randomness or disorder—and this has led to problems in theoretical geomorphology—but it can be readily understood if we look at the changes in phase of water from solid (ice) to liquid and vapour. This sequence increases the vibrations of the molecules (i.e. molecular randomness) and is reflected in an increase in entropy. For melting ice the entropy is higher than for boiling water. In effect, heat and work are not completely interchangeable; rocks cooling after exposure to the sun might release amounts of heat which, if converted to potential energy, would allow them to roll uphill. They do not, because heat cannot be spontaneously converted to work without other changes taking place; in many systems, friction uses up energy. We can say (as a formulation of the Second Law of Thermodynamics) that any spontaneous natural event implies an increase in total entropy.

Friction and resistance

Friction has been mentioned several times, but what actually is it, and further, how can it be measured? Intuitively, we know that a brick will slide more easily on a sheet of ice than on a table-top; the force required, and therefore the friction, is less. However, the force will be the same if that brick is placed flat or upon its end. That is, friction is only partly dependent on the area of contact; it is also proportional to the mass of the object. It can also be seen that the material properties probably have an important part to play in determining frictional resistance. There are two types of friction, both important in geomorphology: sliding and rolling. *Sliding friction* can be of two kinds, *static friction* (μ), the friction which has to be overcome to initiate motion, and *dynamic friction* (μ'), the force needed to keep one surface sliding over another. Rolling friction always has a smaller value than sliding friction and is the force resisting the rolling of a body across a surface. As examples, consider the initiation of movement of scree debris on a slope. Once started it will continue to move some distance. Here the static friction is related in practical terms to the angle of friction, which is the angle whose tangent is the coefficient of friction. Suppose a block is placed on a slope whose angle is increased until movement just starts. This critical angle is the angle of friction. However, the case of a scree slope is not quite so simple, because, once started, rolling friction will come into play, though by how much will depend upon the shape and the angularity of the particles. As the coefficient of rolling friction is less than the dynamic friction it will be the latter that is important in scree-slope development. Rolling friction is more important when considering the continuous shear stress applied by a stream to particles just rolling along the bed.

Failure in soils and rocks

In many problems dealing with slope angles and failure, reference is made to the 'angle of internal friction' or, more loosely, the 'friction angle', though it is more complicated than simply the angle at which a block moves on a plane surface. It is denoted by the symbol ϕ and usually associated with the Mohr-Coulomb failure criterion:

$$\tau_f = c + \sigma_n \tan \phi \tag{2.15}$$

where τ_f is the shear strength of the substance at failure, c is the cohesion (measured in N m^{-2}) and σ_n is the normal stress (in N m^{-2}). Figure 2.7A shows a block of soil subjected to stress. The applied shear stress causes strain, and energy is required to move the particles away from and around each other. The resistance to the shear stress can be considered to be of two types; a cohesion, which is due to inter-atomic forces between clay mineral particles, and a frictional component related to both static and dynamic friction (as well as 'interlocking' of grains) which is attributable to the granular nature of the materials. Grains characterize materials ranging from fine sand to large boulders and recent papers by Statham (1976) and Carson (1977) which discuss some of the complex interactions between material properties, slope angles and scree-slope development, well illustrate this behaviour. Silts represent an intermediate material showing both granularity and cohesion, and present problems of their own.

The behaviour can best be studied by means of shear stress–strain diagrams. Figure 2.7B shows how packing of the soil affects the shape of the curve and the peak shear-strength. Curve (*i*) is typical for densely packed granular material, (*ii*) for very loosely packed grains. If a clay block has previously been sheared in two, the sliding friction between two blocks results in what is often called the residual strength, which plots as the flat portion of such a graph. For any one soil, a whole series of curves can be found for different packings, i.e. densities, at the same normal load (σ_n). We already know from Figure 2.5A that a normal load can be applied to a body as well as a shear stress. In a soil this is the equivalent of increasing the depth that we are considering; the shear stress at failure increases as σ_n increases. This can be seen if τ is plotted against σ_n as in Figure 2.7C. The intercept, c, is the cohesion, the gradient is ϕ and so the equation of the line is that of the failure criterion. For clayey soils, ϕ can often be taken as zero, i.e. no slope on the line; for granular soils there may be no intercept, though granular interlocking can give an 'apparent cohesion'. This simple predictive theory is usually extended to include the concept of 'effective stresses'. Stresses imposed on a soil mass are borne by its structure, grain-to-grain contacts or by the clay structure (Figure 2.7A); but even in a dense clay there are pore spaces and, when these contain water, some of the load is taken by this water. If the normal stress is increased in

A

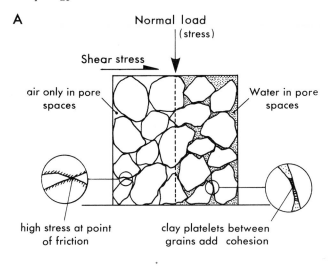

Normal load
(stress)

Shear stress

air only in pore
spaces

Water in pore
spaces

high stress at point
of friction

clay platelets between
grains add cohesion

B

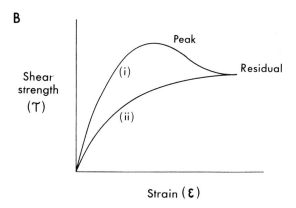

Shear
strength
(τ)

Peak

(i)

Residual

(ii)

Strain (ε)

C

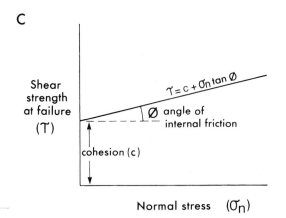

Shear
strength
at failure
(τ)

$\tau = c + \sigma_n \tan \varnothing$

$\varnothing$ angle of
internal friction

cohesion (c)

Normal stress (σ_n)

a partially saturated soil, the shear strength decreases, the amount being related to the presence of the water in the pore spaces; hence we talk of *pore-water pressure* (denoted by u). This pressure is developed if the drainage conditions of the soil do not allow the water to move freely out of the pore spaces. The pore-water pressure may increase so that at a critical pressure value the soil may fail. We can speak therefore of undrained loading, a process that may be associated with mudflow formation (e.g. Hutchinson, Prior and Stephens, 1974) when small flows rapidly load the soil at the upper end so that high pore-water pressures cannot dissipate but increase. Clays are particularly prone to failure by the development of high pore-water pressures, perhaps due to high-intensity rainfall, because of their low permeability. Under periglacial conditions, high pore-water pressures can develop between the surface and the permafrost table, and gelifluction can occur on low-angle slopes (Hutchinson, 1974). How this comes about may be seen by considering a modified equation that takes account of effective stresses. The effective normal stress σ'_n is given by $\sigma'_n = \sigma_n - u$; thus,

$$\tau_f = c + (\sigma_n - u) \tan \phi \qquad (2.16)$$

To show that effective stresses are considered, this is often written as $\tau'_f = c' + \sigma'_n \tan \phi$. An increase in u effectively decreases the value of σ_n, so that for determined material values of c and ϕ, a high pore-water pressure decreases the value of the shear strength at failure. Various equations exist whereby slope angles at failure can be related to c and ϕ, and critical pore-water pressures can be predicted from these. Both soil engineers and geomorphologists need to be able to measure pore-water pressures. This is done with devices known as piezometers which are of various types but respond, often quickly, to changes in the pore-water pressure at a given depth.

We have looked at one type of failure model (there are others that need not concern us here) and the importance of effective stresses by examining soil behaviour in a simple way. The manner in which shear strength is developed and in which different soils react to natural conditions is a subject of much investigation by engineers. In recent years, geomorphologists have been taking an approach to slope problems much more akin to this than the traditional study of slope-profile development. In reality both types of investigation are required.

Rock masses need to be examined in a similar manner. However, in some respects rock mechanics is even less developed than soil mechanics. This is partly because of observational problems, the time taken by a rock mass to respond to changes in stress and also because of practical difficulties in testing when much larger stresses are needed. There are other complications; whereas soils are relatively homogeneous and isotropic on a macro-scale, this is rarely true of rocks. Cleavage, stratification and folding all introduce anisotropy, and joints of varying size make rocks much less homogeneous than might be thought. Indeed, micro-fractures in rocks, around mineral grains for instance, may be critical in determining the ultimate strength of a rock. Pore-water pressures are still important but again difficulties

Figure 2.7A: A generalized view of a block of soil. The pore spaces may be filled with air only, when the load (normal stress) is taken by the intergranular contacts; or they may be filled with water when the load is taken by the water with a corresponding pore-water pressure. Usually a soil is partially saturated but failure can occur if the pressures approach that of the normal load. Intergranular contacts result from friction at asperities but cohesion comes from clay-mineral chemical bonding at a very small scale.

B: The peak strength of a soil can be achieved for a densely packed soil (curve i); after failure, the residual value results but for a loosely packed soil (ii) no peak occurs.

C: A typical plot achieved from a test to obtain the shear strength of a soil; as the normal stress (i.e. load) increases, so does the shear strength.

arise because of long relaxation times and the resulting problems of measurement. Creep of rock is certainly happening wherever there are free rock faces as part of a process that has been called pressure release. Failure of cliffs does occur, though it is very difficult to predict 'when' or 'how big'. It seems clear that geomorphologists need to look at these problems in order to be able to understand such things as valley development better than they do. Scheidegger (1970) gives a more comprehensive treatment of this topic.

Behaviour of fluids

A previous section showed that the mechanical behaviour of a Newtonian fluid (such as water) can be represented by a stress–strain diagram (Figure 2.7B); here the response of the substance is to flow when a *shear* stress is applied and the larger this stress the greater the rate of flow or deformation. The gradient of the curve is the *dynamic viscosity* of the fluid and is related to its molecular composition. Gases have molecules vibrating more rapidly than water and consequently their dynamic viscosity is less. Substances possessing larger molecules than water (such as syrup) generally have higher viscosities but this can be reduced by heating. Viscosity is therefore temperature-dependent and is quoted in poiseuilles at a given temperature. The poiseuille, designated Pl in SI units, is equal to $1 \, N \, s \, m^{-2}$ or $kg \, m^{-1} \, s^{-1}$. Frequently, it is more convenient to measure viscosity as *kinematic viscosity*. This is the dynamic viscosity divided by the density of the fluid, the units being $m^2 \, s^{-1}$. The use of this will be discussed in the next chapter in relation to the friction encountered when fluids flow in pipes or channels.

Just as solid objects exhibit friction when they slide over each other, so does a fluid sliding on a solid. This is an important phenomenon in both meteorology and stream flow. The friction is called drag and can be related both to the nature of the river bed (for instance) and to the nature of the fluid motion. Allied with this is the way in which particles become entrained in the fluid (the physics is similar for both air and water); the fluid flow has a certain amount of energy that can be dissipated by providing the shear stress to allow motion and, if sufficient, to move material on the bed or even to erode the bed. Similarly, the energy level—which specifies the 'critical boundary shear stress' for initiation of particle movement—also controls deposition and sedimentary structures. Chapter 3 develops this further; a more detailed analysis is given by Allen (1973) and a classic work on wind-blown sand is that of Bagnold (1941).

Ice flow

The last type of stress–strain relationship to be considered is the flow of ice. Figure 2.6B has already suggested that ice behaves in a way intermediate between that of a perfectly plastic body and a Newtonian fluid. This is indeed the case; when ice flows a viscosity can be calculated but at very low stresses creep only will occur. At some levels of shear stress, in the region of 0·5 to 1 bar, flow starts to occur and this is the equivalent of the yield point of a perfectly plastic body. With ice, however, there is strictly no 'point' at which this occurs. On the other hand ice, like plastic substances, will fracture, as demonstrated by the formation of crevasses. The form of the line shown for ice in Figure 2.6B is

$$\dot{\varepsilon} = B\tau^n \qquad (2.17)$$

where $\dot{\varepsilon}$ is the strain rate, τ is a shear stress as before, B is a temperature-dependent constant and n a power usually in the range 2 to 4. Note that if $n=1$ we have a Newtonian fluid, and if n is equal to infinity we have perfect plasticity. Using this relationship (Glen's flow law), it is possible to calculate the maximum depths of crevasses (d) on low-angle slopes by using the equation.

$$d = \frac{2\dot{\varepsilon}^{1/n}}{\rho g B} \tag{2.18}$$

where ρ is the density of ice and g the acceleration due to gravity. The numerical values obtained using equation 2.18 agree well with observed values of about 30 m. At depths greater than 30 m the ice flows to inhibit crevasse formation. Another way of looking at this behaviour of ice is to consider a block 30 m high. This will produce a basal stress by virtue of its mass, producing flow to give a strain rate of about 10 per cent per year. Other aspects of the flow law and ice behaviour are discussed in chapter 3.

Flow through porous media; permeability

Another type of flow important in many geomorphological situations is that often known as flow through porous media. Again there is a resistance to flow, this time provided by the intricate network through which water must pass to move, for example, from the ground surface to the water table; lateral and upward movements are similarly affected. Thus, hill-slope throughflow, formation of duricrusts, some aspects of glacial erosion and slope failure all involve this phenomenon. If a material (such as a clay) provides a resistance to fluid flow, high pore-water pressures can result. As has already been seen, this may upset the stability of a slope and so it is important to be able to measure the ease of water movement.

In simple terms, the permeability of a substance (k) is related to the discharge (Q) by:

$$k = Q\mu / A(\Delta P / L) \tag{2.19}$$

where μ is the dynamic viscosity of the fluid, A is the cross-sectional area of the specimen, and ΔP is the applied pressure difference (i.e. the driving force) across the length L. As water is frequently dealt with, the viscosity term is often disregarded.

The units of permeability are $m\,s^{-1}$, though it is common to find another unit (the darcy) used, especially in oil engineering. Its equivalent value can be obtained by substituting centimetre/gram/second units in equation 2.19. Another way of expressing the permeability is as in Darcy's original experiments:

$$Q = \frac{k(h_1 - h_2)A}{L} \tag{2.20}$$

where the pressure difference per unit length $(h_1 - h_2)/L$ is the hydraulic gradient (i) and is related to the pressure (sometimes called 'head') differences in a permeable bed between points distance L apart.

Permeability is thus a frictional effect, a resistance to flow, and is measured as a bulk property of the rock or soil. Various refinements of Darcy's law have been made, especially in order to quantify the 'tortuosity' of passages between particles, a complex relationship involving size distribution, shape, orientation and packing of particles. These factors need to be combined with permeability to give adequate understanding of frost-heave phenomena (Chapter 6) and it has already been pointed out that periglacial gelifluction is related to effective stresses and hence permeability. An appreciation of periglacial activity in the British Isles can thus be seen to require a knowledge of a range of material properties.

Some observational problems

It should be clear by now that modern process geomorphology is closely related to physics and chemistry in that the components of the landscape are governed by their laws. On the other hand, the difficulty of observation of many complex processes more frequently places the geomorphologist near to the engineer in his approach to problems concerning slope stability, water flow and so on. Indeed, our understanding of geomorphological processes owes a lot to engineering theory. Yet it should be remembered that the engineer is frequently concerned with relatively short-term answers (i.e. expediency) while the geomorphologist needs to look at the landscape with an eye on its long-term evolution. Climatic change is one obvious way in which time perspectives may differ. Chandler (1970), comparing some present-day stable slopes on Lias Clay in the East Midlands of England with relict failed slopes, concluded that past (probably periglacial) conditions had resulted in failure on rather lower slope angles than occur at present. In general, however, very little is known about such changes in the effectiveness or rate of action of processes which reflect changes in energy available through time.

As an example, it has been shown by Grove (1972) that climatic changes in central Norway between 1550 and 1880 AD produced an increase in the frequency of rockfalls, avalanches, floods and mudslides compared with both previous and subsequent rates. Now this must reflect variation in energy available rather than in the processes *per se*. In some cases the amount of work done (the 'work rate' in terms of fluvial bed transport) may increase with increasing energy, transport of stream solutes being one such case. Quite frequently there is an energy threshold below which there will be no activity: lack of bed-material movement below the critical boundary shear stress is a good example. The overall geomorphological efficiency can therefore be gauged by statistical methods. On the other hand, the frequency of operation of some processes may be very low and as such might even be regarded as geomorphologically insignificant. It may mean merely that, under present conditions, the energy levels (such as groundwater amounts, pore-water pressures) do not achieve the thresholds at which certain processes are triggered. It may need a combination of effects to achieve this. It seems likely that one direction of progress in geomorphology in the future will be towards a better understanding of such ephemeral processes. In essence, this is a search for critical stress levels in the rocks and soils of the landscape.

3

The nature of fluid motion

C. Embleton, J. B. Thornes and A. Warren

The flow of matter is the response of materials to stress. As explained in Chapter 2, an average stress is the force divided by the area of the body over which the force is being applied, and strain is a measure of the deformation of a body. We define strain as the change in the length of any line element within a body divided by the original length of the line element. If we drop an eraser, a ball of clay, a cube of halite, or a cubic centimetre of honey on the floor, their differences become evident when they hit the floor. The eraser rebounds and bounces about, the soft clay sticks, the halite fractures and the honey spreads out slowly. These differences of behaviour illustrate elastic, plastic, fracture and viscous deformation. Elastic deformation is completely recoverable; viscous and plastic substances cannot reverse the deformation. Viscous substances respond to a shear stress of any magnitude other than zero by flowing and the rate of deformation (flow) depends on the magnitude of the stress. In plastic substances a threshold strength has to be passed before deformation occurs. In this chapter, we shall consider the flow of the three most important fluids in the field of geomorphology—water, ice and air.

Water flow

Viscosity

If we imagine a viscous substance being deformed, a plot of deformation rate (strain rate) against stress will yield a straight line of constant gradient. Water is a typical Newtonian viscous fluid, and has a dynamic viscosity (p. 36) of about 0.01 poise ($0.001 \text{ kg m}^{-1}\text{s}^{-1}$). (Viscosities for some other materials are given in Table 3.1). Viscosity may also be considered at the molecular level as the interference per unit time between parallel layers of fluid flow. This is the *kinematic viscosity* and, as noted in Chapter 2, is equal to the dynamic viscosity divided by the mass density. The state or behaviour of water in open channel flow is governed basically by the effects of viscosity relative to the inertia of the flow. Inertial resistance is

Table 3.1 Dynamic viscosity of some geomorphological materials (in poises)

Glacier ice	$10^{13}-10^{14}$
Flowing lava	$6.5 \times 10^3 - 7.5 \times 10^3$
Mudflow	$2 \times 10^3 - 6 \times 10^3$
Solifluction	10^3
Debris flow	7.6×10^2
Water (20°C)	10^{-2}

Sources: Leopold, Wolman and Miller (1964); Johnson (1970)

a function of mass density; the larger the mass density of a fluid, the greater the force required to produce a specified acceleration in a specified volume.

Laminar and turbulent flow

For fluids of very small viscosity, inertial forces are much greater than viscous forces. The ratio of inertial forces to viscous forces determines the type of flow that occurs. The inertial forces increase with velocity. If the velocity is increased, it will be observed that a streak of dye in the water, coherent, straight and practically of constant width at low velocities, becomes highly distorted. The flow under the first condition is said to be *laminar*; under the second, *turbulent*. In laminar flow, transfer of momentum from the faster to the slower moving parts of the flow occurs at the molecular level.

The ratio of inertial forces to viscous forces determines whether the flow is laminar or turbulent, and is represented by the *Reynolds number*

$$R = \frac{vL}{\upsilon} \qquad (3.1)$$

where v is the velocity (m s^{-1}), L is a characteristic length, such as the depth of flow (m), and υ is the kinematic viscosity $(\text{m}^2\,\text{s}^{-1})$. Being dimensionless, the Reynolds number can be a measure of any size of system and is independent of the units of measurement. When R increases, turbulence sets in; when R is small, frictional forces prevail and flow is laminar. The zone separating the two types of flow is rather large in streams compared with pipes, for which the measure was originally designed. The lower critical Reynolds number depends on channel shape and is in the vicinity of 500–600. The upper boundary can be regarded as about 2000. In most open channels laminar flow occurs rarely (except near the boundary, see p. 45, and on hillslopes, as described on p. 50). The fact that the surface stream appears smooth and glossy does not indicate that flow is laminar.

With small depths the value (R_{crit}) at which transition flow sets in may be very low, especially where the bottom is steeply sloping and rough, and gravity waves develop. The ratio of flow velocity to the speed of movement of gravity waves is a characteristic of the state of flow. It is represented by the *Froude number* (F). Since a gravity wave is propagated at a celerity $C = \sqrt{gh}$ where g is acceleration due to gravity and h is depth, then

$$F = v / \sqrt{gh} \qquad (3.2)$$

Gravity waves which occur in shallow water result from a momentary change in the local water depth, which creates waves that exert a weight or gravity force. When $F > 1$, the flow is said to be *supercritical*; when $F < 1$, it is *subcritical*. The two criteria, the Reynolds number and the Froude number, are used to discriminate between various types of flow so that the class of flow is determined for a given depth and velocity. The distribution of types is shown in Figure 3.1.

Channel flow

Channel flow may be classified in terms of spatial and temporal variations as well as instantaneous conditions at a point. It is measured as the product of the average water speed and cross-sectional area:

$$Q = Av \qquad (3.3)$$

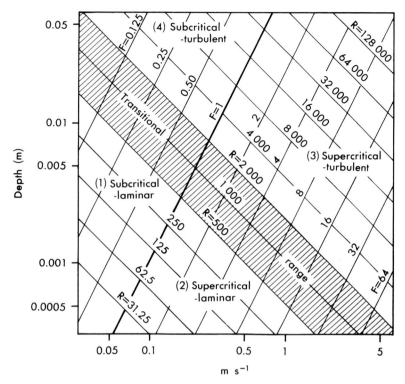

Figure 3.1 Class of flow according to velocity and depth (based on Ven T. Chow, 1959).

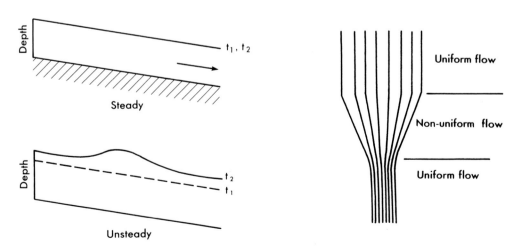

Figure 3.2 Diagram to illustrate flow types in time and space.

The flow is said to be continuous if, for a set of sections

$$Q = A_1 v_1 = A_2 v_2 = \ldots A_n v_n \qquad (3.4)$$

where the subscripts represent the different measuring stations. The property of continuity is very useful and equation 3.4 is the equation of continuity for steady flow. If the measured velocity at a station remains constant over the period of observation it is said to be *steady*; if it fluctuates, the flow is said to be *unsteady*. If a reach is observed at an instant in time and it varies in velocity from place to place then the flow is *non-uniform*, whereas a spatially constant velocity characterizes uniform flow. These different types of flow are illustrated in Figure 3.2. For the purposes of engineering, spatially varied flow and unsteady flow may be regarded as varying gradually or rapidly (Chow, 1959). This enables simplifications to be made in the treatment of complex flow conditions. The flow of water in rivers is usually unsteady and non-uniform, but if change with time is slow, as in very big rivers, unsteady flow may be approximated by steady flow.

   Unsteady flow is very difficult to model analytically. In stream channels it is recognized by water waves which are propagated through the channel. The most familiar of these are flood waves. Indeed a water wave in a river channel generally means any change of discharge, velocity or depth with time. In river channels only two types of unsteady flow are generally considered: (*i*) discontinuous, or rapidly varying unsteady flow such as progressing surges, bores and depressions and (*ii*) gradually varying unsteady flow, such as progressing long waves.

   The stability of flow depends on channel shape as well as flow conditions. Using the Manning resistance equation (equation 3.9), the *Verdnikov number* yields the relationship between depth–width ratio and the Froude number. The stability criterion is then given as

$$F < 3/2 + 3(h/W) \qquad (3.5)$$

where F is the Froude number, h the depth and W the width. When this criterion is exceeded the flow is unstable and perturbations are likely to grow in amplitude (Liggett, 1975). For flow over a plane under the same assumptions and for an infinitely wide channel the Froude number must be lower than 1·5 for stable flow.

Turbulence

The general assumptions made so far are that the main flow is down the channel and that the water can be treated as large fluid 'packages'. It is also implied that lines representing average velocity lie more or less parallel to the banks and the formulae mentioned above refer to time averages. Superimposed on these time averages are other less general flow types, representing complex, secondary motions. In recent years the development of the hot-film anemometer (Blinco and Simons, 1974) has enabled the velocity in the water to be measured in three coordinate directions that match the older comparable investigations of turbulence in the atmosphere. Some fluctuations of velocity at different depths in the channel are shown in Figure 3.3. Kalinske (1943) carried out measurements at three levels in rivers and examined the frequency of deviations of velocity from the average. He found that the magnitude of the fluctuations increases from the surface downwards. This means that flow is more irregular near the bed and that the relative intensity of turbulence at different depths is similar for different sections, though the spread of values is quite large. Kalinske concluded that the relative intensity of turbulence depended on the roughness of the bed and on the Reynolds number. Near the actual bed Einstein and El Samni (1949) used pressure rather than velocity fluctuations as a measure of turbulence and this is true of recent work. Their results show that pressure distributions near the boundary must be skewed rather than normal.

The turbulent motion in streams varies in scale and is usually evaluated in terms of the autocovariance of the velocity or shear stress. Prandtl (1925) argued that in turbulent flow there are small bodies of liquid that have a fairly independent motion over a certain distance, until they lose their identity by mixing with the surrounding turbulent medium. This representative distance is called the mixing length. Yalin (1971) theoretically derived the autocovariance function of the turbulence process which represents the fluctuations caused by the largest eddies. He found high autocorrelations at lengths equal to six times the flow depth and these were interpreted as being partially responsible for pools and riffles on the bed (see p. 44).

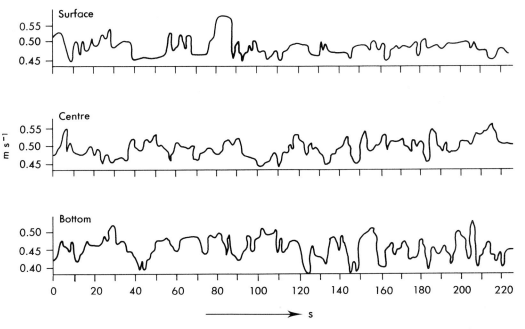

Figure 3.3 Fluctuations of velocity at different depths in a channel (after Blinco and Simons, 1974).

A most important group of secondary currents are the cross-currents which occur in river bends. Secondary currents have also been observed in straight as well as curved channels so the basic instability of turbulent flow is probably involved. Yalin argued that the behaviour of the largest macroturbulent eddies would be systematically affected by a permanent discontinuity in the channel, and this would occur in the same and reverse kind of disturbance, occurring alternately at equal interval lengths along the direction of flow. If turbulence is not fully developed, the wavelength of this periodicity will depend on roughness and the Reynolds number. The expected pattern of secondary circulation under this model is shown in Figure 3.4. Beneath it is the pattern observed by Hey and Thorne (1975). The latter should be compared with the model proposed by Wilson (1973). He suggested that it is possible to rationalize the single-spiral flow at the meander apex with a double-spiral condition at the point of inflexion. In order to achieve this, secondary flows are dissipated and new

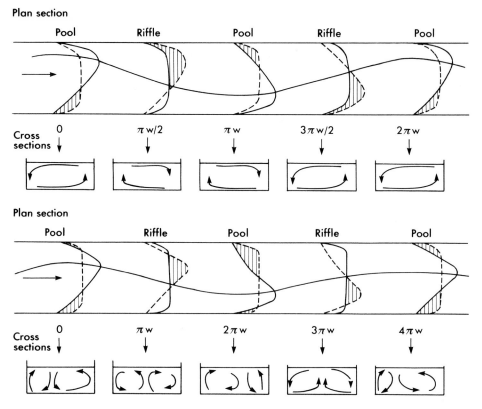

Figure 3.4 The pattern of flow in bends according to different authors; *w*=channel width, π=3·14159...

ones generated at the apex of the meander bend. The results obtained by Hey and Thorne show a twin-spiral system with convergence of surface flow at the meander apex. This means that on either side of the bend there must a zone of zero secondary circulation where old cells are dissipated and new ones generated. Bridge and Jarvis (1976) also report the existence of a second smaller spiral on the bend in addition to a single spiral over the whole point bar. The distribution of secondary currents has important implications for the distribution of transverse water-surface slopes and of pressure and shear on the bed. These are discussed in Chapter 7. The initiation of cross-currents in straight channels has also been discussed by Einstein and Li (1956), Einstein and Shen (1964) and Shen and Komura (1968).

The boundary layer

Returning now to the cross-sectional distribution of velocities (Figure 3.5), it will be observed that the velocities are higher in the centre of the flow and decrease towards the boundary. In other words, a velocity gradient is set up at right-angles to the direction of macroscopic flow. Water in immediate contact with a rigid body adheres to it and is therefore stationary relative to the surface. If a plate is set vertically in the middle of a stream parallel to the flow, the velocity of flow increases from the surface outwards, very rapidly at first, then more slowly, until the velocity of the free-flowing water is attained at a certain distance from the

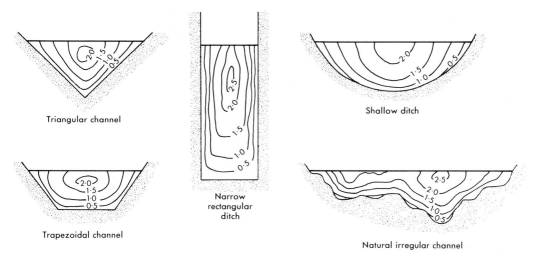

Figure 3.5 A typical pattern of cross-sectional velocities (in feet per second).

surface. The zone in which the flow is appreciably retarded by the viscosity of the water and friction against the surface is known as the boundary layer.

The flow in the boundary layer may be either laminar or turbulent. If the flow is laminar and uniform and the depth of the boundary layer is known, the effect of boundary roughness on the velocity is fairly easy to calculate. Beyond the boundary layer for steady uniform flow, there will be no variation in velocity and the form of the velocity profile will be essentially parabolic.

As the velocity increases, the inertial forces acting on the boundary layer become so large relative to the viscous forces that turbulence sets in. Under these circumstances (without sediment) the distribution of velocity becomes approximately logarithmic with depth. The turbulent boundary layer is divisible into three sub-layers with quite different flow structures, the laminar sub-layer, the buffer layer and the fully turbulent part of the boundary. The laminar sub-layer is next to the bed and banks and is very thin compared with the other two. The thickness of this sub-layer (δ_{sub}) is given approximately by

$$\delta_{\text{sub}} = \frac{5v}{v*} \text{ where } v* = \sqrt{\frac{\tau_0}{\rho}} \tag{3.6}$$

That is, it is related to the velocity (v) and the boundary shear velocity ($v*$); the latter in turn is a function of the boundary shear stress (τ_0) and the density of the water (ρ). The boundary shear stress is the pull of the water on the wetted area, equal to the weight of water multiplied by depth and channel slope. Three different types of flow are normally defined. When the bed roughness projections (K) are less than δ_{sub}, the flow is *aerodynamically smooth*; with K up to 14 times δ_{sub} the flow is *transitional*, and with K greater than 14 times δ_{sub} the flow is said to be *fully turbulent*. These results are for sand-grains and somewhat different figures may be quoted by other workers. Einstein (1950) and Keulegan (1938) define fully turbulent flow when $K > 5\delta_{\text{sub}}$, where δ_{sub} is the theoretical thickness given by

$$\delta_{\text{sub}} = 11 \cdot 6\, v/v* \tag{3.7}$$

Sometimes the boundary layer separates from the boundary and enters the flow, being separated from the walls by separation bubbles or 'vortices' (Fig. 3.6A) which are important for erosion. Separation occurs whenever a fluid stream of suitably large Reynolds number encounters a sufficiently adverse pressure gradient or expands over a sharp edge. In three dimensions the rollers and vortices produced by the separation phenomena turn out to be quite complex (Allen, 1970a). The particles of water beyond the separation point actually turn backwards as shown schematically in Figure 3.6B. The layer of discontinuity between the two flows is rolled up to the accompaniment of large and small eddies. The tendency for the boundary layer to separate is greater when the boundary layer is laminar than when it is turbulent, because in turbulent flows there is a much more intense interchange of fluid between adjoining layers with different velocities.

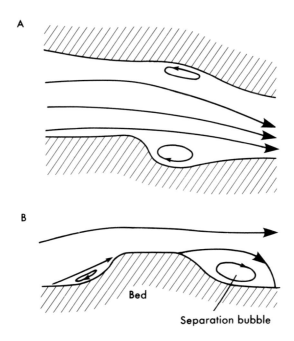

Figure 3.6 Vortices produced by flow separation (after Allen, 1970a).

An important consequence of the separation is the occurrence of *form resistance*, which may be thought of as macroscopic roughness. The resultant of the various normal forces acting on the boundary layer creates drag and dissipates energy. Streamlining (by erosion) reduces this drag. Sundborg (1956) claims that it is 'justifiable to regard the flowing water in its entirety as a fully developed turbulent layer, of a thickness practically the same as the depth of water, and the morphological processes as determined by the interaction of the bottom surface and the water's state of flow—the variations in the state of turbulence, the viscous shear and the form resistance being dependent on the shape of the bottom, and the grain size of the material there.' When flow occurs in an open channel, resistance is encountered by the water as it flows downstream and this resistance is counteracted by the gravitational forces acting on the body of water in the direction of motion. A uniform flow will be developed

Table 3.2 Typical values of the Manning roughness coefficient for natural channels

Streams	Min.	Normal	Max.	Flood Plains	Min.	Normal	Max.
Small, clean; Lowland, straight	0.025	0·030	0·033	Pasture, no shrubs, short grass	0·025	0·030	0·035
Small, weedy; deep pools	0·075	0·100	0·150	Cultivated mature field crops	0·030	0·040	0·050
Small mountain streams, cobbles with large boulders	0·040	0·050	0·070	Shrubs, medium-dense with foliage	0·070	0·100	0·160
Large streams (top width >33 m)	0·025	...	0·060	Heavy woodland with flood stage reaching branches	0·100	0·120	0·160

Source: Ven te Chow (1959)

if resistance is balanced by gravitational forces. About 1769, Chezy developed an equation using this proposition, relying on the assumption that the flow per unit area of stream bed is proportional to the velocity and that the effective component of the gravitational force causing the flow must equal the total force of resistance by uniform steady flow. This leads to the equation:

$$v = C\sqrt{rs} \tag{3.8}$$

where v is velocity, r is the hydraulic radius $= A/(2w+d)$ where A is the total area and w and d are width and depth; s is channel slope and C is the Chezy resistance factor. A similar development is the Manning equation:

$$v = \frac{1}{n} r^{2/3} s^{1/2} \tag{3.9}$$

where n is the Manning roughness coefficient. This is a complex though widely used coefficient. It is highly variable and depends on many factors such as the size and shape of grains on the bed, vegetation, channel irregularity and stage and discharge. Some characteristic values are given in Table 3.2 from Chow (1959).

These equations reflect our continuing ignorance of the turbulent process. They were first derived for uniform, steady flow, although their use has become common in non-uniform and unsteady flow. The primary difficulty in predicting the frictional resistance still lies in estimating the resistance coefficient for a natural situation. Figure 3.7 shows the variations in Manning's n obtained by Baltzer and Lai (1968). The large scatter at low Reynolds numbers is to be expected since high accuracy in the data is required under this condition. Unfortunately any error in the resistance coefficient leads to an error in velocity or flow of the same magnitude. If some of the flow occupies part of the floodplain this will also complicate the assessment of frictional resistance (Toebes and Sooky, 1967).

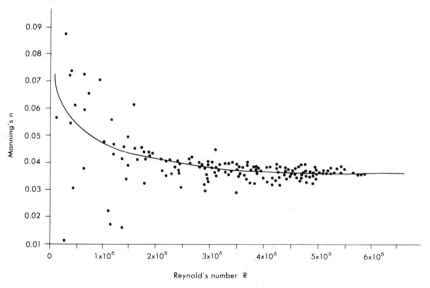

Figure 3.7 Variations in Manning's n for a natural channel (Baltzer and Lai, 1968).

Another widely used, and in many respects more satisfactory, dimensionless friction coefficient is the Darcy–Weisbach factor (f);

$$f \propto grs/v^2 \qquad (3.10)$$

(notation as in 3.8). This expression was developed for pipes but is also used for open channel flow. Leopold, Wolman and Miller (1964) relate it to particle size in the expression

$$\frac{1}{\sqrt{f}} = 2 \log \frac{d}{D} + \text{constant}$$

$$\qquad (3.11)$$

where D is the diameter equal to or larger than 84 per cent of the bed size material and d the depth of flow. In the Brandywine Creek, Wolman (1955) found that the constant equals unity. A plot of f against the Reynolds number shows that, under tranquil flow, the Darcy–Weisbach factor plots as a linear function of R and in transitional or fully turbulent flow the rate of decrease of f with increasing R is much smaller, under some conditions becoming negligible.

Energy in a stream channel

The total *energy* at any particular point in the channel is determined essentially by the elevation, the pressure head and the velocity head. The pressure head is due to the depth of water, $d \cos \alpha$, where d is the depth and α the channel bed slope, and the velocity head is $v^2/2g$. For channels with a very low slope, the total energy at a section is:

$$H = Z + d + kv^2/2g \qquad (3.12)$$

where k is a coefficient allowing for an unequal velocity distribution and Z is the height of the bed of the channel above a given datum. This energy has to remain constant down channels, so that

$$Z_1 + d_1 + v_1^2/g = Z_2 + d_2 + v_2^2/g \qquad (3.13)$$

which is the Bernouilli energy equation.

Specific energy is the energy per kilogram of water at any section with respect to the channel bed. This is therefore only a function of flow depth. The specific-energy curve is a plot of specific energy against depth; for a channel of low slope it appears as in Figure 3.8. There is a point at which specific energy is at a minimum. When the depth of flow is greater than this critical depth, the velocity of flow is less than the critical velocity for the given discharge, so that the flow is sub-critical. When the depth of flow is less than the critical depth, the flow is super-critical. Thus critical flow is also that state of flow at which the specific energy is at a minimum for a given discharge. If a channel suddenly changes slope (and velocity) to a much flatter one, the specific energy conditions may change quite rapidly. Such a situation is frequently marked at the water surface by subduction of the upstream water beneath a standing wave called the *hydraulic jump*.

Flow on hillslopes

In shallow flows such as occur on hillslopes, the conditions are somewhat modified. Efforts have been applied in three particular directions. First, attempts have been made to compare the flow on hillslopes with that in river channels. Secondly, there has been an effort to understand the effects of rainfall on the flow characteristics, and finally there have been investigations on the distances over which the flow remains stable. The latter may have particular

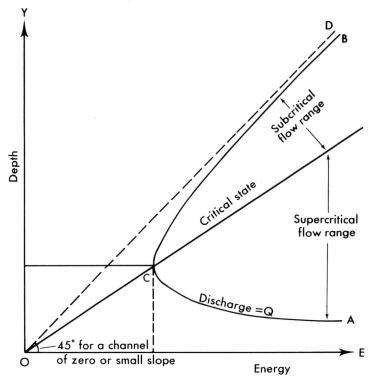

Figure 3.8 The specific energy diagram for a flow in an open channel.

geomorphological importance for it is inferred that the form properties adjust directly to the flow conditions. The latter view has been advocated for many years (e.g. King (1953) and Schumm (1962)) but is only recently being fully evaluated.

Early work in this area was initiated by Horton and others (Horton, Leach and Van Vljet, 1934), who studied the basic behaviour of laminar sheet flow essentially through the rating curve:

$$Q = kd^m \tag{3.14}$$

where Q is the discharge, d the flow depth and k and m constants for a given set of conditions. They argued, using the Manning equation, that for fully turbulent flow m should be equal to 5/3 and, for fully laminar flow, equal to 3. The most extensive theoretical investigation was that carried out by Keulegan (1944), who employed a mathematical analysis to obtain a complete solution to spatially varied discharge over a sloping surface. Comparable studies have been carried out by Woo and Brater (1962) and recently by Muzik (1974). Emmett (1970) carried out both laboratory and field experiments and extended his work to cover the likely effects of overland flow on hillslope geometry. In the laboratory, the Reynolds numbers of flows were found to be in the range between 1500 and 6000. Flows over smooth surfaces, both turbulent and laminar, were found to have exponents in the rating equation 3.14 very close to those predicted by Horton *et al*. With rough surfaces the values of m were less than those expected theoretically and this was attributed to a more complex and more rapid increase in depth of flow due to roughness. Variations of flow depth with different slope conditions were also found to be quite strong. With increasing slope, depth approaches infinity hyperbolically. The Stanton–Moody diagram, plotting Darcy–Weisbach friction against Reynolds number, showed a rapid drop in resistance between 2000 and 5000. The data plot higher than expected values for smooth flow in both parts of the curve, but this is attributable to the failure to make the flume completely smooth.

In the field, greater depths of runoff were experienced. The higher f values reflect both the magnitude of relative roughness and the character of runoff, but the f values were poorly correlated with ground slope. A most interesting result was that the downslope change in roughness was approximately zero, indicating that relative roughness must be increasing

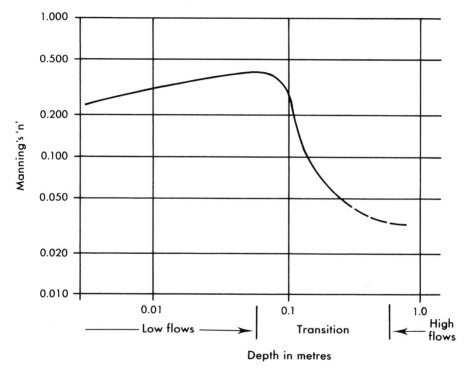

Figure 3.9 The effect of vegetation on roughness (as measured by Manning's *n*) according to the experiments of Palmer (1946).

in that direction, perhaps due to variations in micro-relief. Field data showed a tenfold increase in resistance on natural plots when compared with laboratory surfaces.

Most work shows that the friction factor is appreciably larger when the runoff is being affected by falling rain, though Savat (1977) found that this effect diminishes when the degree of turbulence of the flow increases or the slope becomes steeper. Emmett (1970) used the term *disturbed flow* to cover conditions with falling rain. The most important effect of falling rain appears to be to retard the flow and increase the depth for a given discharge. The increase in depth in Emmett's experiments was appreciably above the 17 per cent measured by Parsons (1949): 50 per cent for a Reynolds number of 200 to about 65 per cent for a Reynolds number of 1000. Here, the isolated effect of the type of artificial rainfall used in the experiments was approximately double the friction factor over that for flows without rainfall.

The surface roughness varies enormously when vegetation is involved. Parsons's experiment compared measured depths and those expected on the basis of theory. The ratios were 1·0 for a smooth surface, 1·5–2·8 for bare soil and 10·00 for bluegrass. The influence of the shape of vegetation on open channel flow was demonstrated by a series of laboratory experiments conducted by Palmer (1946). Figure 3.9 shows a typical result. In the low flow range the roughness is consistently high since the vegetation cannot bend and yield, due to the low velocities corresponding to the low depths. In the transitional range, the hydrodynamic drag acting on the individual elements of vegetation begins to cause bending, resulting in a reduction in roughness which should reach a minimum in the high flow range where the high velocities have bent the vegetation completely. The apparent roughness measured by Palmer depends not only on the vegetation characteristics but also on the additional loss of energy induced by the rhythmic motion or vibration of the usually flexible vegetation elements. In the case where the vegetation protrudes through the water layer and consists of rigid elements such as trees, the roughness of the ground between the trees must also influence the overall apparent roughness, as shown by Petryk and Bosmajian (1975).

Ice motion

By far the greater part of the ice on the Earth is glacier ice; this section will exclude other types of ice such as lake ice or icicles, whose crystallographic structure and behaviour differ somewhat from those of glacier ice. Glacier ice is a polycrystalline substance derived mainly from the consolidation of snow crystals. Its density is about $0·9 \, g \, cm^{-3}$ and the transition from 'firn' or old compacted snow, with a maximum density of about 0·82, comes about by gradual expulsion of air bubbles and increase in crystal size. The hard blue ice of glaciers is mostly air-free and relatively impermeable; it will often have taken several decades to form and the crystals may be up to several centimetres in length. Its hardness varies with temperature (1–2 at 0°C, about 4 at −40°C), and its crystals often show a preferred orientation, which is a secondary characteristic acquired through deformation and flow.

The behaviour of glacier ice under stress has been the subject of extensive research, both in the laboratory and in the field, and in the last 25 years there have been major advances in our understanding of this aspect of glaciology. The behaviour is complex, and may be studied on three levels: the deformation of a single ice crystal, the deformation of a polycrystalline aggregate of ice, and the flow of an actual glacier or ice sheet.

Ice crystals

Because of its tetrahedral atomic structure, ice is a substance that can crystallize hexagonally. The atoms are arranged in paired layers of hexagonal lattice structure, and deformation under stress takes place principally by gliding along these layers, parallel to the basal plane of the crystal. Only very low stresses are required to initiate this deformation, which Weertman and Weertman (1964) interpret in terms of dislocation theory. It is suggested that irregularities in the crystal structure, arising during uneven crystal growth, allow layers of atoms to glide over each other more easily than in a perfect crystal. With a constant applied shear stress, the rate of deformation (strain rate) increases initially with time but after a few hours settles down to a steady rate. This behaviour, known as creep, is analogous to the response of certain metals to stress.

Polycrystalline ice

Geomorphologically, the behaviour of an aggregate of ice crystals is of much more direct interest since this is the substance of which glaciers are composed. Because of the interlocking of ice crystals in an aggregate, the response of polycrystalline ice to stress differs somewhat from that of a single crystal. Laboratory studies by Glen (1955, 1958), Steinemann (1958), Rigsby (1960) and others since, have shown the existence of three distinct stages of creep. When a specimen is first subjected to stress, the strain rate initially decreases with time. This may be contrasted with the response of a single ice crystal, whose creep is not obstructed by adjacent crystals packed around it, as in glacier ice. As stresses in actual glaciers act over long periods, this first stage of 'transient creep' occurring over a few hours is unimportant and will not be considered further. After a few hours, provided the stress level is not much below 1 bar (10^5 N m^{-2}), the strain rate settles to a steady value ('secondary' or 'quasi-viscous' creep). If the stress is now increased beyond about 4 bar, the creep rate falls initially and then accelerates to a higher rate ('tertiary' creep), a change probably associated with recrystallization as the crystals become aligned in a more favourable direction for creep. The crystal sizes of samples subjected to these higher stresses are found to be smaller than those deformed by lower stresses.

Several processes contribute to creep, including gliding of individual crystals on the basal planes, recrystallization and the migration of crystal boundaries. Barnes *et al.* (1971) show that, at moderate stresses, creep processes are related closely to ice temperature. Below $-8°C$, the behaviour is dominated by basal gliding; between $-8°C$ and $-1°C$, creep is associated with a liquid phase at the grain boundaries and grain-boundary sliding; while near $0°C$, pressure melting and regelation (see p. 63) are the chief processes.

About 1955, Glen showed in a series of experiments that a power-flow law could closely describe the behaviour of ice under stress. The relationship proposed was of the form

$$\dot\varepsilon = B\tau^n \tag{3.15}$$

where $\dot\varepsilon$ is the strain rate (see p. 28) and τ the applied shear stress, B is a constant depending on temperature, and the exponent n is greater than unity. In his experiments, Glen (1955, 1958) used a range of shear stresses from 1 to 11 bar, and temperatures from $-0.02°C$ to $-12.8°C$. It should be noted that the stress range includes values much higher than those encountered in glaciers, that the tests were performed at temperatures slightly below to well below pressure melting point, and that for the lower stress levels, the tests were not continued long enough to reach a steady state. Over the range that he tested, Glen found values for the exponent n of about 3.2. Steinemann (1958) used a stress range of 0.7 to 20 bar, a temperature range of $-1.9°C$ to $-21.5°C$, and suggested that values of n increased with increasing stress, varying from 1.85 to 4.2. These and other experiments failed to give consistent values either for B (the constant sensitive to temperature) or for n. Part of this failure was because a steady state of deformation was not always achieved in the tests, and part because the samples of ice used in different experiments were structurally different. Nevertheless, it has become apparent that a simple power-flow law is not adequate to describe the rheology of ice over a large stress range. Meier (1960) and Mellor and Smith (1967) have proposed a two-term flow law of the form

$$\dot\varepsilon = A\tau + B\tau^n \tag{3.16}$$

to fit a range of data including not only laboratory tests but also measurements on actual glaciers. Some workers have recently looked more closely at the response of polycrystalline ice to very low stresses. All emphasize the great difficulty of testing its behaviour under small stresses because of the time-consuming nature of the experiments, especially at low temperatures. At $-10°C$, for instance, a stress of 0.5 bar produces a strain rate of less than 0.001 year^{-1}, and it will be apparent that even experiments extending over as long as a year may be much too short to give reliable results. Yet the vast bulk of glacier ice is being subjected

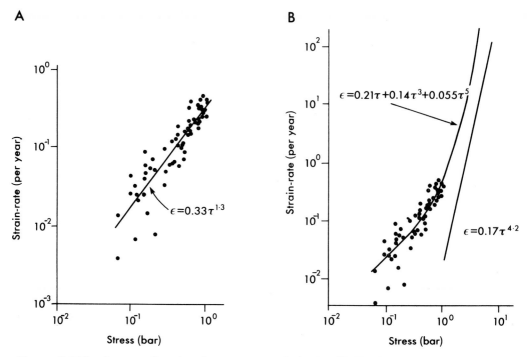

Figure 3.10A: A power-flow law for temperate glacier ice (Colbeck and Evans, 1973). **B**: Comparison with Glen's power-flow law with $n=4.2$. Also shown is a best-fit curve of polynomial type (Colbeck and Evans, 1973).

to shear stresses of 1 bar or less. Mellor and Testa (1969) conducted tests at 0.43 and 0.093 bar, with a temperature of $-2.06°C$ and found that a power law with $n=1.8$ represented a good fit. Colbeck and Evans (1973) recently reported the results of 90 experiments using compressive stresses of less than 1 bar, in which large blocks of glacier ice ($0.52 \times 0.13 \times 0.13$ m) were tested in a tunnel in the glacier from which the ice was obtained. This enabled the tests to be conducted at pressure melting point. Assuming a power-flow law, the best fit of the data by least squares gives $B=0.33$ and $n=1.3$, but there is in fact a wide scatter of points (Figure 3.10). A better form of flow law to fit the data is of the polynomial type suggested by Lliboutry (1969), in which

$$\dot{\varepsilon} = A\tau + B\tau^2 + C\tau^4 \qquad (3.17)$$

Colbeck and Evans find a good fit in the following expression:

$$\dot{\varepsilon} = 0.21\,\tau + 0.14\tau^3 + 0.055\tau^5$$

They argue that this form of flow law is superior to Glen's for glacier ice at pressure melting point and at realistic stress levels.

Glacier flow

Glaciers and ice sheets represent a temporary storage of water in solid form on the Earth's surface: they are integral parts of the hydrological cycle. It is estimated that three-quarters of the Earth's fresh water is at present locked up in glaciers, but it is water that is constantly in transport through the glacier system (Figure 3.11) between the input in the form of

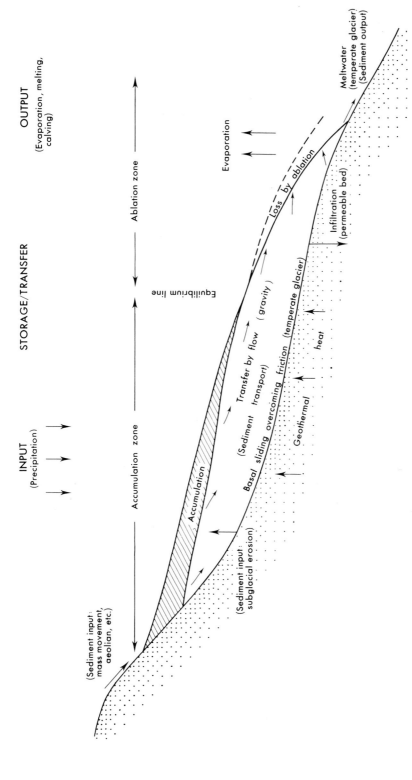

Figure 3.11 Glaciers as open systems. Flows of sediment and water through the system are generally indicated, but storage (e.g. sub-glacial sedimentation, englacial water bodies) is omitted.

precipitation and the output mostly in the form of meltwater. The energy that drives the glacier system is principally solar and gravitational: solar energy governs the precipitation feeding the glacier, and gravity is the force causing downhill movement. Geothermal energy also affects the glacier system by causing basal melting beneath some types of glacier (see p. 296). The energy of a glacier system is dissipated by friction with the bedrock floor, internal friction of the moving ice, the transport of rock debris, and by loss of heat externally.

Before dealing with the ways in which glacier ice moves over an irregular and immobile surface, it is useful to summarize some facts about surface and internal velocity distributions.

Surface movements

Rates of surface motion vary from zero (stagnant ice) to tens of metres a day. Highest values are associated with relatively short-lived surges, during which speeds of over 100 m day^{-1} have been recorded (e.g. by Thorarinsson, 1969). Rates of ice discharge cover an equally large range: Variegated Glacier, Alaska, 20 km long, has a discharge of about 0·2 m^3 s^{-1} at the equilibrium line; Jakobshaven Isbræ, in contrast, pours out over 400 m^3 s^{-1} on the west coast of Greenland, while the calving ice-shelves of Antarctica lose in total about 30 000 m^3 s^{-1}. Measurements of surface velocities are most accurately obtained by repeated theodolite or geodimeter survey of stakes in the ice, and less accurately (for short periods) by photogrammetry. Lack of bedrock outcrops in the case of ice sheets or large ice caps is a problem in establishing trigonometrical or photogrammetric control; traversing may then have to replace triangulation and, in the ultimate, recourse to repeated astronomical fixes may be necessary though, compared with triangulation, this method is relatively inaccurate.

In a glacier flowing down a channel of uniform cross-section and slope, velocity will be greatest at the equilibrium line separating the accumulation zone upstream and the ablation zone downstream, for here the volume of ice being discharged will be greatest. Downstream, as ablation progressively reduces the volume of flowing ice, velocities will diminish, approaching zero at the snout. Friction with the valley sides retards the motion towards the edges of a glacier, so that transverse velocity profiles are commonly parabolic, the centre-line velocities often being four or five times greater than velocities near the margins. Some glaciers exhibit very rapid transverse increases in velocity close to the margins, with little differential surface movement in the central core: this is termed plug flow or Blockschollen movement and is typical of many surging glaciers or glaciers below ice falls.

Changes in velocity along or across a glacier surface are of as much interest as absolute values of velocity. These changes or 'velocity gradients' are expressed as strain rates, which may be positive ('extending') or negative ('compressing'):

$$\dot{\varepsilon} = \frac{1}{l} \frac{\Delta l}{\Delta t} \tag{3.18}$$

where l is the initial distance between two markers, and Δl is the change in their separation over time t. Because strain rates for glaciers are so small, the unit of time is normally a year. To determine all the surface strain-rate components at a place involves setting up a diamond pattern of four stakes with one diagonal lying approximately along the line of flow, and a stake where the two diagonals cross. The sides and diagonals are measured with steel tape and the measurements repeated after any desired interval of time (a few months or a year, say). Nye (1959) gives details of the computation of surface strain-rate components from these data.

Vertical components of flow must not be ignored, though they are much more difficult to measure precisely. Velocity vectors have a downward component in the accumulation area and an upward component towards the snout in the case of a simple valley glacier flowing down a uniform gradient.

Temporal variations in rates of glacier flow have long been recorded. Evidence is accumulating to show that seasonal variations in flow are closely related to the amount of meltwater reaching the base of a glacier and helping to lubricate the sliding process. This applies only to temperate and sub-polar glaciers. In the case of the former, the ice throughout the glacier is at pressure melting point and meltwater exists at the base of the glacier throughout the year. Its greater abundance in summer (from ice melt, at the surface, or from summer rainfall) is likely to be the fundamental cause of faster motion at this time. Some sub-polar glaciers are frozen to bedrock in the winter or even for most of the year, but summer meltwater may find its way to the base for a short time and give rise to sudden acceleration of movement. Cold glaciers frozen to bedrock do not appear to exhibit seasonal variations in rate of flow. Short-interval variations in velocity over periods of hours, days or weeks, are also known, particularly near the margins of glaciers. One possible cause is a form of stick-slip motion at the ice–rock contact owing to irregularities or large boulders jammed here; meteorological conditions are also relevant, for periods of heavy rain or warm spells can add suddenly to the quantity of basal meltwater.

Internal movement

Englacial flow rates can only be measured by means of artificial tunnels (or, rarely, natural meltwater stream tunnels) or drill-holes. The latter must be cased and their deformation

Plate II Hanging glacier, Athabasca, Alberta. (C.E.)

by the surrounding flowing ice measured by inclinometer at fairly long intervals of time (usually a year or more). Examples of detailed surveys of englacial movement by drill-holes include the work on Athabasca glacier, Alberta (1966–67) (Plate II) and Lower Blue glacier, Washington State (1957–61) (Raymond, 1971; Shreve and Sharp, 1970); surveys using arti-ficially-dug tunnels include those on Skautbreen, Norway (McCall, 1960) and Meserve glacier, Antarctica (Holdsworth and Bull, 1970). Natural subglacial cavities have been util-ized beneath the Glacier d'Argentière, France, to provide extremely detailed records of basal

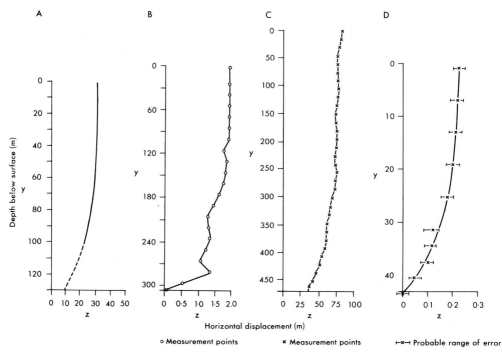

o Measurement points   × Measurement points   ⊢×⊣ Probable range of error

Figure 3.12 The deformation of vertical boreholes in four temperate glaciers: **A**: Jungfraufirn (Perutz, 1950), over 14 months; **B**: Malaspina glacier (Sharp, 1953), over 13 months; **C**: Salmon glacier (Mathews, 1959), over 1 year; **D**: Saskatchewan glacier (Meier, 1960), over 2 years. Note that the vertical and horizontal scales differ in each case. Direction of ice flow from left to right. Many of the minor irregularities in the curves for **B** and **C** may have no significance, being within the limits of instrumental error. In **D**, probable errors are indicated for each measurement, and the line is the curve of best fit.

ice movement (Vivian and Bocquet, 1973). Figure 3.12 illustrates some of the results obtained from boreholes. In general, these show that velocities decrease exponentially from the surface to the base of the glacier. It is important to note that most of our data on englacial movement relate to temperate glaciers, and that the accuracy of englacial velocity measurements and strain rates is far less than that of corresponding surface measurements because of the limited precision of inclinometry.

Flow laws for polycrystalline ice have already been described. Glen's power-flow law approximates very well to the internal deformation of actual glaciers. Rates of closure of tunnels have been measured, for instance on the Z'Mutt glacier, Switzerland, and Skaut-breen, Norway, yielding values for the exponent *n* of about 3. However, tunnels in the more actively moving ice of ice falls, such as that on Austerdalsbreen, Norway, have shown faster rates of closure than expected because of the large longitudinal stresses present in such parts of a glacier. Partly reflecting the fact that the constant B in Glen's equation is sensitive to temperature (a fall from zero to $-15°C$ decreases the flow rate by an order of magnitude), cold glaciers flow more slowly and tend to possess slightly deeper crevasses because the ice can support greater shear stresses and deforms less readily. In the upper layers of Meserve glacier, Antarctica (Holdsworth and Bull, 1970), a power-flow law with $n = 1·6 \pm 0·1$ provides a good fit to the data from borehole deformation, but near the base of the glacier, higher values of n ($= 4·3 \pm 0·5$) were needed to describe tunnel-closure rates.

Theoretical models of internal velocity distribution have been proposed by Nye (1965) for channels of varying cross-sectional form. Assuming a parabolic cross-profile for the channel, isothermal ice, a simple power-flow law and no basal sliding, Figure 3.13A shows the relative distributions of velocity and shear stress in a channel whose width is four times its depth. For comparison Figure 13B shows the actual distribution for two cross-sections of Athabasca glacier (Raymond, 1971) which differs markedly from Nye's predicted behaviour, principally because of the high rates of basal sliding that this glacier exhibits. A more complex model (Reynaud, 1973) allowing for basal sliding and for subglacial water pressure (Fig. 3.13C) shows much better agreement.

Another parameter that has to be taken into consideration is the longitudinal velocity gradient. The discussion of internal deformation so far has mostly assumed that the glacier is flowing down a channel of constant inclination. In reality, glaciers flow over irregular floors and, probably, accentuate irregularities by the erosional processes described in Chapter 8. Often, valley segments with relatively gentle gradients alternate with steeper sections down which the glacier pours as an ice fall (Plate III). At the foot of an ice fall, the reduced angle of slope will cause the glacier to move more slowly and therefore thicken or 'compress'; conversely, extension and thinning will occur where velocity gradients increase. Nye in 1952 described a simple model of the slip-line fields associated with longitudinal extension and compression based on a plastic flow law. Figure 3.14 shows a further development of these ideas (Nye and Martin, 1968) in relation to a glacier's potential ability to excavate rock basins along the floor of a valley. Using a more realistic power-flow law, the theoretical slip-line boundaries become zones of high shear-strain rate.

As more and better data on englacial movement become available, it is becoming possible to improve the models of glacier flow. Glen's power-flow law provides a good model of ice behaviour over a limited range of stress, though values of the exponent n vary from near 1 to more than 4. Whether more complex flow laws of polynomial type will provide a better approximation to reality is not yet known with certainty. Furthermore, our understanding of the flow of cold glaciers still lags behind that of the flow of temperate glaciers because of the continuing paucity of field data for the former.

Basal and side slip

The contribution of basal sliding to total glacier motion varies from zero (in the case of cold glaciers frozen to bedrock) to over 90 per cent (in the case of temperate glaciers with large quantities of basal meltwater under high subglacial pressure). The figures in Table 3.3, however, should be regarded as only approximate for three reasons: first, error occurs in establishing the relationship between sliding velocity at a point A on the glacier bed and velocity at a point B on the glacier surface, such that B is vertically above A; secondly, the figures refer to observations at the sites of boreholes and tunnels and may not be truly representative of conditions in the surrounding area of the glacier bed; and thirdly, basal sliding is irregular over time. Illustrating the last point, Vivian and Bocquet (1973) showed how

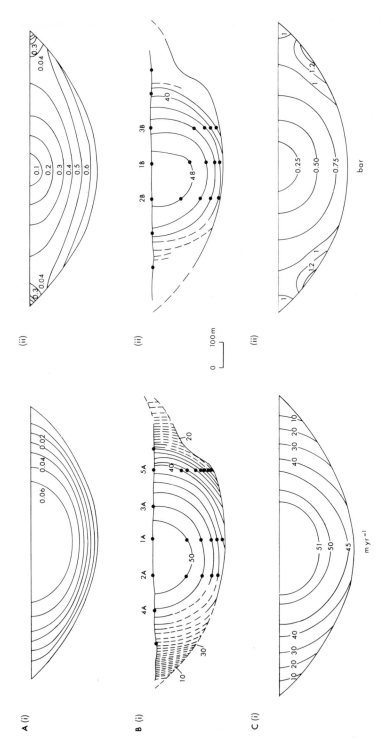

Figure 3.13A: Theoretical distribution of velocity (i) and shear stress (ii) for a parabolic channel whose width is four times its depth. Data are plotted in dimensionless form (Nye, 1960).

B: Actual distribution of longitudinal velocity (m year⁻¹) in two cross-sections of Athabasca glacier (Raymond, 1971).

C: Theoretical distribution of velocity (i) and shear stress (ii) for a channel as specified in **A**, according to Reynaud (1973).

the sliding velocity measured over 1 year in a cavity beneath the Glacier d'Argentière, France, ranged from zero to more than 9 cm hour^{-1}, including one period of complete halting for 3 days.

Irregular jerky movements also characterize the side slip of many glaciers. Surface side-slip measurements do not involve construction of boreholes or tunnels and data are therefore more readily obtainable, but rates of side slip are no useful guide to rates of basal sliding.

Plate III Austerdalsbreen, Norway. The main valley glacier (foreground) is fed by twin ice-falls from the Jostedalsbreen (plateau ice cap, skyline). Ice flow measured in the left-hand ice-fall varies from over 5 m day^{-1} at the top to about 2 m day^{-1} near the bottom where the flow is strongly compressive. Near the point of the photograph, velocity has further diminished to about 0·3 m day^{-1}. Thickening of the glacier and its increase in width below the ice-falls is associated with the excavation of a rock basin in the subglacial valley floor. (C.E.)

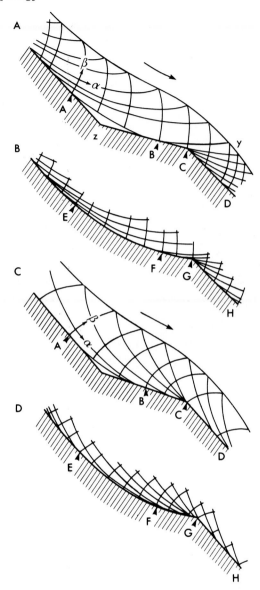

Figure 3.14: Slip-line fields (after Nye and Martin, 1968): **A**: compressing flow, before erosion; **B**: compressing flow, after erosion; **C**: extending flow, before erosion; **D**: extending flow, after erosion.

Figure 3.13B shows how high rates of basal slip for Athabasca glacier diminish to quite small values at the edge of the glacier surface.

The behaviour of cold ice in regard to basal and side slip is of great interest, though data are still insufficient to establish precise laws. Totally cold ice frozen to bedrock (as in the Camp Century borehole, Greenland) cannot exhibit differential movement at the actual ice–bedrock junction, though there will be shearing in the lowest ice layers which increases rapidly upwards away from the bedrock floor. There are, however, some cold ice masses whose bases are just at pressure melting point (e.g. probably beneath Byrd Station, Antarctica) where basal sliding is theoretically possible, although it may be inhibited by obstacles projecting up into the cold ice above; and there are some polar ice masses where cold ice is thought to overlie a substantial layer of temperate ice, beneath which basal sliding may take place as for wholly temperate glaciers. The thermal régime of glaciers thus has very important implications: its control of basal sliding will directly affect total glacier motion and there are also fundamental consequences for subglacial erosion and deposition that will be dealt with in Chapter 8.

Table 3.3　Proportion of total glacier movement contributed by basal sliding (per cent)

Glacier/ice sheet	Thermal régime	Per cent basal sliding	Reference
Thule, Greenland	Cold	0	Goldthwait, 1961
Meserve Glacier, Antarctica	Cold	0	Holdsworth and Bull, 1970
Jungfraujoch	Cold	0	Haefeli, 1963
Jungfraufirn	Temperate	26	Perutz, 1950
Salmon Glacier, BC	Temperate	55	Mathews, 1959
Østerdalsisen, Norway	Temperate	65	Theakstone, 1967
Athabasca Glacier, Alberta	Temperate	81–87	Raymond, 1971
Blue Glacier, Washington	Temperate	90	Kamb and LaChapelle, 1964
Skautbreen, Norway	Temperate	90	McCall, 1952

Two main processes of sliding are thought to operate beneath temperate glaciers. Regelation slip (Kamb and LaChapelle, 1964) involves local melting of ice in areas of high compressive stress (as on the upstream faces of bedrock obstacles) and the flow of the resulting meltwater around the obstacle to refreeze on the low-pressure downstream side. Kamb and LaChapelle (1964) and others have studied the process in the laboratory and in the field. Souchez *et al.* (1973) and Vivian and Bocquet (1973), studying cavities beneath the Glacier d'Argentière, France, have confirmed that interstitial water can and does refreeze at the base of a temperate glacier in accordance with change of pressure. Observations over 3 years in one cavity showed refrozen ice some centimetres thick on some visits and melting on others. Weertman (1957, 1964) has shown that the theoretical rate of sliding permitted by the regelation mechanism is proportional to the size of the bedrock obstacles, their conductivity and the basal shear stress:

$$S_a = \frac{CD}{\rho HL} \tau_B r^2 \tag{3.19}$$

where S_a is the sliding rate, D is the coefficient of thermal conductivity, H is the heat of fusion (80 cal g^{-1}, ρ is the density of ice, τ_B is the basal shear stress, L is the obstacle size (length down-glacier), and r is a roughness factor (the down-glacier spacing of the obstacles divided by their length). As the obstacles increase in size, the sliding rate is reduced because of the decreasing efficiency of heat transfer through the obstacle. Obstacles larger than about 0·1 m will probably reduce regelation slip to negligible amounts. Furthermore, the process is inapplicable to cold glaciers frozen to bedrock.

The second mechanism of basal sliding is that of creep, or the deformation of ice in accordance with applied stress as described in the previous section. As the stress increases around obstacles, the strain rate rises, and therefore we have a mechanism that, unlike regelation slip, will in theory allow ice to flow over and around large obstacles. For simplicity, it is assumed that there is a power relation between shear stress and strain rate (Glen's law), and also that the hydrostatic (normal) pressure at the base of the ice is sufficient to prevent sub-glacial cavity formation. Weertman (1957) has then shown that the sliding velocity S_b by this process is given by:

$$S_b = BL\left(\frac{\tau_B r^2}{2}\right)^n \tag{3.20}$$

where B is the temperature-sensitive constant in the flow law, n is the flow-law exponent and other notation is as before. In the case of cold glaciers, only equation 3.20 will apply, but in the case of temperate glaciers where both regelation slip and creep may operate, the speed of sliding S, when $S_a = S_b$, will be given by:

$$S = 2^{1-n/2}\tau_B{}^{n+1/2}r^{n+1}\sqrt{\frac{BCD}{\rho H}} \tag{3.21}$$

If $n=3$, this simplifies to $(1/\sqrt{2})\tau_B{}^2 r^4 \sqrt{BCD/\rho H}$, in which it will be seen that the sliding velocity is proportional to the square of the basal shear stress and the 4th power of the bed roughness. The latter is then a very sensitive parameter, which is unfortunate because, being invisible, it is very difficult to assess realistically. In 1964, Weertman modified the relationship to allow for a more realistic spectrum of bedrock obstacle sizes and to allow for some heat passing through the ice around the bedrock obstacles.

So far the possible existence of sub-glacial cavities has been ignored in the case of the creep mechanism. This is a significant omission, and Lliboutry (1959, 1964, 1968) has made 'sub-glacial cavitation' (not to be confused with cavitation erosion, p. 285) the basis of his sliding theory for temperate ice. Sub-glacial cavities will immediately reduce the friction; moreover, if they are filled with meltwater, as is likely beneath a temperate glacier, the water pressure may serve further to reduce the basal pressure, as well as helping to maintain the cavities. The effective basal pressure will be equal to $(\rho gh - p)$ where ρgh is the 'normal' hydrostatic pressure and p is the sub-glacial water pressure. Other things being equal, smaller values of the term $(\rho gh - p)$ will give rise to enhanced basal sliding. This argument applies irrespectively of whether the sub-glacial water is concentrated in cavities (as Lliboutry considers most likely) or whether it mostly occurs as a relatively thin basal water film (Weertman). Zero values of $(\rho gh - p)$ will represent the situation when the glacier floats on the basal water and virtually loses contact with the bed; such conditions may be approached in cases of catastrophic glacier advances (Lliboutry, 1964; Vivian, 1966; Weertman, 1962). Values of $(\rho gh - p)$ will be greatest at the level of the englacial water table and will diminish towards the deeper parts of the glacier bed: Raymond (1971) considers that this may help to explain the great lateral variations in basal sliding beneath Athabasca glacier, from as much as 87 per cent near the centre to less than 4 per cent near the margins.

Basal meltwater beneath temperate glaciers may accumulate from three sources:

1. meltwater of sub-aerial origin (melting snow, melting glacier surface ice, precipitation) finding its way by englacial and sub-glacial routes to the base of the ice;
2. ice melted by geothermal heat (average $0 \cdot 5 \text{ cm}^3 \text{ cm}^{-2} \text{ year}^{-1}$ but increasing up to five-fold in regions of abnormally high heat flow such as Iceland); and
3. ice melted by basal sliding and internal deformation (heat of friction).

Many observations (see p. 25) have suggested that basal meltwater, acting as a lubricant, is an important factor in the sliding of temperate glaciers. Weertman (1964, 1966) considers that the thickness of a basal water layer will be critical in this connection. If the bedrock is impermeable, the thickness (d) will be a function of the hydraulic gradient and the amount of meltwater (Q) being discharged in the layer:

$$d = \left(\frac{Q\,12\mu}{\rho g \sin \alpha} \right)^{1/3} \tag{3.22}$$

where μ is the viscosity of water and α is the ice surface slope. Boulton (1972a) discusses the correction factor necessary for permeable bedrock, a factor that has been too often neglected.

It is clear that the basal sliding of glaciers and ice sheets is a complex process depending on many variables that are either difficult to define or difficult to measure in practice. Inevitably the theoretical models of Lliboutry, Nye, Weertman and others are greatly simplified compared with reality. A major problem is that of defining a realistic bed form (Lliboutry, 1968), which will require more quantitative field studies of glacier beds exposed in tunnels or following recent ice retreat. Another aspect for which data are still meagre is the distribution and nature of occurrence of basal meltwater and the water pressures encountered. At present, our understanding of basal sliding lags behind that of the internal deformation of moving ice, which is unfortunate geomorphologically since it is the former that is most relevant to understanding the sub-glacial processes of erosion and sedimentation.

Air motion

Air and water are both fluids, and many of the principles applied to their flow are the same. Air is a mixture of gases, is compressible, and has low viscosity and density. In place of the well defined upper surface of a shallow liquid, at which waves may form, the atmosphere has vertical (and horizontal) density gradients which may contribute to the growth of gravity waves. The same is true in the oceans.

All geomorphologically effective winds are turbulent; they have high Reynolds numbers so that inertial forces are much more important than viscous ones (p. 40). Much of the turbulence is highly irregular, but regular patterns do occur for short periods and geomorphological interest focuses on these because it is they that seem to determine the shape of regular surface forms. The forms persist beneath irregular flow by interacting with it and creating their own regular flow patterns. The discussion of flow patterns here cannot be very analytical. A more rigorous, if idiosyncratic, discussion is found in Scorer's books (1958, 1977).

Turbulent or secondary flow always involves rotation. This is measured as 'vorticity', a vector whose intensity can be related to angular velocity. Vorticity in the atmosphere is generated in many geomorphologically relevant ways: by the rotating Earth; by heating and convection; by topographic deflection; and by shear over the ground. Two or more of these can, and frequently do, operate together.

The Coriolis effect, produced by the Earth's rotation, is reflected in aeolian landforms when, as is often the case, they are aerodynamically rougher than surrounding surfaces. The geostrophic wind, well above the surface, blows in a direction determined by the balance between the pressure gradient force and the Coriolis or deviating 'force' (Figure 3.15). The surface wind is held back by friction, and, in the absence of any drag from overlying layers of air, this surface wind then deviates towards the pressure-gradient direction. The rougher the surface, the more the deviation, which is leftward in the northern hemisphere and rightward in the southern. The geomorphological effect of this is discussed on page 343.

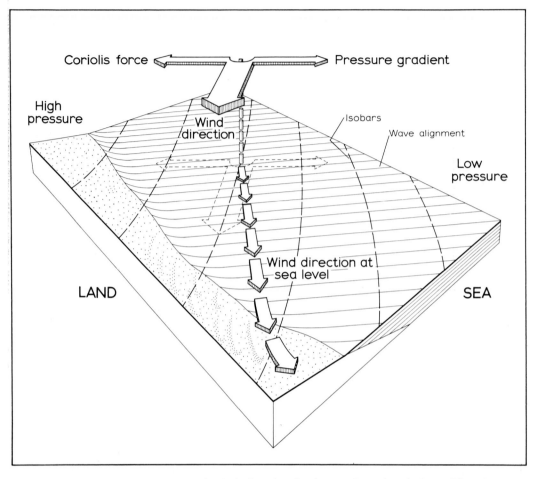

Figure 3.15 The determination of wind direction in the northern hemisphere. The change at the coast is explained in Chapter 10.

Some meterologists maintain that the vertical component of the Coriolis force helps to generate large vortex rolls of the type more commonly thought to be associated with convection (see below). Wipperman (1969) further maintained that such flow had an additional wave-like motion at right-angles to the vortex rolls and of a longer wavelength. He found evidence for both types of flow from aerial photographs of dunes. Unlike convective vortices or the deviation produced by friction, Wipperman thought his vortices would tend to be about 14° to the right of the geostrophic wind.

In the planetary boundary layer (up to about 6 km), vorticity generated by convection has a widespread impact on aeolian landforms. The motion is roughly analogous to the controlled laboratory experiment in which a viscous fluid in a tank is heated from below and cooled from above. If there is too much heat to be dissipated by molecular motion, the fluid density at the base of the vessel decreases and its buoyancy causes it to rise. This motion, combined with compensatory downward movement, forms a system of hexagonal or square-topped 'Bénard' cells. The cells are smaller in shallower fluid. In a stratified and often chaotic atmosphere, convective motion is much more complicated, but regularity does sometimes appear. Simple cellular motion happens when there is little horizontal wind; this is distorted into long vortex rolls roughly parallel with the flow when there is a wind (Figure 3.16). The Coriolis force or shear may add to the effect (see above and below). Regular patterns are picked up when clouds form on top of the rising air, these being known as 'cloud streets' (Kuetner, 1971). The vortices are commonly twice as wide as they are high and spaced between 2 and 8 km apart. These regular arrangements only occur when there is gentle heating, as in the morning, and they disperse in the chaotic convection of the heat of the day. The rolls are oriented usually to the left of the cloud-base wind in the northern hemisphere. Pairs of more persistent vortices occur downwind of hot humps (hills). Hanna (1969) has collected a mass of information linking vortex flow to longitudinal dunes.

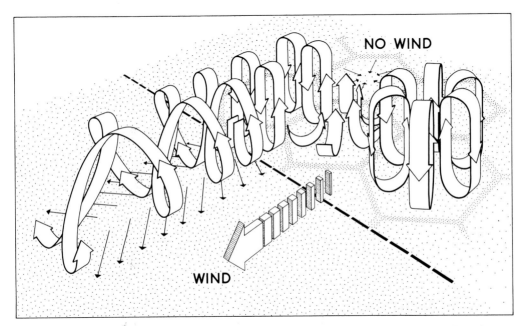

Figure 3.16 The formation of cellular air motion above a heated plain and its distortion to vortical flow by a wind.

Wave-like motion is another form of rotatory flow in the atmosphere. Scorer (1977, pp. 163–84) shows that there are three main controls of such motion: the density stratification (acting with gravity), the velocity gradient (shear) and the heights of discontinuities. The velocity gradient is often the most important control of wave amplitude. A common case of wave development is when strong vorticity and a vertical velocity component are imparted by flow over a large ridge. Standing lee waves form, whose amplitude and wavelength are related to the size and shape of the obstacle, steep-sided ridges giving the strongest waves. In intense cases rotors, with reverse flow, develop under the waves. Hidy (1967, pp. 125–7) quotes work on the famous Owens Valley waves in the lee of the Sierra Nevada in California where it was found that at very small Froude numbers (p. 40) flow was irregular; at $F \simeq 0.1$ three standing waves formed and at $F \simeq 0.2$ one intense wave developed with a rotor. Intense upward movement of dust has been observed in these flows. The most developed lee waves, according to Scorer (1977, pp. 201–2), occur when there is downward flow off a plateau, or when a series of ridges manages to amplify waves downwind. In aeolian geomorphology,

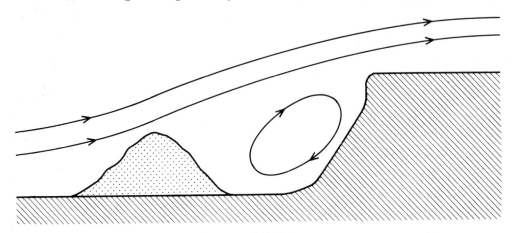

Figure 3.17 The bolster of air formed when a wind blows over an escarpment. The dune in the diagram should be compared with the map in Figure 10.17.

such waves may initiate a regular dune pattern which can then reproduce itself downwind in less regular flow (see p. 338).

The study of flow around obstacles has several other lessons for aeolian geomorphology. It is well known, for example, that flow up and over an escarpment creates a standing rotor in a 'bolster' of air beneath (Figure 3.17). This can help to generate fore-dunes or echo-

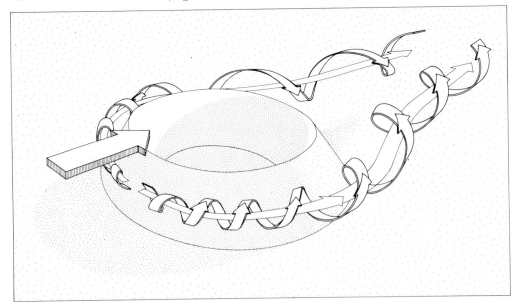

Figure 3.18 The development of vortices around a Martian crater (after Greely *et al.*, 1974).

dunes as shown in Figure 10.17. Such a model has been used to explain the form of 'obstacle dunes' around buildings and around craters and volcanoes on Mars (Greeley *et al.*, 1974; Figure 3.18). In the Martian case, the rotors apparently swing round the smooth slopes and escape downwind as parallel vortices. With different intensities of flow, different lee-dune patterns can be generated experimentally, each with a Martian analogue. The well defined lee depressions behind hills in the Sudanese fixed dunes argue for some

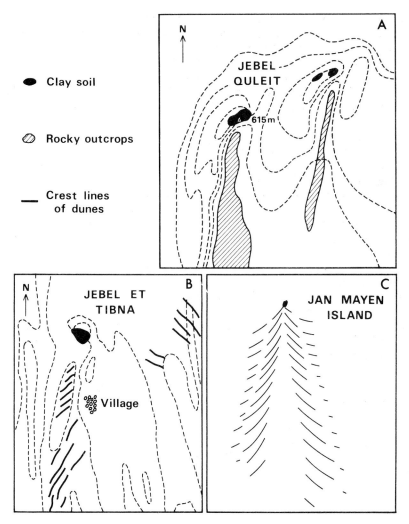

Figure 3.19A: A lee dune ridge with superimposed wake-like minor dunes trailing downwind of a small hill in the central Sudan.

B: Two depressions each bounded by twin dune ridges in the central Sudan.

C: The wake formed in the lee of Jan Mayen Island in the North Atlantic and made visible in a cloud cover. The wind was blowing from the top right (after Scorer, 1977).

such vortical flow (Figure 3.19), but some lee-dune patterns suggest a fan-shaped wake similar to that behind a boat or as shown in cloud patterns in the lee of a hill (Figure 3.19; Scorer, 1977, pp. 194–9).

Of more widespread interest is the flow pattern in the lee of smaller obstacles, such as dunes themselves. Allen (1968) has made an extensive study of such flows as they relate to sub-aqueous ripples, and many of his findings are relevant to aeolian dunes. The way in which interaction between a dune and the flow generates vorticity is discussed at length on pp. 338–9.

Shear on the bed is the most important generator of vorticity near the ground (less than 10 m up). It also contributes to larger-scale motion as discussed above. Just above ground level the wind velocity never rises above zero and above this point there is a sharp velocity gradient in the boundary layer (Figure 3.20). As velocity increases, this zone becomes shallower. In turbulent flow the energy is carried down from the main flow by eddies, the most intense transfer being in the steepest velocity gradient near the bed. The turning of fast over slow flow imparts vorticity to these eddies. The simplest view of this vorticity is to envisage line vortices with axes across the flow, rather like rollers or a layer of spaghetti (Figure 3.20; Scorer, 1958, p. 139). Such a pattern would not retard the flow, i.e. would not abstract energy from it, and since observation shows that eddying indeed abstracts energy, the actual pattern must be much more complex, with vortices aligned at different angles. Scorer and others envisaged that three-dimensional disturbances, perhaps small longitudinal vortices in the boundary layer, twist, distort and even 'knot' the spaghetti and break it up into smaller, thinner, more energetic irregular units. The beginning of such a pattern is sketched in Figure 3.20. The mean motion itself would stretch and thin out the vortices aligned diagonally to it. The eddies start at a large scale and dissipate to smaller and smaller sizes, until they lose their energy to viscous molecular motion (heat) when they reach about 0·25 mm diameter. There must always be a laminar boundary layer of about this size just above the surface.

The outstanding characteristic of these motions is their irregularity. It is hard to discover regularly occurring patterns, although longitudinal vortices do commonly appear. Above all, turbulence is dispersive. While sand grains carried in such flow usually have too much momentum to be dispersed upward, a saltating flow of sand (p. 332) is fanned out horizontally downwind. Dust, however, is rapidly dispersed upward into the atmosphere and so is carried far downwind (p. 331). Shear-generated, near-ground turbulence therefore leaves little pattern on the ground, except perhaps in longitudinal ripples on ventifacts (p. 329) and in the transient form of sand streamers (Verstappen, 1968).

The importance of the boundary layer to aeolian geomorphology lies in the energy that shear imparts to the bed for it is this that accomplishes erosion. The generalizations used to understand the enormous complexity of this process are based on the work of Prandtl.

Faster winds have steeper velocity profiles and therefore greater shear. For turbulent flow this means that

$$\tau = \rho \ (Km + \mu)\frac{d\bar{v}}{dz} \tag{3.23}$$

where τ = shear, ρ is the density of the air, $\bar{v}$ is the mean velocity over a period and z its height of measurement; Km is an expression of the 'Reynolds stress', i.e. the amount of energy dissipated by turbulence, and μ is the dynamic viscosity (p. 36). In turbulent flow $Km \gg \mu$ so that μ can be ignored. Km varies with height, buoyancy and surface roughness (the controls of turbulence).

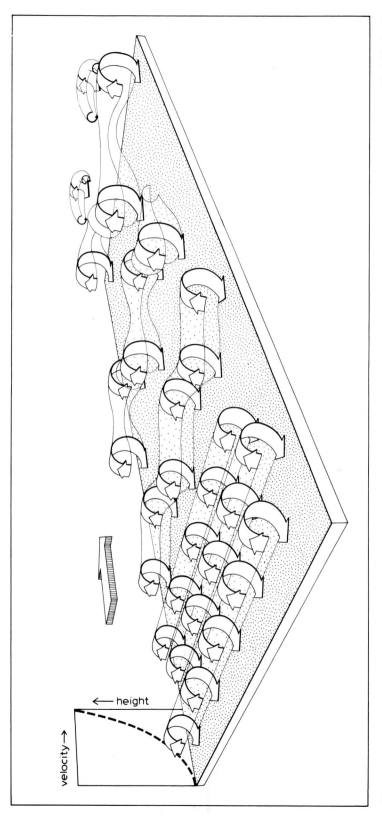

Figure 3.20 A hypothetical view of the growth of turbulent eddies at the ground surface. The rolls are broken up, by some unspecified kind of longitudinal disturbance in the air motion, into small roll-like units. A diagrammatic velocity profile is shown on the left.

Equation 3.24 can yield the expression τ/ρ which is commonly transformed to another quantity, known as the 'shear velocity', V_*, the relationship being

$$V_* = \sqrt{\frac{\tau}{\rho}}$$

V_* is therefore related to $d\bar{v}/dz$ and Km becomes a virtual constant, known as 'the von Karman constant', K, which is usually taken as 0·41. So

$$V_* = 0\cdot41\frac{d\bar{v}}{dz} \tag{3.24}$$

Near the bed, even in non-adiabatic conditions (Nunn, 1966), the velocity profile is fairly close to a straight line on semi-logarithmic paper (log z against linear $\bar{v}$), with zero velocity at some height above the bed z. Hence V_*, and therefore shear, can be defined even more simply:

$$V_* = \frac{K\bar{v}_a}{\log z_a/z_o} \tag{3.25}$$

where $\bar{v}_a$ is the mean velocity over a period at height z_a. The interactions of shear with a mobile bed are discussed in Chapter 10.

4
Weathering

D. Brunsden

Weathering is the response of materials within the lithosphere, to conditions at or near its contact with the atmosphere, the hydrosphere and the biosphere (after Reiche, 1950; Keller, 1957). Three groups of interrelated processes are normally identified. *Physical weathering* is the mechanical breakdown of rock without any contributory chemical alteration, decomposition or mineralogical adjustment of rocks to a given chemical environment. *Biological weathering* is the contribution to or removal of minerals and ions from the weathering environment and physical changes due to growth or movement.

Figure 4.1 shows the relationships between these processes, their controlling variables and the responses. The following general tendencies should be noted:

1 A movement toward a more stable state where mineral assemblages, formed under conditions of high pressure and temperature within the earth, adjust to the new environmental conditions, atmospheric pressure and lower temperatures at the Earth's surface.
2 Irreversible changes from a massive to a clastic or plastic state as breakdown proceeds (Polynov, 1937) (columns 1–6).
3 Changes in volume, density, grain size, surface area, permeability, consolidation and strength (column 4).
4 The formation of new minerals, aggregates and solutions (columns 5–6).
5 The resistance of original minerals (columns 5–6).
6 The transfer, translocation, dispersion, aggregation and concentration of minerals and salts (column 7).
7 The preparation of rock for subsequent erosion and transport by geomorphological processes.
8 The production of new land surfaces and deposits.

Factors controlling rock weathering

The nature and rate of weathering depend on several sets of factors including climate, the properties of the rock-forming materials and the biosphere. These combine through time to yield the basic properties of the weathering mantle. The five basic factors of soil formation according to Jenny (1941) are climate (cl), topography (r), parent material (p), organisms (o) and time (t) which combine to yield the soil properties (s).

$$s = f(cl, r, p, o, t) \tag{4.1}$$

s is regarded as the dependent variable and the factors as the independent variables.

Climate and weathering

Weathering processes are heavily dependent on the prevailing climatic conditions (Table 4.1) but it is difficult to isolate, at any scale, the nature of the climatic parameters that control

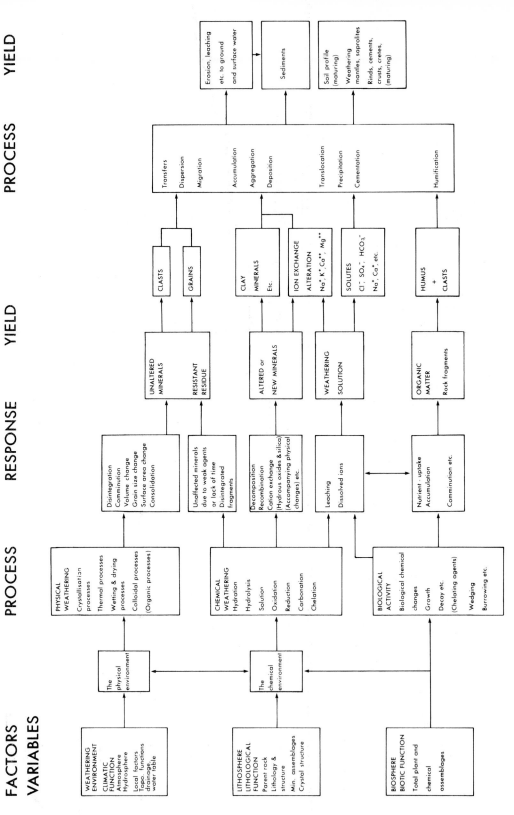

Figure 4.1 General diagram of the factors and processes of weathering.

the main weathering processes. Usually it is assumed that landscape is climatically controlled, and yet the relationships between climate, process and landform are imperfectly understood. There are few studies of the effect of a specific climatic parameter on a given weathering process. Moreover, many climatic effects are indirect (e.g. through vegetation) and the exact relationships are difficult to discover and quantify. Usually it is only possible to relate generalized climatic parameters (e.g. mean annual rainfall) to broad concepts of process (e.g. mechanical weathering). However, the most important controls are undoubtedly the availability of water and temperature conditions.

The intensity, frequency and duration of precipitation–evaporation events are important in weathering studies since chemical reactions depend on a supply of water for many essential processes (hydration, hydrolysis, solution)—to leach and remove soluble constituents and thereby enhance further reactions, to control the concentration of hydrogen ions, organic content and oxidation-reduction conditions, and to control the distribution of clay minerals and the zonation of the weathering mantle.

Temperature conditions, especially absolute values, seasonal and diurnal variations, control the frequency and magnitude of freeze–thaw, and partly control wetting, drying and salt-crystallization cycles. Chemical weathering is active in moist areas, where temperatures are high, since an increase in temperature often accelerates chemical reactions. In drier areas, high temperatures decrease vegetation cover and chemical weathering because of the increase in evaporation. Physical weathering is, however, accelerated by the increase in wetting–drying and salt-crystallization cycles. Similarly, low temperatures in moist areas inhibit some chemical reactions and increase freeze–thaw and the development of ground ice. It is worth emphasizing, however, that, in high latitudes and at high altitudes, chemical processes can still be active and may even account for a major part of the lowering of a land surface (Rapp, 1960). This is particularly true of processes of reduction, solution and hydrolysis.

Climate–weathering relationships at the macro-scale
It is possible to construct world-wide distributions of weathering based *only* on the assumptions that the rate of chemical weathering increases with temperature and water availability, and that the physical processes increase at lower temperatures. The parameters chosen are generalized from available world data and most commonly use mean annual temperature and mean annual precipitation, but they do provide an initial attempt to quantify the relationships between climate and weathering (Figures 4.2, 4.3, 4.6; Table 4.2) (Peltier, 1950; Tricart and Cailleux, 1955; Büdel, 1963; Leopold, Wolman and Miller, 1964; Strakhov, 1967; Wilson, 1968; Pedro, 1968).

Climate–weathering relationships at the meso-scale
It is also possible to refine the general weathering characteristics identified on the world-wide scale. The methods involve, first, the development of more sophisticated climatic parameters, particularly to make allowance for seasonality and evaporation; secondly, the use of indices related to leaching and the acidity of the weathered mantle; thirdly, the calculation of indices of the regional effect of frost action, usually on building-stones, and fourthly, the zonation of the mantles or soils of major land units in terms of clay content, clay mineralogy or the mobility of elements.

Weinert's precipitation ratio Weinert (1961, 1964, 1965) used potential evaporation (E, Meyer formula), total precipitation during the warmest month (P), (J, January) and annual precipitation (P_a) to produce a precipitation ratio (N) that could be used to explain the location

Table 4.1   Generalized weathering characteristics in four climatic zones (various sources).

	Processes	Grading
GLACIAL/ PERIGLACIAL	Frost very important. Susceptibility to frost increases with increasing grain size. *Taiga*, fairly high soil leaching. Low rates organic matter decomposition, 1:1 clays can form. Iron mobility varies but generally low clay formation. *Tundra*, low prec., low temp., permafrost—moist conditions, slow organic production and breakdown. May have slower chemical weathering. Algal, fungal, bacterial weathering may occur. Granular disintegration occurs. Hydrolytic action reduced on sandstone, quartzite, clay, calcareous shales, phyllites, dolerites. Hydration weathering common due to high moisture.	Hard rocks breakdown to well-graded gap-graded material. Soft rocks change slowly. Chem. weathering increases plasticity and decreases grain size.
TEMPERATE	Prec. evap. generally fluctuate. Both physical weathering and chemical weathering occur. Iron oxides leached and redeposited. Carbonate deposited in drier areas, leached in wetter areas. Increased precipitation, lower temperatures reduce evaporation. Organic content moderate-high, breakdown moderate. Silicate clays formed and altered. *Deciduous* forest areas—abundant bases, high nutrient status, biological activity moderate-high. *Coniferous* areas, acidic, low biological activity, leaching common.	Grading changes in steps particularly stages, e.g. threshold slopes. Plasticity increases in late stages by decomposi
TROPICAL ARID/SEMI ARID	Evaporation exceeds precipitation. Rainfall low. Temperature high. Seasonal. Organic content low. Physical weathering, salt weathering, granular disintegration, dominant in driest areas. Thermal effects possible. Low organic input relative to decomposition. Slight leaching produces 2:1 clays and $CaCO_3$ in soil. Sulphates and chlorides may accumulate in driest areas. Increased prec. and decreased evap. toward semi-arid areas and steppes yield thick organic layers, moderate leaching and $CaCO_3$ accumulation.	Coarse-grained but good sorting by transport agencies produces poorly gra deposits. Becomes finer-grained away source. Evaporites cause aggregation. Salts, crusts, detrital sands, gravels, silt and calcareous fragments common.
TROPICAL HUMID	High rainfall often seasonal. Long periods of high temperatures. Moisture availability high. Weathering products (a) removed or (b) accumulate to yield red and black clay soils, ferruginous and aluminous soils (lateritic), calcium-rich soils. Calcareous rocks generally heavily leached where silica content is high, soluble weathering products removed and parent silica in stable products are sandy. Where products remain, iron and aluminium oxides are common. Usually intense, deep weathering, iron and alumina oxides and hydroxides predominate. Clay minerals of 2:1, mixed and 1:1 lattice occur with increasing rain. Organic content high but decomposition high.	Generally well-graded in initial stages. Finer grained and poorly graded with t

of the boundary between disintegration and decomposition processes in South Africa for engineering purposes (Figure 4.4). Weinert calculated:

1 a potential evaporation–precipitation ratio

$$R = \frac{E_J}{P_J} \qquad\qquad (4.2)$$

2 a seasonality index

$$D = \frac{12P_J}{P_a} \qquad\qquad (4.3)$$

Strength	Permeability	Compressibility and consolidation	Rates (mm year^{-1})
Peak shear strength reduced with weathering. Residual strength not obtained. Mode of failure changes from brittle to plastic.	Increases with production of coarse debris. Decreases with further disintegration.	Soil becomes compressible. Preconsolidation effects removed.	Narvik 0·001 Spitsbergen 0·02–0·2 Kärkevagge 0·32–0·5 Alaska 0·04

Marked loss of strength mainly occurs in stages but also in steps during course of weathering.

e.g.s

	c'	ϕ'm
Leighton Buzzard	0	33–43
Belgium Molsand	0	35–50
Exmoor slate	0	42
Pennine Sandstone	0	36
Salop. Mudstone.	0·1	25
London Clay	0·15	20

Generally as related to index properties of parent rock.

As expected from index properties. Preconsolidation effects removed.

Askrigg 0·5–1·6
Limestone 0·083
Austria 0·040
Austria 0·015

Generally high, related to grading. Plasticity characteristics generally unknown.

Generally high for weathered soils near source. Decreases away from parent material. Local variations common.

As expected from index properties. Low near source. Loessic soils may be metastable. CBR values 47% unsoaked 40% soaked (for fans).

Egypt 0·0001–0·0005
Australia 0·6–1
Egypt 0–2

Published work indicates consistent, isotropic. ϕ'm generally in range 25–35. Strength increases with increased grain size. Laterization and cementing increases strength.

e.g.

	c'	ϕ'm
Brazil	0·1–0·5	30–33 sandstone
	0·2–0·4	26–29 gneiss
	0·4–0·5	29–30 basalt

Variable—often low, related to index properties rather than weathering process.

Effects of pre-consolidation removed. Some soils metastable after leaching. Coeff. of consolidation variable but high. Compressibility values average.

e.g. Florida 0·005

where $D>1$=summer rain, $D<1$=winter rain
$D=1$=equal distribution

3 the precipitation ratio

$$N=\frac{12E_J}{P_a} \qquad (4.4)$$

In South Africa, empirical studies have shown that there is a fundamental weathering boundary between $N>5$, where disintegration is the dominant process, and $N<5$ where decomposition is most important. Other relationships showed that hydrous mica is produced where $N>6$; montmorillonite is produced from basic igneous rocks where $N=2-5$, but kao-

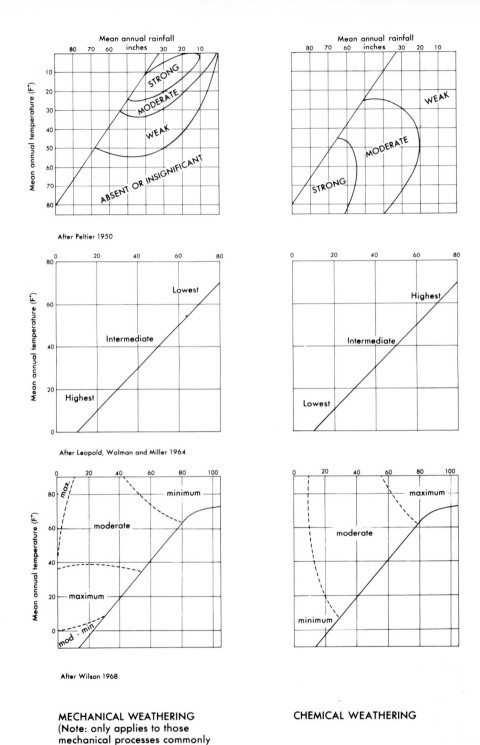

Figure 4.2 Possible relationships between climate and weathering.

Table 4.2 Leaching and acidification indices developed to relate the major soil distributions to climatic parameters

Author	Index	Formulation	Comments & Usage
Crowther (1930)	Leaching factor	$LF = R - 3 \cdot 3T$	Used to assess the influence of rainfall and temperature on soils on a broad zonal basis.
Meyer (in Jenny, 1941)	Leaching quotient	$NS = \dfrac{P}{Sd}$	Widely used by Jenny (1935, 1941, etc.) to distinguish arid-humid boundary of $NS = 200-250$.
Prescott (1949)	Precipitation–leaching effectiveness index	$PE_I = \dfrac{P}{Em}$	Used to relate major soil groups to rainfall and evaporation.
Hallsworth (1952)	(a) Evaporation equation	$E_p = K_1 \dfrac{VP_t(100 - RH)}{100}$	Group of measures to explain the distribution of soils in terms of evaporation, leaching, base release and acidification. Originally applied to basalt soils in New South Wales.
	(b) Leaching potential (water passing through soil)	$Wp = K_2 \dfrac{P}{E_p}$	
	(c) Rate of base release	$Rb = K_3 \, Dw_t$	
	(d) Acidification potential	$Ap = \dfrac{K_2 \, (P/E_p)}{K_3 \, Dw_t}$	
	(e) Acidification index	$A_I = \dfrac{P/E_p}{Dw_t}$	

Notation: R=Mean annual rainfall (cm); P=Mean annual rainfall (mm) (original form used); T=Mean annual temperature (°C); Sd=Mean absolute saturation deficiency of air (mm mercury); E=Average annual evaporation from free water surface; E^m=a value varying from $0 \cdot 67$–$0 \cdot 80$, mean $0 \cdot 73$; E_p= Evaporative power; K_1, K_2, K_3, Constants; VP_t=vapour pressure at temperature t°C; RH=relative humidity; Wp=water passing through soil profile; Rb=Rate of base release; Dw_t=the dissociation of water at t°C (mean ann. temp.); Ap=acidification potential=Balance between water passing through soil and the bases released by weathering.

linite is produced under the same conditions from acid igneous rocks. These figures can only be applied to South Africa but, since these initial results were achieved, the technique has been applied in Hong Kong, parts of Guiana, Idaho and West Germany (Saunders and Fookes, 1970).

Leaching and acidification indices are used to relate the major soil distributions to rainfall, evaporation and leaching (Table 4.2). The purpose of these techniques has been to relate some of Jenny's (1941) factors of soil formation to known soil characteristics. The indices have not been tested over a wide range of weathering mantles nor fully applied in geomorphological studies, and their full value therefore is unknown.

Indices of frost weathering The regional effect of frost weathering has been the subject of similar studies. Perhaps the best known method is that of the American Society for Testing Materials (1971 a, b) which devised an index for the effect of weathering on building-stones. This index is the product of the average annual number of freezing-cycle days (any day in which air temperature passes either above or below 0°C), and annual average winter precipitation (sum of mean monthly precipitation in inches, between and including the normal dates of the first and last killing frosts). The index is used to map the occurrence of negligible (<100), moderate (100–500) and severe (>500) weathering susceptibility to frost action (Figure

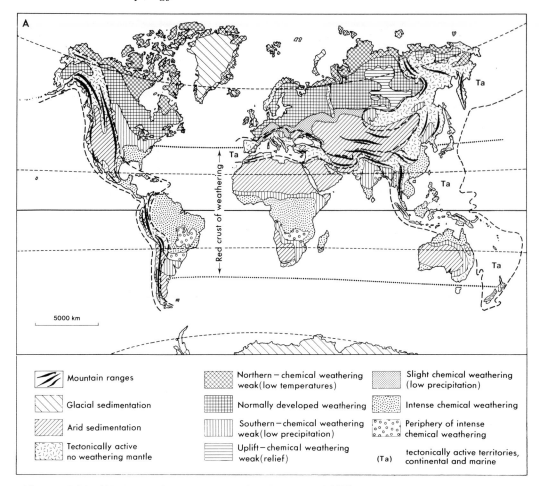

Figure 4.3A: World weathering zones (after Strakhov, 1967).

4.5A). An alternative summary of potential frost weathering in North America is provided by Visher (1945) (Figure 4.5B); unlike Figure 4.5A, this does not take into account water availability.

Zonation based on specific properties of soils   Attempts to relate the effects of specific climatic parameters to weathering stem from Jenny's (1935, 1941) original work on the basic igneous rocks (diorite and gabbro) along a north–south line from New Jersey to Georgia. This showed that the clay content of soils was exponentially related to mean annual temperature but increased linearly with rainfall (Figures 4.6 and 4.7). Later work in Hawaii in alternating wet–dry climates and in wet climates without a dry season (Sherman, 1952) and elsewhere has confirmed that there is a zonation of clay-mineral types with increasing rainfall. Assuming that each climatic area can be considered as a unique environment, it will possess a character-istic clay–mineral assemblage. Some caution is needed, however, since the assemblage itself may not be unique. The principle of equifinality suggests that a given grouping of minerals may be produced by different arrangements of climatic parameters.

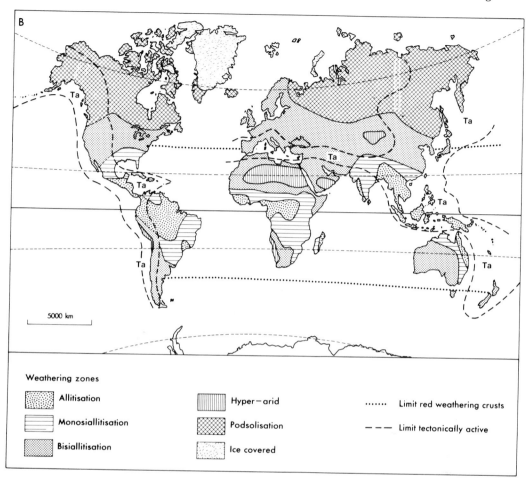

B: Principal weathering zones on a world scale (after Pedro, 1968; Strakhov, 1967; Thomas, 1974). Ta=tectonically active areas.

A full discussion of the relationships between climate and the mobility of various soil elements is beyond the scope of this chapter but the interested reader can find an excellent review in Birkeland (1974). The mobility of iron, the distribution of nitrogen and relative amounts of organic matter, the depth of the zone of carbonate accumulation, variations in colour and mineral etching by solutions all reflect the general effects of precipitation, evaporation and temperature (Figure 4.8). As a result, there is a well defined zonation of weathering mantles, especially in tectonically stable areas, areas far removed from base-level changes or areas that have been least affected by climatic change (Strakhov, 1967).

Climate–weathering relationships at the local scale
Local weathering differences may be a response to micro-climate. At this scale, seasonality, frost pockets, cold-air drainage, aspect and exposure to wind or rapid insolation all cause variations in evaporation, soil temperature and leaching potential. In addition, rainfall and temperature variations with elevation become important in determining the distribution of

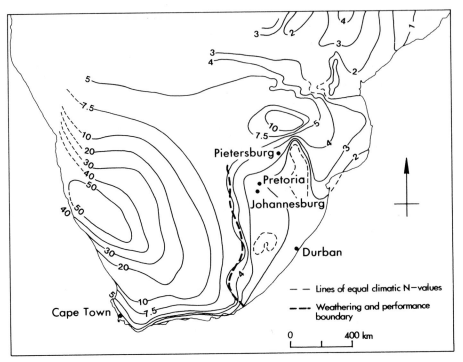

Figure 4.4 Map of Weinert's (1965) precipitation ratio (N) and the location of the boundary between disintegration and decomposition processes in South Africa (see text for fuller discussion).

local weathering processes and rates. They are of particular importance to the pedologist but have received little attention from the geomorphologist except with respect to small features such as tafoni, weathering pits (Plate IV), the production of solifluction debris, asymmetrical slopes or unequal weathering on the sides of buildings and monuments (e.g. Raistrick and Gilbert, 1963).

The hydrosphere and weathering

An alternative and complementary approach to the establishment of climatic weathering environments is to establish the nature of the hydrological zones in which weathering takes place (Table 4.3). A simple threefold division into sub-aerial, sub-surface and sub-aqueous environments is generally applicable to all climates and geological regions (Keller, 1957).

The sub-aerial zone
Sub-aerial chemical weathering is often less obvious in its effects than in the other environments since soluble products are removed and surface crumbling proceeds slowly. Conversely, physical weathering is often clearly visible where rocks are exposed directly to the thermal and moisture changes of the atmosphere. It should not be assumed on such evidence that chemical weathering is less effective in this zone, but it is worth pointing out that there may be a fundamental difference between expansion weathering in the sub-aerial and near-surface zones and a tendency to constant-volume weathering at depth (Ollier, 1965).

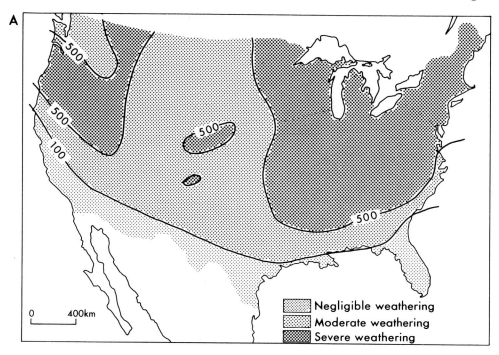

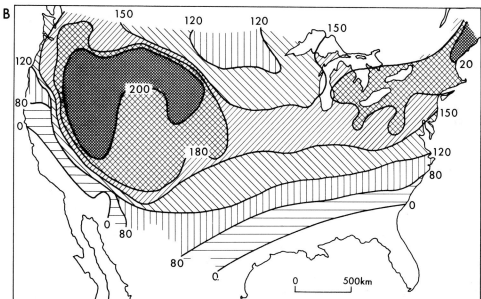

Figure 4.5A: Weathering regions in the USA defined by the ASTM Standard C216 weathering index. The measure incorporates frost frequency and water availability as assessed by average winter rainfall totals (see text).

B: The frequency of freeze–thaw in the USA (after Visher, 1945) on the basis of average number of frost cycles per year.

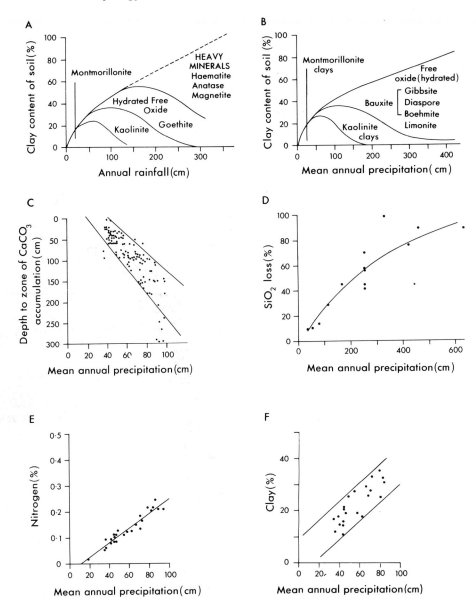

Figure 4.6 Some relationships between weathering changes and precipitation. **A**: Progressive clay-mineral development in Hawaii as a function of annual rainfall, in an alternating wet and dry climate (after Sherman 1952); **B**: The relationship between clay content, mineral type and mean annual precipitation in a continuously wet climate, Hawaii (Sherman, 1952); **C**: Depth of $CaCO_3$ accumulation and mean annual rainfall (Jenny and Leonard, 1934; Russell and Engle, 1925); **D**: Silica loss with mean annual precipitation from a 10 000–17 000 year old volcanic ash, Hawaii (Hay and Jones, 1972); **E**: Nitrogen percentage and mean annual precipitation (Chorley, 1969); **F**: Clay percentage and mean annual precipitation (Jenny, 1941).

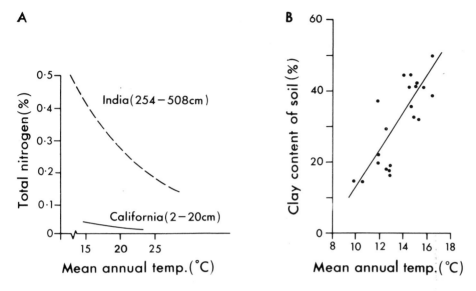

Figure 4.7 The variation of **A** : total nitrogen percentage and **B** : clay content of soil with mean annual temperature (0°C) (Jenny, 1941).

Plate IV Weathering pit in granite on Goatfell, Arran. The diameter is about 40 cm. (C.E.)

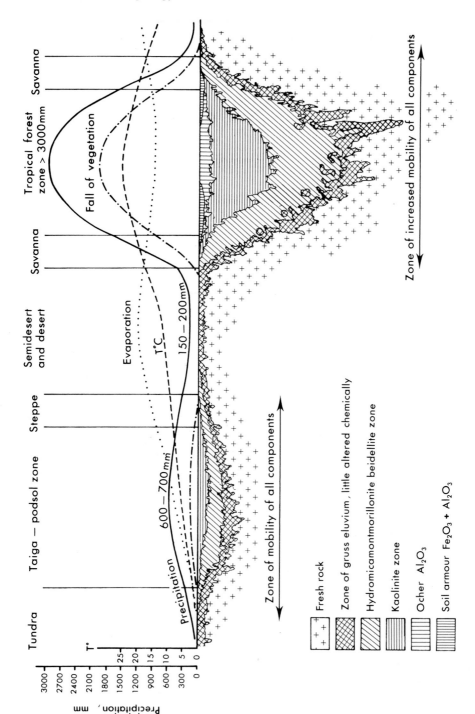

Figure 4.8 Relationships between climate and weathering mantles (after Strakhov, 1967).

Table 4.3 The properties of the main hydrological weathering environments

Macro-environment	Micro-environment	Properties
Atmospheric	(a) Overhead	Closely related to micro-climate of surface. Depends on rainfall constituents as primary source of cations and anions.
	(b) Near the surface	pH variable 4·9 (Av. 5·7) at 25°C. CO_2, SO_2, H_2SO_4 pollutants. Cations Na, K, Mg, Ca; Anions Cl, SO_4.
Sub-surface	(a) Upper soil	Biologically active. Fluctuating temperature, pH, Eh, moisture. Fluctuating conc. metal ions. Al_2O_3, SiO_2 in solution. Frequent change in direction of water movement. High CO_2.
	(b) Below upper soil— solid rock	Biologically less active. Temperature and pH fluctuation less pronounced. CO_2, H_2S, H^+ variable. Low Eh. High conc. of M ions.
Sub-aqueous	(a) Below ground-water table	pH variable. Eh variable. Rain water renewed in permeable rocks. Stagnant conditions possible where waterlogged or impermeable.
	(b) Open fresh water	pH variable. Eh variable. Carbonate and bicarbonate anions available. Ionic conc. lower than marine. Often relatively stagnant.
	(c) Open marine water	pH usually >7. Eh variable. Chloride, sulphate, borate anions available. Ionic conc. high. Often relatively stagnant.

Recently, studies of the chloride, sulphate, sodium, potassium, magnesium and calcium contents of rainwater and of the products of pollution, CO_2, SO_2, H_2SO_4, indicate that surface weathering may also depend heavily on local variations in these substances, especially in coastal, desert and urban areas. Rain is thus a mixed electrolyte with varying constituents and a variable pH (usually 4–9, mean 5·7 at 25°C) and is a primary source of ions which are added to a given weathering system.

Despite the importance of these constituents, however, geomorphologists have mainly focused attention on the nature of the surface forms and on qualitative, intuitive explanations. Tors, weathering pits, tafoni, fluting and etching of limestone and granite and the production of cretes and varnishes are common subjects.

The sub-surface zone
The sub-surface environment is of considerable interest in chemical weathering since it is here, in the zone of circulating and often renewed ground water, that chemical processes are most efficient. The zone, or parts of it, has been variously referred to as the vadose zone,

Table 4.4   Some characteristics of the principal hydrological zones (after Keller, 1957)

		↑	Rainfall pH=4–9 (Av. 5·7) Cations Na, K, Mg, Ca Anions Cl, SO$_4$	Pollutants CO$_2$, SO$_2$, H$_2$SO$_4$
Sub-aerial Zone	Evapotranspiration			↓
Surface			↕ Wetting / Drying	↑ Freeze Thaw ↓ → Runoff
	Percolation zone	↓	Leaching ↓ eluviation	Lateral → through-flow
Sub-surface zone	Capillary zone	↕	illuviation Variable oxidation—reduction	
	Fluctuating Water Table	↑	migration of soluble ions away from the rock face →	
Sub-aqueous zone	Saturated zone	↓	Anaerobic conditions Possible stagnation	→ water movement
	Water base			

aeration zone, oxidation zone, the belt of soil moisture, the capillary zone (fringe) and the belt of water fluctuation. Its most important feature is the rapid exchange of water and air as rainwater percolates toward the water table or evaporates at the surface (Tables 4.3 and 4.4).

In general, the upper layers of the regolith suffer the most intense weathering, for here alternations of water and air, organic life, leaching and other physical processes are most active. At lower levels, water movement is mainly downward with considerable leaching but with precipitation where drying-out occurs. Toward the base, oxidation and reduction are common, especially in the zone of capillary water just above the water table. This zone is particularly important in limestone and may even control the level of cave formation for it is here that fresher unsaturated water is most likely to be available.

The sub-aqueous zone
This may be divided into three environments, beneath the water table, beneath open fresh water and beneath open salt water. Only the first has received much attention by geomorphologists and only this will be considered here. The sub-aqueous zone has been variously termed the zone of saturation, the phreatic zone, the zone of reduction, the belt of discharge and stagnation and the zone of supergene enrichment. The main characteristics are that all voids are permanently filled with water, with the addition of solutions, precip-itates and eluviated particles from above (Table 4.4). Towards the top of the zone, water movement and discharge to springs or rivers allows solutes to be removed, and weathering reactions continue relatively slowly with a tendency to achieve a steady state between the production and the rate of removal of weathered material or solutes. Beneath this, some

parts of the groundwater may be immobile in deep sedimentary or regolith basins. It is often assumed that weathering is very slow in such situations since weathering products are not removed by water movement, but, by such processes as ionic diffusion—the migration of soluble ions away from the rock faces toward areas of lower ion concentration—some weathering may still occur.

Unfortunately, several authors (Cotton, 1945; Reiche, 1950) have indicated that weathering beneath the water table may become so slow as effectively to cease altogether. The idea is partly embodied in some descriptions of the basal surface of weathering (Ruxton and Berry,

Plate V Pseudo-bedding and strong vertical jointing control the form of Great Mis Tor, Dartmoor. The rounded edges of some joint-blocks are probably due to subaerial weathering, though sub-surface deep weathering has also been suggested. (J. Gerrard)

1959) and used by Linton (1955) in his explanation of the characteristic form of tors (Plate V). However, the argument depends on the assumption that weathering is coincident with the vadose zone and with oxidation, and ignores the importance of hydrolysis and reduction. Sharp weathering contacts between weathered and unweathered rock really depend on the position of the groundwater base rather than the groundwater surface (Ollier, 1969).

Evidence that weathering can occur at or beneath the water table is provided by many records of very deep weathering (Ollier, 1969; Thomas, 1974), by the apparent control of cave formation in limestone by water level (Sweeting, 1950), and by the nature of oxidized-reduced mineral zones (Bateman, 1950) but the exact relationships are not always clear. In tropical areas the processes are complex and related particularly to the fluctuating levels of the water table. The main changes are to feldspars and the ferromagnesian minerals by ionic migration (Nye, 1955; de Lapparent, 1941; Lelong and Millet, 1966; Lelong, 1966).

Lateral movement of water is of fundamental importance to the nature and rate of weathering in both the sub-surface and sub-aqueous zones (Table 4.4). At the surface, weathering products are relatively easily removed by mechanical processes or in solution. In the near-surface layers, water movement is slow—perhaps a maximum of $5\,m\,hr^{-1}$—and therefore solid particles are less likely to be removed. The size of sediment moved is limited by the size of the voids and the strength of the electro-chemical forces that resist detachment. Although intuitively it might be expected that translocation of clays, soil grains and colloids will take place, direct evidence is often lacking and assumptions have to be made from observations of increased clay content with depth or distance downslope. The latter could, of course, be caused by overland flow.

Solutional removal is at a maximum in the near-surface environment and most soluble products may be removed. This condition persists until deep into the weathering mantle and is only really limited by water stagnation. Generally speaking, ions are preferentially removed from the upper horizons by leaching and travel laterally or vertically in the direction of water flow, downslope and across the soil and weathered mantle horizons, eventually to reach the rivers unless they are redeposited on the way. Net lowering of the landscape therefore follows water movement and the dissolved load of ground water and rivers may give an indication of the efficiency of these processes, except where solution products are fixed by other mechanisms or recycled, as in some tropical regions.

Topography and weathering

The topography of an area has a profound effect on the nature and rate of weathering processes. Slope angle is a major control on the balance between the rate of production of weathered rock and the rate of removal by denudation processes. Erosion generally increases with slope angle to a point where the accumulation of a deep chemically weathered mantle is impossible. Relief and slope steepness together have a marked effect on the rates of surface runoff, infiltration and throughflow. The areas most favourable for chemical weathering are those of moderate relief, gentle slopes, good drainage, unimpeded throughflow and rapid replacement of vadose water.

Similar controls are exerted by position on the slope. The term 'toposequence' (Ollier, 1969) may be used to describe the downslope sequence of soil types or weathering mantles. This may be a simple distribution caused by variations in drainage or soil thickness, or complex and dependent on variable parent materials, relict deposits and erosional discontinuities. Usually there are typical toposequences for a given area. Position affects the downslope increase of soil moisture, organic content and clay content. Weathering of primary minerals tends to increase downslope with a positive feedback mechanism in which more clay holds more moisture which in turn enhances chemical processes. This idea is the basis of many geomorphological arguments concerning the origin of scarpfoot-pediment breaks-of-slope and flared slopes (Ruxton and Berry, 1957; Twidale, 1976). It is also a partial explanation of the change in properties and slope angles which accompanies the progressive weathering of slope materials from talus through taluvium to colluvium (Carson, 1969, 1971b).

Orientation and slope shape in plan are other important factors since they affect the microclimate, soil temperature and rate of evapotranspiration. These in turn partially control soil-moisture variation, the distribution of organic matter, pH and depth to base saturation and thus weathering reactions.

Parent material and weathering

Rock properties

The influence of parent material on weathering may be considered at several scales. The large-scale forms of a landscape will show the effect of differential weathering depending on the relative resistance of the rock (lithology) and the macro-structure. Rock structure and disposition, bedding planes, joints and faulting strongly control patterns of water movement and the movement of soluble products. The relative resistance of a particular rock type to weathering, however, is more affected by lithology and texture as measured by such properties as cleavage, porosity, water absorption, saturation coefficient, grain size, and crystal size, shape and perfection. Rocks with a coarse grain size tend to weather rapidly due to the large void space and permeability (Table 4.5). In other cases this is partly balanced by the greater surface area which accompanies decreasing grain size, so that many fine-grained rocks are very susceptible to weathering. Crystal shape is important since platey fragments weather more easily than more solid forms, and perfect crystals tend to be more

Table 4.5 Approximate average porosities and relative permeabilities for common rock types

Rock type	Porosity (per cent)	Relative permeability
Granite	1 ⎫ Igneous	1
Basalt	1 ⎭	
Shale	18	5
Sandstone	18	500
Limestone	10	30
Clay	45	10
Silt	40	—
Sand	35	1100
Gravel	25	10000

After Spicer (in Birch *et al.* (1942), Kessler & Sligh (1940), Leopold *et al.* (1964))

resistant than irregular, incomplete or fractured forms. Porosity, water absorption and the saturation coefficient, a measure of the water absorbed per unit time as a fraction of available pore space, have been found of particular value in assessing rock durability against both chemical and mechanical weathering and the suitability of stone for building purposes. Other relevant properties include volumetric expansion as a measure of insolation weathering; the thermal conductivity and diffusivity of the rock in relation to thermal weathering or crystallization; and tensile strength as a measure of resistance to crystal-growth pressures.

Based on these properties, it is possible to propose a general rock-stability sequence:

Quartzite, chert>granite, basalt>sandstone, siltstone>dolomite, limestone (Birkeland, 1974)

The silicate minerals

At all scales the mineral composition of the rock mass is probably the most important factor. Since silicate minerals make up about 90 per cent of all rock-forming minerals in the lithosphere, discussion of parent materials as a factor in rock weathering must concern itself with their properties. This has been recognized and fully discussed in many modern texts on weathering and therefore only a basic outline is provided here. For greater detail, reference should be made to Keller (1957), Loughnan (1969), Garrels and Christ (1965) and Garrels and Mackenzie (1971).

The structure of minerals may be thought of as lattices in which atoms are arranged in repeated, regular, three-dimensional forms. Stable arrangements are usually achieved by linking oppositely charged ions, the positive cations balancing the negative anions to give electrical neutrality. Dense packing, which ensures that each ion is surrounded by ions of opposite charge, ensures stability due to ionic bonding. Covalent bonding where adjacent atoms share electrons is also present but specific to the atoms concerned.

The silicate minerals have a very stable structure based on the SiO_4 tetrahedron. The bonding between oxygen and silicon is covalent with ionic bonding of the cations and anions in varying proportions. The number of anions bonded to a given cation is termed the coordination number. Thus fourfold coordination is known as tetrahedral, sixfold as octahedral, etc.

Table 4.6   Classification of silicate minerals (after Loughnan, 1969)

Aluminosilicate structures		No. oxygens shared with adjacent silica tetrahedra	Example
Name	Structure		
Neosilicates	Discrete tetrahedra (SiO_4)	None	Olivine
Sorosilicates	Two tetrahedra (Si_2O_7)	1	
Cyclosilicates	Tetrahedra in hexagonal rings and stacked in columns (Si_6O_{18})	2	Beryl
Inosilicates	Single chains, unidirectional (Si_2O_6)	2	Pyroxenes
Inosilicates	Double chain (Si_4O_{11})	2 and 3	Amphiboles
Phyllosilicates	Sheetlike, hexagonal (Si_4O_{10})	3	Micas
Tectosilicates	Interlinked network of tetrahedra (Si_4O_8)	4 (all)	Feldspars, Feldspathoids and quartz

Polymerization, or a reaction between two or more molecules of the same compound to form larger molecules with the same empirical formula but with a multiple of the molecular weight is also used to give a division of the silicate minerals. The division given by Loughnan (1969) is summarized in Table 4.6.

Although the basic tetrahedral structure of silica is strong and chemically stable, some of the combinations provide weaknesses that allow weathering to proceed with varying efficiency. First, internal structure controls cleavage (e.g. the phyllosilicates) which facilitates mechanical break-up, the access of water and the subsequent solution of the cations (Loughnan, 1962). Secondly, the cations (Al^{+++}, Fe^{+++}, Fe^{++}, Mg^{++}, Ca^{++}, Na^+, K^+) form weak links in the structure. Sodium and potassium, for example, are easily neutralized by water. Thus the lower the interlocking and covalent bonding, and the greater the combination of cations, the greater the possibility of weathering. Resistance increases with an increase in the silicon:oxygen ratio and decreases with a substitution of Si^{++++} ions. For example, calcic-plagioclase is less resistant than sodic-plagioclase since Al^{+++} replaces every other Si^{++++} in the former but only every fourth Si^{++++} in the latter.

Many of these characteristics, which exert a strong influence on the course of weathering, are believed to originate in the conditions of formation of the minerals from a rock melt. Silica tetrahedra appear to link more easily at higher temperatures and Al ions also more readily enter the structure. It is therefore widely accepted as a general rule that the minerals that crystallize first, i.e. at higher temperatures, are less stable than those that crystallize at lower temperatures (Bowen reaction series; Bowen, 1928).

Several attempts have been made to establish a stability series for the rock-forming minerals. Goldich (1938) studied the persistence of minerals in soils following the weathering of granite, gneiss, diabase and amphibolite to yield a now classic sequence (Figure 4.9). A

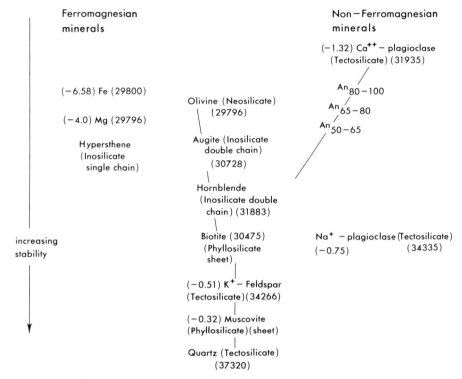

Figure 4.9 Stability series of some common rock-forming minerals, together with Gibbs's free-energy values for weathering reactions, bonding energies and silicate-structure classification.

study of the persistence (frequency of occurrence) of heavy minerals in sedimentary rocks of increasing age yielded very similar results (Pettijohn, 1941) and the sequence received partial confirmation from Reiche (1943) who obtained a series from a calculation of a weathering-potential index based on chemical composition. This work clearly shows that the weathering-stability sequence is the reverse of the crystallization series (Tables 4.7 and 4.8). Attempts to find reasons for this distribution have, however, been less successful. Keller (1954) calculated the bond energies for cation-oxygen bonds and the 'energies of formation' for the silicate minerals. These increase in the expected manner but do not correlate with the Goldich sequence if the cation linkages are included. Similarly, the use of 'energy indices' (Gruner, 1950) and the degree of packing (Fairbairn, 1943) have yielded anomalous results. Curtis (1976b) reports, however, that a possible answer may lie in the formational free energy

Table 4.7   Weathering sequence of clay-sized minerals in soil

Primary	Gypsum	(also halite)	most susceptible
	Calcite	(also dolomite)	↑
	Olivine-Hornblende	(also diopside)	
	Biotite	(also chlorite)	
	Albite	(also anorthite, microcline)	
Secondary	Quartz	(also christobalite, muscovite)	
	Hydrous mica—Inter		
	Montmorillonite	(also beidellite)	
	Kaolinite	(also hallosite)	
	Gibbsite	(also boehmite)	
	Haematite	(also goethite, limonite)	↓
	Anatase	(also ilmenite)	least susceptible

(Jackson *et al.*, 1948).

Table 4.8   Persistence orders, bonding energies, energy indices, packing indices and Gibbs free energy values for some common minerals

Persistence order (Pettijohn, 1941)	Mineral	Bonding energies (Keller, 1954)	Energy indices (Gruner, 1950)	Packing index (Fairbairn, 1943)	Gibbs's free energy values Kcal g atom^{-1} (Curtis, 1976b)
—	Gehlenite	26 890	—	5·3	—
—	Spinel	28 008	—	—	—
—	Forsterite	29 796	1·28	5·9	−4·00
22	Fayalite	29 800	1·28	5·7	−6·58
11	Epidote	30 020	1·46	—	—
—	Idocrase	30 360	—	—	—
—	Akermanite	30 391	1·24	—	—
5	Biotite	30 475	—	5·5	−
17	Augite	30 728	1·35	5·9	—
14	Topaz	31 712	1·55	—	—
9	Staurolite	31 823	1·59	6·7	—
—	Nepheline	31 860	1·30	—	—
—	Almandine	31 878	(1.40)	—	—
12	Hornblende	31 883	1·45	5·7	—
—	Anorthite	31 935	1·44	—	−1·32
20	Diopside	32 052	1·35	5·9	−2·72
—	Dickite	32 165	—	5·9	—
—	Tremolite	32 284	1·45	—	−2.24
—	Enstatite	32 344	1·40	—	—
—	Analcite	32 410	1·53	3·7	—
−2	Muscovite	32 494	1·55	5·5	−0·32
—	Talc	32 516	1·56	5·6	—
—	Pyrophyllite	32 558	1·73	—	—
18	Sillimanite	32 957	1·55	6·2	—
—	Orthoclase	34 266	1·48	5·0	−0·51
—	Albite	34 335	1·49	—	−0·75
—	Quartz	37 320	1·80	5·2	—

and the free-energy changes (the sum of the free energies of formation of all the reaction products, minus the sum of the free energies of the reactants, in standard states), which accompany decomposition, for these agree well with persistence observations and explain many anomalies (Table 4.9).

Table 4.9 Free energies of formation and standard free-energy changes of reaction for some common minerals, together with sample weathering equations with primary minerals as reactants and formational free-energy values for reactants and products. A worked example is also provided (*all from Curtis, 1976b*)

Mineral	Formula	Free energies of formation Reaction Products $\Delta G°f$ Kcal mol^{-1}	Standard free-energy change of reaction $G°r$ Kcal mol^{-1}	Gram atom values to correct for relative no. of atoms $G°r$ Kcal g atom^{-1}	
Fayalite	Fe_2SiO_4	329·7	52·7	−6·58	
Forsterite	Mg_2SiO_4	491·9	44·0	−4·00	
Clinoenstatite	$MgSiO_3$	394·4	20·9	−2·98	magnesium
Diopside	$CaMg(SiO_3)_2$	752·8	38·1	−2·72	silicates
Authophyllite	$Mg_7Si_8O_{22}(OH)_2$	2716·3	137·2	−2·49	
Tremolite	$Ca_2Mg_5Si_8O_{22}(OH)_2$	2779·1	123·2	−2·24	
Anorthite	$CaAl_2Si_2O_8$	955·6	23·9	−1·32	
Low Albite	$NaAlSi_3O_8$	884·0	23·1	−0·75	feldspars
Microcline	$KAlSi_3O_8$	892·8	15·9	−0·51	
Muscovite	$KAl_3Si_3O_{10}(OH)_2$	1330·0	17·3	−0·32	mica

Phase	Formula	Formational free energies for reactants and products stable in weathering environment $\Delta G°_F$ (Kcal mol^{-1})
Hydrogen ion	$H^+(aq)$	0·0
Water liquid	H_2O (l)	56·7
Magnesium ion	Mg^{2+} (aq)	108·9
Quartz	$SiO_2(s)$	204·7
Calcium ion	Ca^{2+} (aq)	132·2
Kaolinite	$Al_2Si_2O_5(OH)_4(s)$	904·0
Sodium ion	Na^+ (aq)	62·5
Potassium ion	K^+ (aq)	67·7
Oxygen gas	O_2 (g)	0·0

examples

$Mg_2SiO_4(s) + 4H^+(aq) = 2\,Mg^{2+}(aq) + 2H_2O(P) + SiO_2(s)$
$CaMg(SiO_3)_2(s) + 4H^+(aq) = Mg^{2+}(aq) + Ca^{2+}(aq) + 2H_2O(l) + SiO_2(s)$

Example of calculation of Gibbs's free-energy value for the weathering equation in kcal mol^{-1}
$Mg_2SiO_4(s) + 4H^+(aq) = 2Mg^{2+}(aq) + 2H_2O(l) + SiO_2(s)$
using $\Delta G°_r = \Delta G°_f$ (products) $- \Delta G°_f$ (reactants)
$\Delta G°_r = \Delta G°_f(2\,mg^2) + \Delta G°_f(2H_2O) + \Delta G°_f(SiO_2) - \Delta G°_f(Mg_2SiO_4) - \Delta G°_f(4H)$
$\Delta G°_r = 2(108·9) + 2(56·7) + 204·7 - 491·9 - \text{zero}$ (from tables)
$\quad = 535·9 - 491·9$
$\quad = 44·0\,\text{kcal mol}^{-1}$

Other parent materials
This discussion has centred on the weathering of silicate materials due to their predominance in surface materials. It is important, however, to recognize that some parent materials, especi-

ally the sedimentary rocks, contain lower amounts of these minerals and may depend for their behaviour on the presence of material which has already gone through or been produced by a weathering process. The percentage and type of clay minerals or cementing agents, and the type of residual minerals or rock clasts, are obvious determinants of the rate and type of process that will occur.

Limestone is a special case since it possesses a high solubility and is usually distinguished by the development of special geomorphological forms that are always strongly associated with weathering, such as solutional grooving, karren, swallow-holes and caves. The dominant minerals are carbonates (calcite, $CaCo_3$) and dolomite $(CaMg(Co_3)_2)$. These minerals also occur as calcareous muds, silts, sandstones, marls or as a metamorphic rock, marble. In general, calcite weathers more readily than dolomite, though Picknett (1972) has shown that magnesium in small quantities increases the rate of solution.

The process of solution is complex and more fully discussed on p. 108. As solution plays a large part in the removal of limestone, it is necessary to note here that the dominant controls are the temperature, the CO_2 content, the pH (Trombe, 1952; Picknett, 1964) and the mobility of the water passing through the limestone outcrop, for these control the chemical equilibria of the reactants and products which in turn determine the maximum solubility of the limestone (a full discussion is provided in Garrels, 1960). The actual amounts dissolved depend on the achievement or non-achievement of equilibrium and the related rates of solution. Thus, although the chemical environment is the main determinant of the process, the nature of the parent material itself is important in that the lithology, texture, porosity, jointing and other structural features largely control the surface area available to water, the rate of water movement and hence the rate of solution.

The biotic factor

The relationship between weathering and the biome establishes several distinct weathering processes. The general effect of this factor can be summarized in terms of the physical and chemical changes to the weathering environment.

The term 'physical effects' refers to the breakdown of rock and the transfer and mixing of particles. Physical breakdown by burrowing and wedging increases the surface area of rock which is exposed to subsequent chemical attack and assists in the penetration of moisture and air. Mixing facilitates weathering by creating new contacts between water, air and minerals. The importance of this process should not be underestimated. Nye (1955), for example, reports that earthworms can deposit $50t\,ha^{-1}\,year^{-1}$ and termites $1\cdot25t\,ha^{-1}\,year^{-1}$ under Nigerian forest conditions. Similarly it would be unwise to neglect the effect of man who, by disturbing vegetation, cultivating soils and mechanically moving weathered material, has an enormous but unknown effect on weathering.

The chemical effects of the biome are often complex. Macro-flora are responsible for the cycling and accumulation of elements and nutrients by taking them up, through their roots, sometimes from depths up to 30 m. Leaf leachates, important for cheluviation processes, create complex compounds and mobilize a wide range of elements. Plant roots may have acid reactions or exchange reactions with metal cations; the production of humic and fulvic acids enhances decay and the solubility of some elements. Humic acids in particular combine with clay minerals, silica and calcium, and are important in chelation (p. 114). In some cases bacteria and algae can have important chemical effects on minerals. The most important group of bacteria are chemotrophic. These are active in waterlogged and reducing conditions, where they produce sulphides but also oxidize iron, manganese and other minerals, and assist in the solution of silica. Nitrogen-fixing bacteria are common to most soils and certainly help to control soil pH and levels of organic matter.

Algae utilize mineral nutrients for growth and are known to be responsible for the mobilization of iron and manganese, producing desert varnishes by concentrating oxides at the surface (Scheffer *et al.*, 1963; Parfenova and Yarilova, 1965). They should perhaps be regarded as a near-surface process in areas where moisture or high humidity persists longest.

Gross changes to the chemical environment include

1 the increase of soil moisture due to shading and the water-holding capacity of humus;
2 the decrease of evaporation but, on the other hand, the increase in transpiration;
3 reduction of temperature by shading, and reduction in wind velocities;
4 increase in the CO_2 content of the soil atmosphere by respiration; this increases the carbonic-acid content and lowers pH, which in turn enhances certain reactions;
5 protection of the ground surface from erosion, so that the balance of production and removal of weathered products is controlled and a deep-weathering environment is promoted.

Rates of rock weathering

The durability of a rock depends on the rate at which its constituents break down and the intensity and duration of the weathering processes operating on the rock. Measurement of weathering processes involves rather tentative procedures since the changes are often slow, variable in time and space, and take place within rocks and on crystal surfaces that are inaccessible.

Changes in datable geomorphological landforms or mantles
A common technique is to make qualitative or semi-qualitative assessments of changes that occur to datable geomorphological surfaces. These involve descriptions of the change in appearance or surface texture of boulders incorporated in such features as river terraces or alluvial fans; the formation of various thicknesses of soil on newly exposed rock, volcanic debris, glacial tills or landslide detritus; the loss of individual minerals or the decrease in the amounts of specific constituents; the gain of new minerals such as clays or the increase in concentration of resistant, usually heavy, minerals.

Some results are summarized in Table 4.10. It should be noted that the data range from qualitative statements, such as 'significant change in 80 years', to more precise measurements such as '58 mm of soil developed in 1000 years' or 'all $CaCo_3$ content (5%) leached out in 124 years'. Such figures are difficult to interpret in terms of landform change but it is possible to use them to indicate the orders of magnitude of changes that occur in various environments and thus to calibrate theoretical models. A general conclusion is that highly weathered mantles may require prolonged periods of environmental stability to produce significant soil depths or complete rock decomposition, a fact that should be considered by those who would allow such deposits to develop in interglacial periods in temperate regions!

Qualitative measures of the degree of decomposition or disintegration
Where data on the exact age of a rock surface or the precise rate of breakdown are lacking, some system of relative degree of change may be used. As an example the following scale was used to measure changes to tombstones in New England (Rahn, 1971):

1 Unweathered.
2 Slightly weathered; faint rounding of corners of letters.

Table 4.10   Weathering changes to datable geomorphological surfaces and measurement of chemical changes.

Author	Method	Rate
LEACHING OF CaCo₃		
Tamm, 1920	Leaching of $CaCO_3$	All removed (0·5%) in 124 years
Tamm, 1920	Leaching of $CaCo_3$	All removed (0·5%) in 124 years
Salisbury, 1925	Leaching of $CaCo_3$	All removed (6·3%) in 280 years
Hissinck, 1938	Leaching of $CaCo_3$	All removed (9·4%) in 300 years
LOSS OF CONSTITUENTS		
Schreckenthal, 1935	General properties	Significant in 80 years
van Baren, 1931	Soil analysis	SiO_2 decreased 6·4% Al_2O_3 increased 2·05% Fe_2O_3 increased 1·04% FeO increased 0·44% CaO increased 0·72%
Parsons *et al.*, 1962,	Soil analysis	Steady state org. matter in 550 years Steady state org. matter in 1000 years
Dickson & Crocker, 1953	Soil analysis	Steady state of some constituents in 1200 years
SOIL DEVELOPMENT		
Hay, 1960	Soil depth	6 feet soil in 4000 years Clayey soil 1·5–2 feet/1000 years Glass decomposed 15 g/cm²/ 100 years
Ruxton, 1966	Depth of weathering profiles	58 mm/1000 years
Leneuf & Aubert, 1960	Leaching rate of Si, Ca, Mg, K, Na	Ferrallitization of 1 m granite 22 000–77 000 years
Haantjens & Bleeker, 1970	Nature of mantles	Skeletal soil—5000 years Immature—5000–20 000 years Mature—20 000 years
Prendall, 1962	Duricrusts	9 m duricrust in 10^6 years
WEATHERING OF SURFACE STONES		
Birkeland, 1974	Boulder breakdown	Weathered surface in 10^5–10^6 years Decomposition in $>10^5$ years Difference in mineral ratios 10^4 years

3 Moderately weathered; rough surface, letters legible.
4 Badly weathered; letters difficult to read.
5 Very badly weathered; letters almost indistinguishable.
6 Extremely weathered; no letters remaining, rock scaling.

Assuming a constant rate of change and a hundred-year standard, the use of this sequence yielded a rock-durability sequence that agreed with the elevation of the rocks in the natural landscape.

mantles. The measures range from qualitative statements to accurate

Area	Comments
L. Ragunda, Sweden	From top 10 inch soil under pine heath
L. Ragunda, Sweden	From top 25 inch, mixed forest
Sand dunes, Southport, UK	Dunes dated by tree rings and maps
Reclaimed polders, Holland	Newly-exposed seafloor
Tyrol moraines	Soil acidity, silt and clay increased
Indonesia, soils developed on debris derived from eruption of Krakatoa	Pyroclastics, 45 years old 'considerable weathering'
Willamette Valley, Oregon, Iowa	CaCo$_3$ and pH required 10–300 years for steady state depending on initial values and amount of leaching
St Vincent, W.I.	Volcanic ash to clayey soil
Hydrographers Volcano, Papua	650 000 years old
Ivory Coast, forest zone	Granite, two mica
New Guinea	
Uganda	Granites
Colorado	Generalized from many studies. Surface of tills of Early to Pre-Wisconsin age.

Some other examples are given in Table 4.11. Generally such methods are imprecise with respect to time and are of most use when establishing durability sequences, to describe weathering zones, or to relate to quantitative measurements of rock strength or other properties and thereby extrapolate the measured values to other sections.

Lowering of rock surfaces
More exact measures of surface weathering can be obtained by recording the amount of rock lost from a surface. Although these methods give an impression of precision they suffer

Table 4.11   Some examples of weathering-grade classifications

Author	Class	Description
BASED ON FRIABILITY		
Ollier, 1965	1	Fresh, a hammer tends to bounce off the rock
	2	Easily broken with a hammer
	3	The rock can be broken by a kick*, but not by hand
	4	The rock can be broken apart in the hands, but does not disintegrate in water
	5	Soft rock that disintegrates when immersed in water
Melton, 1965	1	No oxidation stain, visible alteration or weakening to fresh fragments
	2	Surface stained and pitted, interior: no visible change
	3	Surface deeply pitted, thick weathering rind, interior: some stains, fragment broke after repeated hammering, definitely weaker than fresh rock
	4	Partially decomposed throughout, still cohesive, could be broken in hand or by light blow of hammer
	5	Thoroughly decomposed, not able to resist rough handling or being dropped
BASED ON SOIL MATURITY		
Shaw, 1928	Raw soil	—
	Young soil	Slightly weathered
	Immature soil	Only moderately weathered
	Semi-mature	Already considerably weathered
	Mature	Fully weathered
	Stage	
van Baren, 1954	Initial	Unweathered parent material
	Juvenile	Weathering started, much unweathered
	Virile	Easily weathered minerals decomposed, clay content increased
	Senile	Decomposition in final stage, only resistant minerals survive
	Final	Soil development complete and weathered out under prevailing conditions
BASED ON ZONAL CHARACTERISTICS (Granite rocks)		
	Zone	
e.g. Wilhelmy, 1958	1	Red-yellow loam (soil)
	2	Structure lost in upper part but not below
	3	Decomposed with rounded corestones
	4	Less weathered, blocks, angular, locked, some weathered bands
Soil zone		
	Zone	
e.g. Ruxton & Berry, 1957	I	Residual debris, structureless sandy clay, clayey sand
	II	Residual debris, rounded, free corestones
	III	Angular, locked corestones, residual debris
	IV	Partially weathered, minor residual debris along structural planes

* Ollier recommends we keep our boots on.

Table 4.11—*contd.*

BASED ON DEGREE OF DECOMPOSITION
Fookes and Horswill, 1969

Term	Grade	Abbreviation	Soils (i.e. soft rocks)	Rocks (i.e. hard rocks)
true residual	VI	Rw	The material is completely changed to a soil of new structure and composition in harmony with existing ground surface conditions	The rock is discoloured and is completely changed to a soil with the original fabric completely destroyed
completely weathered	V	Cw	The material is altered with no trace of original structure	The rock is discoloured and is externally changed to a soil, but the original fabric is mainly preserved; the properties of the soil depend in part on the nature of the parent rock
highly weathered	IV	Hw	The material is mainly altered with occasional small lithorelicts of original soil; little or no trace of original structure	The rock is discoloured: discontinuities may be open and the fabric of the rock near to the discontinuities is altered; alteration penetrates deeply inwards but lithorelicts are still present
moderately weathered	III	Mw	The material is composed of large discoloured lithorelicts of original soil; separated by altered material	The rock is discoloured; discontinuities may be open and surfaces will have greater discolouration with the alteration penetrating inwards; the intact rock is noticeably weaker, as determined in the field, than the fresh rock
slightly weathered	II	Sw	The material is composed of angular blocks of fresh soil, which may or may not be discoloured; some altered material starting to penetrate inwards from discontinuities separating blocks	The rock may be slightly discoloured; discontinuities may be open and have slightly discoloured surfaces; the intact rock is not, as determined in the field, weaker than the fresh rock
fresh	I	Fr	The parent soil shows no discolouration, loss of strength or any other effects due to weathering	The parent rock shows no discolouration, loss of strength or any other effects due to weathering

from the disadvantages that the exact form of the original surface is not always known and the measurements are only of the physical manifestations that have occurred to the surface. Chemical changes, solutional losses and internal changes to void space are not discovered. In addition, it is necessary to assume a constant weathering rate if average figures are to be assessed. The measurements are in the nature of uncontrolled experiments since the intensity and character of the responsible processes are largely unknown.

The data in Table 4.12 allow some general conclusions to be drawn. Rates of surface lowering seem to be slower in arid, periglacial and high-relief areas, but higher in humid temperate and warm temperate coastal conditions. The figures are, however, very variable; of great value would be experiments which exposed different rock types to the same controlled weathering conditions and the same rocks to different environments. Such experiments are

Table 4.12 Some records of surface-weathering rates

Author	Surface	Location	Rock type	Period (years)	Result	Average mm yr^{-1}
Geikie, 1880	Tombstones	Edinburgh	Sandstone	200	Little	0·051
			Slate	90	Engraving clear	0·085
			Marble	90	Crumbling	0·102
Goodchild, 1890	Tombstones	Yorkshire	Kirkby Stephen lmst.	500	Based on 1 inch	0·106
			Tailbrig lmst.	300	of weathering	
			Penrith lmst.	250	in period	
			Askrigg lmst.	240		
Barton, 1916	Ancient structures	Aswan	Granite	4 000	Fresh	0·0009—
		Giza	Granite	5 400	Flaking 0·5–0·8 cm	0·0015
Emery, 1960	Great Pyramid	Giza	Hard grey lmst.	1 000	Little	0·01–0·02 in part
			Soft grey lmst.	1 000	1–2 cm pits	0·2
			Grey shale	1 000	20 cm niches	0·2 (average over whole surface)
			Yellow limey shale	1 000	Rubble	1·32
Akimtzev, 1932	Kammetz fortress	Ukraine	Limestone	230	30·48 cm	
Sweeting, 1960	Erratic block	N. England	Carb. lmst.	12 000	30–50 cm	0·025–0·042
Sweeting, 1960	Glacial striae	N. England	Carb. lmst.	13	3–5 cm	2·2–3·8
Sweeting, 1960	Runnels in glacial surface	N. England	Carb. lmst.	13	15 cm	11·5
Bauer, 1962	Bare surface	Austrian Alps	Limestone	1 000	9–12·5 mm	0·009–0·0125
Bauer, 1962	Under soil and dwarf pine	Austrian Alps	Limestone	1 000	28 mm	0·028
Hodgkin, 1964	Jetty intertidal	Norfolk Is.	Limestone	16	28 mm	0·5–1·0
Ollier, 1969	Coastal notch	Point-Perm., Aust.	Limestone	—	—	1·0
Emery, 1941	Inscriptions	La Jolla, Calif.	Limestone	—	—	0·5
Kaye, 1959	Inter-tidal notch	Puerto Rico	Limestone	155	—	1·0
Dahl, 1967	Glacial surface	Narvik	Granite	10 000	—	0·00105
Rapp, 1960	Glacial surface	Spitsbergen	Lmst. chert	70	—	0·02–0·2
Rapp, 1960	Glacial surface	Spitsbergen	Sandstone	10 000	—	0·34–0·5
Rapp, 1960	Glacial surface	Kärkevagge	Micaschist, garnet mica, lmst.	8	—	0·04–0·15
Trudgill, 1976 (Using micro-erosion meter)	Coastal rock	Aldabra	Low magn. calcite	<1	—	0·35
			High magn. calcite	<1	—	0·51
			Well cemented algal lmst.	<1	—	0·09
Trudgill, 1976	Lmst. pavement	Co. Clare	Poorly cem. limestone	—	—	0·11
						0·003–6·3

now usually carried out using micro-erosion meters (Dahl, 1967; High and Hanna, 1968) or topographs (Boorman and Woodell, 1966) which permit accurate records. On the other hand, these results are highly localized and require a carefully designed sampling scheme before they can be related to a wider area. They are most useful where it is necessary to distinguish the rates on different lithologies over short periods and on a short time-scale.

Experiments using rock tablets
An extension to the methods of measuring surface lowering is to use rock tablets cut to a standard size, weighed, placed in a particular weathering environment for an experimental period and then reweighed. The weight loss can be related to other measurable variables such as pH, organic matter or permeability and used to establish comparison with other environments or rock types (Table 4.13). From Table 4.13B it is clear that it is possible clearly to distinguish the effect of lithology; the role of climate is, however, far from clear.

Table 4.13A   Surface lowering of limestones under various soils and vegetation types (after Trudgill, 1976)

Rock	Soil type	Vegetation	Area	Results, mm yr^{-1}
Limestone	Rendzina	Calcicole	UK	0·003–0·006
Limestone	Calcareous brown earth	Calcicole	UK	0·001–0·002
Limestone	Acid brown earth	Calcifuge	UK	0·008–0·03

B   Surface lowering of two limestones in temperate and tropical environments (after Trudgill, 1976)

Climate	Rock	Rate, mg yr^{-1}
Temperate	Carboniferous limestone	0·3
Temperate	Aldabra limestone	9·2
Tropical	Carboniferous limestone	0·1
Tropical	Aldabra limestone	2·7

Rates determined from water-chemistry data
A useful method of estimating the rate of chemical weathering is to analyse the water-chemistry data available from rivers, or on a smaller scale, to measure the dissolved ion content of throughflow water (Weyman, 1970), soil drainage from field drains (Perrin, 1965), small instrumented catchments (Cleaves *et al.*, 1970) or cave systems. There are many sources of variation, including the variable accuracy of the data, the uncertainty of the source of the dissolved material, the need to assume that the erosion is evenly distributed across the catchment area, and the assumption that the weathering system is in equilibrium with the environment and not rapidly changing from some unknown, unspecified conditions. Other sources of variation are the variability of rock type in the compared catchments, the wide range of catchment area and the unknown residence time of the water when it is in contact with its weathering materials.

Published figures such as those given by Strakhov (1967) (Table 4.14) must therefore be treated with reserve but it should be possible, as more data become available, eventually to set up controlled experiments to eliminate the worst errors. At present the figures show

Table 4.14 Annual chemical denudation for various rivers of the world estimated from water-chemistry data (after Strakhov, 1967)

	Rivers	Chemical denudation t km^{-2}
Northern rivers, temperate and cold climates	**Mountain rivers**	
	Neva	10·0
	Yenisei	11·4
	Luga	17·0
	Narova	17·7
	Onega	20·0
	Ob	12·2
	W. Dvina	25·0
	Mezen	16·0
	N. Dvina	48·0
	Plains rivers	
	Kolyma	5·5
	Yana	3·9
	Pechora	17·0
	Indigirka	11·0
	Amur	10·1
	Yukon	22·0
Rivers of temperate to hot subtropical and tropical climates	**Mountain rivers**	
	Kura	23·4
	Amu Darya	78·1
	Av. S.E. Asia	93·0
	Terek	125·0
	Rion	209·0
	Samur	270·0
	Sulak	290·0
	Rivers rising in mountains or uplands	
	Kuma	9·0
	Kalaus	5·0
	Amazon	13·0
	La Plata	18·0
	Mississippi	28·4
	Kuban	35·0
	Plains rivers	
	Dnieper	17·0
	S. Bug	14·0
	Don	22·0
	Ural	15·0
	Volga	32·5
	Dniester	39·5

Plate VI Deep-weathered gneiss, Dhankuta, Nepal. The development of rills and gullies by raindrop impact and overland flow has been accelerated by forest clearance and poor land management. (J.C. Doornkamp)

that chemical denudation does occur in all environments, not least in arctic-alpine areas, though in general chemical erosion seems to be most efficient in hotter mountainous areas (Plate VI).

The only extensive attempts to eliminate some the many sources of variability are studies of limestone denudation where rock-type has been held reasonably constant, the data have been corrected for catchment area, allowance made for variations of water hardness with temperature and CO_2 supply, and the data grouped for climatic regions. A full review of these problems and data is given by Smith and Atkinson (1976) (Figure 4.10). The calculation of chemical denudation (weathering products?) requires long records (30–50 years) of solution load and stream discharge. The procedure involves establishing the relationships between these two variables using least-squares analysis and the construction of a flow-duration curve. The overall solution loss is then calculated as the average of the instantaneous rate of removal of soluble products for each discharge class, weighted for the time over which discharge occurs.

For limestone regions few such lengthy records exist and therefore calculations are based on the general equation

$$X = \frac{\overline{Q}}{A} \cdot \frac{\overline{l}}{10^6 \, r} \cdot \frac{1}{n} \qquad (4.5)$$

where X=mean erosion rate (m^3 km^{-2} year^{-1})

$\overline{Q}$=mean annual runoff (m^3 year^{-1})

$\bar{l}$=mean water hardness (mg l^{-1} CaCO$_3$) or (CaCO$_3$+MgCO$_3$)

r=bulk density of limestone (g cm^{-3})

A=catchment area (km^2)

$\dfrac{1}{n}$ fraction of catchment area occupied by limestone

Several manipulations of this equation are used to overcome the paucity of data, to generalize, or to extrapolate from other areas. Common changes involve estimating Q/A from other gauged streams in the area; estimating Q from mean annual precipitation and mean annual potential-evaporation records; omitting MgCO$_3$ from the calculation of total hardness, and assuming a common value of 2·5 g cm^{-3} for the bulk density of the rocks, all of which introduce sources of error.

The most commonly used variant of the equation is that attributed to Corbel (1957) where

$$X=\frac{4ET}{100}\cdot\frac{1}{n} \tag{4.6}$$

where E=mean annual runoff (decimetres) (for $\overline{Q}/A$)

T=hardness

$\dfrac{1}{n}$ proportion of limestone in the catchment

Results from such calculations thus incorporate several possible sources of error but they do yield a useful idea of the orders of magnitude of chemical denudation (Figure 4.10). The figures only partially agree with the total chemical-denudation figures calculated by Strakhov (1967) (Table 4.14) and are often in disagreement with the figures for surface lowering of measured rock surfaces (p. 99) but these discrepancies are probably accounted for by the uncontrolled nature of the former data and the selective nature of the latter. In particular, it is difficult to compare surface-lowering data which do not measure internal losses, and total denudation figures which do not specify exactly where the erosion is taking place.

Weathering processes

Chemical weathering

Since this book is mainly concerned with geomorphological considerations, the accounts that follow are kept brief and do not go deeply into chemical, thermodynamic or kinetic principles. For the interested reader there are several modern summaries of these aspects in Keller (1957), Marshall (1964), Garrels and Christ (1965), Loughnan (1969), Garrels and Mackenzie (1971) and Curtis (1976a). The most important processes are solution, carbonation, hydration, hydrolysis, fixation, chelation, oxidation and reduction. In all of these, water plays a greater or lesser part, so that we will begin by examining its nature and behaviour.

The role of water

Water molecules consist of one oxygen and two hydrogen atoms covalently bonded. The molecule is dipolar since the bonding is asymmetrical. The hydrogen atoms have a net posi-

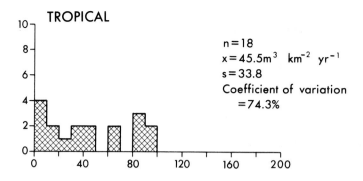

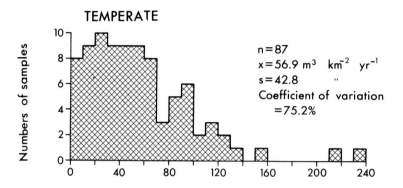

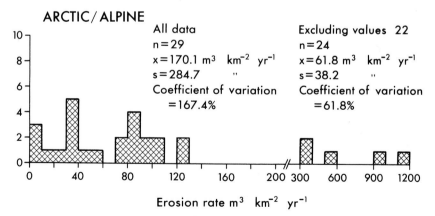

Erosion rate m³ km⁻² yr⁻¹

Figure 4.10 Limestone erosion rates in m³km⁻²yr⁻¹ for three climatic zones, corrected for catchment area and for variations in water hardness with temperature and CO_2 supply (after Smith and Atkinson, 1976).

tive charge and the oxygen atoms a double negative. This structure favours ionic bonding, through the hydrogen atoms, into tetrahedral groups of four molecules, and this provides the properties of high tensile strength, surface tension, adhesion and capillarity. The molecules that are connected by hydrogen bonds are also separated by unbound molecules that move freely and therefore allow flow. Molecules can exchange easily between molecule clusters so that ionic bonds are constantly changing. This movement is related to temperature, the number of bonds increasing at lower temperatures. Dipolar water molecules are able to orient themselves in an electrical field, attach themselves to other oppositely charged ions and therefore weaken the ionic bonding of other molecules. Water is thus a very effective solvent.

Weathering environments are controlled closely by the amount of water actually held in the soil mantle. Water is held by adhesive forces with organic and inorganic material, and cohesive forces with other water molecules. The thinner the film of water, the more tightly it is held; thicker films are more mobile. Since weathering occurs whenever water is present, the amount of water is significant mainly in the sense of flushing or removal of products. If this does not happen, ionic concentrations in thin films of capillary water may approach saturation which inhibits further weathering.

Solution
Recent studies of erosion rates of continents or catchments in various climates and on individual lithologies have revealed that present-day denudation is significantly aided by the removal of solids in solution (Table 4.15). Solution may therefore be regarded as an essential process by which minerals break down as they dissociate into their component ions. These are removed from the mineral structure and transported away in the environmental medium. The figures (Table 4.15) demonstrate a high rate of calcium and bicarbonate removal from the northern continents. Silica, however, seems to be more mobile in Africa than elsewhere. This situation is possibly explained by the gross climatic and organic differences between these regions.

In most environments solution is aided by the ability of water to move through the mantle and by its overall availability. It is thus heavily dependent on the environmental factors— temperature, redox potential (p. 115) and pH (p. 112) of the solvent. The solubilities of some common minerals are shown in Figure 4.11.

The so-called special case of the solution of limestone, however, deserves further mention, for whereas the solution of ions derived from the silicate rocks leads to breakdown and the production of new minerals or detrital clasts, limestone is often very pure and leaves little residue. Thus further solution is not inhibited by the production of insoluble clays and hydroxides.

The two main minerals involved in limestone weathering are calcite ($CaCO_3$) and dolomite ($CaMg(CO_3)_2$). These are both soluble particularly in water containing carbonic acid derived by the solution of carbon dioxide in water, thus:

$$CO_2 + H_2O \rightleftharpoons H_2CO_3 \tag{4.7}$$

$$CaCO_3 + H_2CO_3 \rightleftharpoons Ca^{++} + 2HCO_3^- \tag{4.8}$$

or $$CaMg(CO_3)_2 + 2H_2CO_3 \rightleftharpoons Ca^{++} + Mg^{++} + 4HCO_3^- \tag{4.9}$$

The solubility of carbon dioxide depends on its concentration in the air (partial pressure), and the temperature of the solution. At higher partial pressures, more CO_2 can be dissolved; for a constant concentration, but higher temperature, less can be dissolved (Picknett, 1964, 1972).

It follows that calcite and dolomite are removed when carbon dioxide is available, as con-

Table 4.15 Chemical composition, dissolved load and solution denudation for the continents

Continent	Total dissolved load ×10^{14} g	Solution denudation t km^{-2}	Composition (p.p.m.)											
			NO$_3$	Fe	HCO$_3$	SO$_4$	Cl	Ca^2	Mg^2	Na	K	SiO$_2$	Total
N. America	7·0	33	1	0·16	68	20	8	21	5	9	1·4	9·0	142
S. America	5·5	28	0·7	1·4	31	4·8	4·9	7·2	1·5	4	2	11·9	69
Asia	14·9	32	0·7	0·01	79	8·4	8·7	18·4	5·6		9·3	11·7	142
Africa	7·1	24	0·8	1·3	43	13·5	12·1	12·5	3·8	11	—	23·2	121
Europe	4·6	42	3·7	0·8	95	24	6·9	31·1	5·6	5·4	1·7	7·5	182
Australia	0·2	2	0·05	0·3	31·6	2·6	0	3·9	2·7	2·9	1·4	3·9	59
World total	39·3												
World average	—	—	1	0·67	58·4	11·2	7·8	15	4·1	6·3	2·3	13·1	120

From Garrels and Mackenzie (1971) and Livingstone (1963) (Figures in p.p.m. from Loughnan, 1969)

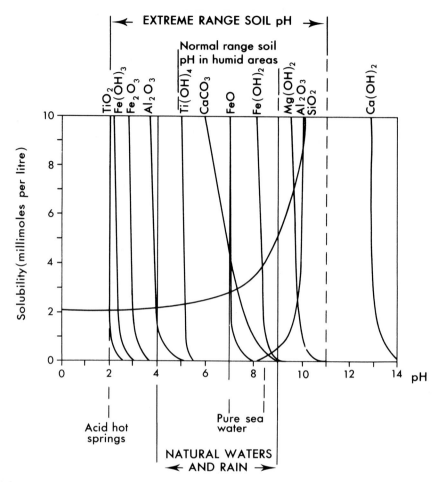

Figure 4.11 Solubility against pH for some products of chemical weathering. Some pH limits of natural environments are also shown (Loughnan, 1969).

trolled by temperature, but it is also necessary to take into account the time that the solution is in contact with the rock itself. As long as the water is in contact it can continue to dissolve limestone until the reactions involved attain equilibrium. If the water leaves the limestone mass before this, it will be unsaturated with calcium carbonate (aggressive) and the dominant control of solution is the rate of water flow. If water movement is slow (say in rock pores), the saturation value may be the appropriate measure.

It is necessary to clarify the meaning and importance of the term 'aggressivity'. The bicarbonates in solution are prevented from precipitating as carbonates as long as there is a certain quantity of free carbon dioxide in solution—the equilibrium CO_2 content. If there is more carbonic acid than this, the excess is aggressive CO_2 (Rodier, 1960) which will permit the further solution of carbonates. It is therefore important to know the various rates of water flow involved as well as the extent to which equilibrium has been achieved. For example, in permeable high-relief or high-rainfall areas, the flow rates should be important, but in low-relief, stagnant or poorly-drained regions, saturation will be dominant. The time taken for equilibrium to be achieved will vary from hours for slowly moving films in soil or rock pores, to several days for freely moving water in cave systems. Turbulent water in swallets or short conduits rarely reaches equilibrium. The hardness of water leaving a system will therefore depend heavily on the physical characteristics of that system.

It should also be borne in mind when assessing hardness values that the hardness of water is due not only to calcium and magnesium carbonates but also to sulphates, chlorides and nitrates. The former are termed temporary or carbonate hardness, the latter permanent hardness. Non-carbonate hardness may account for a considerable proportion of the denudation of limestone terrains and great care should be taken in applying such figures to geomorphological studies of water chemistry (Table 4.16).

Table 4.16 Principal salts causing hardness (after Douglas, 1969)

Total hardness	Non-carbonate permanent	Carbonate temporary
Calcium hardness	$CaSO_4$ $CaCl_2$ $Ca(NO_3)_2$	$CaCO_3$
Magnesium hardness	$MgSO_4$ $MgCl_2$ $Mg(NO_3)_2$	$MgCO_3$

Readers who are interested in these complex phenomena should read the detailed accounts in Garrels (1960), Trudgill (1976), or Smith and Atkinson (1976). A further discussion on hardness and aggressivity of water, together with methods of measurement, is given by Douglas (1969).

Carbonation
Carbonation is the reaction of carbonate and bicarbonate ions with minerals and is a common process, in association with hydrolysis, in the breakdown of feldspars and carbonate minerals. Carbon dioxide which is abundant in the atmosphere and soil air is absorbed by water to form carbonic acid (H_2CO_3). This facilitates the base-exchange process since it ionizes to form H^+ ions and HCO_3^- ions. As discussed earlier, calcium carbonate and magnesium carbonate are removed without the production of a residue (except from resistant materials). In the case of feldspars, however, the hydrogen ions combine to produce a clay residue of aluminium silicates but the bicarbonate ions commonly form soluble products which are removed. For example, potassium carbonate is removed from the weathering of orthoclase feldspars thus:

$$2KAlSi_3O_8 + 2H_2O + CO_2 \rightarrow Al_2Si_2O_5(OH)_4 + K_2CO_3 + 4SiO_2 \qquad (4.10)$$

Hydration
Hydration is the combination of an element or compound with water, the minerals actually absorbing the water into the mineral lattice. The process seems to be important in four ways.

First, some minerals such as gypsum ($\rightleftharpoons$anhydrite) and ferric oxide (to goethite) are readily hydrated to form new minerals, usually with an accompanying volume change. The process is reversible and the resultant flexing may lead to rock disintegration.

e.g. $$CaSO_4.2H_2O \rightleftharpoons CaSO_4 + 2H_2O \qquad (4.11)$$

Secondly, some clay minerals, e.g. montmorillonite, possess wide lattices that permit the easy ingress of water molecules and are characterized by swelling during wetting and drying cycles. The sodium-rich Mancos shales of Colorado will increase in volume by up to 60 per cent under free swelling conditions and repeated wetting can cause soil creep. Similar swelling takes place in montmorillonite clays at the head of the Gulf of Suez, even under quite arid conditions, and can cause cracking of overlying materials.

Thirdly, salts trapped in pore spaces can undergo rapid and repeated (even diurnal) hydration. The resulting pressures (see p. 28) can fracture rock if they exceed the tensile strength of the rock or heave individual crystals from the rock face to cause granular disintegration.

Finally, hydration, by taking water deep into mineral structures, prepares the way for further processes of hydrolysis and solution. Thus by 'reaching places that other agents cannot reach', weathering is greatly enhanced.

The hydrogen-ion concentration (pH) and hydrolysis
The relative concentration of the hydrogen H^+ and hydroxyl OH^- ions determines the soil reaction or acidity of a solution. This is expressed as the pH, the logarithm to the base 10 of the reciprocal of the hydrogen-ion concentration, measured in grams per litre. Neutral pH $(=7)$ is therefore 0·0000001 (1×10^{-7}) g l^{-1} of water of free H^+ ions. (The reciprocal is 1×10^7 whose log is 7.) Values above or below this standard are used to express alkalinity or acidity.

The concentration of H^+ ions is provided by several processes including the polarization of the water molecule at crystal edges and dissociation into hydrogen and hydroxyl ions; oxidation of sulphides to yield sulphates and sulphuric acid; the dissolution of CO_2 with water to give carbonic acid, especially from the soil atmosphere; the production of acids by humus and biotic activity; the osmotic exchange of H^+ ions and nutrients, such as Ca^{++}, Mg^{++}, K^+ cations, by plant roots; and the absorption of cations at the surfaces and edges of clay minerals.

The concentration of H^+ ions is of fundamental importance in chemical weathering. The most important summary points are that the decomposition of minerals is heavily dependent on hydrolysis. H^+ ions are small, mobile, carry large electrical charges and readily penetrate mineral lattices so that mineral breakdown can proceed rapidly. High concentrations allow the formation of complex clay minerals, the variety of which is closely controlled by the exact pH level. Thus, for example, neutral pH leaves Al_2O_3 as a residue with SiO_2 partly soluble. At alkaline pH, Al_2O_3 is very soluble and SiO_2 partly soluble so that both can be removed. Some common mineral solubilities with respect to pH are shown in Figure 4.11, on which pH values for typical soil and water environments are also plotted.

The weathering of the silicate minerals is accomplished readily by the process of hydrolysis in which the cations of a mineral are replaced by the H^+ ions of water and the OH^- ions combine with them to form solutions. It involves the chemical decomposition of a substance by water, the water itself also breaking down. In hydrolysis, water is therefore a reactant and not merely a solvent and the H^+ ion enters the structure of the weathering products. The actual chain of reactions is known to be complex and probably step-like from one product to another, but there is no doubt that hydrolysis is a most active process to which coarse-grained and feldspar-rich rocks such as granite are particularly vulnerable.

Hydrolysis is often accompanied by an increase in the pH of the solution. A suitable measure is the hydrolysis reaction or abrasion pH of a crushed mineral. For example, it can be demonstrated that, when alkali-rich minerals are ground up in pure water, the production of fresh mineral–water interfaces leads to the release of alkali cations and a rise in pH. For any mineral there is a limit beyond which further grinding does not increase the pH value and a state of equilibrium is reached. The feldspars generally have high abrasion pH indices, such as orthoclase feldspar (pH8), oligoclase (pH9) and albite (pH10). The micas range from pH7 for muscovite to pH8 for biotite. The carbonates yield pH8 for calcite and pH9 for dolomite, and quartz has a lower value at pH7. Of course, in the natural weathering environment these high values are modified by the production of H^+ ions from sources such as rainwater or decayed organic matter, and by the leaching-out of some of the cations, but the index does provide a comparative measure of the initial hydrolysis reaction.

The products of hydrolysis can be leached out altogether, remain in solution, take part in cation exchange or be part of the crystal lattice of new clay minerals. Silica and aluminium have complex reactions. Silica is slightly soluble at all pH values but increases markedly at pH 9 due to the ionization of H_4SiO_4. Aluminium is not very soluble in the pH range of natural waters (Figure 4.11) and generally forms clay minerals and hydroxides.

Finally, it is worth noting that hydrolysis causes expansion and contraction of silicate structures. The increase in volume leads to pronounced physical changes to the rock including disintegration of the surface and a loss of the original textural appearance.

Ionic potential and leaching processes
Ionic potential (Z/r) is the ratio of the charge of an ion (in valence units, Z) to its radius (r) in ångström units $(10^{-10}\,m)$ (Cartledge, 1928). Ions in solution attract water molecules in proportion to the factor Z/r. Some elements such as the cations Na^+, K^+, Ca^{++}, Mg^{++} are in true ionic solution in weathering, and possess low ionic potentials of $Z/r<3.0$. Elements with high potentials, $Z/r>9.5$, form complex soluble anions with oxygen, and include silicon and nitrogen. In an intermediate position are such elements as aluminium and iron which possess Z/r values of 3–9.5 and tend to be precipitated as hydrolysates during the hydrolysis process (Gordon and Tracey, 1952). This measure therefore provides some idea of the likely behaviour of ions during leaching.

Leaching is widely regarded as a fundamental factor in weathering, the degree of leaching being controlled by the climate, topography and parent material. Within a weathering mantle it varies with the zonation of the profile, there being pronounced differences between the vadose (free percolating water) zone and the phreatic (beneath the water table) zone. In the soil profile it is largely responsible for the processes of illuviation and eluviation; it is controlled both by the position of the water table and the ability of the soil to transmit water by lateral throughflow.

The rate of mineral breakdown and the nature of the secondary products are directly related to the amount of water passing through the weathering mantle. Weathering reactions are enhanced or retarded by the removal of soluble products which either accumulate in the lower horizons, are precipitated as crusts or removed altogether as part of the chemical denudation of the landmass. Eventually even the most resistant minerals can be removed and thus leaching should be regarded as the most fundamental process of weathering.

Leaching can be estimated using one of the leaching factors (Table 4.2, p. 79) but a good idea of the relative characteristics can also be obtained by measuring the infiltration capacity of the soil, the amount of throughflow from different depths, or by leaching experiments and permeability tests in the laboratory. The mobility of many of the common cations under leaching conditions has been summarized by Loughnan (1969). These are shown in Table 4.17.

Cation exchange
Many clays and colloids (suspensions of particles of less than 2 μm in diameter) are of laminated form with large surface areas. They are usually regarded as possessing negative

Table 4.17   The mobility of the common cations in relation to leaching for pH values in the normal range of 4–9 (after Loughnan, 1969)

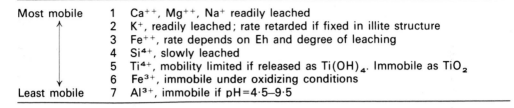

Most mobile	1	Ca^{++}, Mg^{++}, Na^+ readily leached
	2	K^+, readily leached; rate retarded if fixed in illite structure
	3	Fe^{++}, rate depends on Eh and degree of leaching
	4	Si^{4+}, slowly leached
	5	Ti^{4+}, mobility limited if released as $Ti(OH)_4$. Immobile as TiO_2
	6	Fe^{3+}, immobile under oxidizing conditions
Least mobile	7	Al^{3+}, immobile if pH$=4\cdot5$–$9\cdot5$

electrical charges and hence repel each other in solutions. They try to achieve neutrality by absorption of positively charged cations such as calcium (Ca^{++}), hydrogen (H^+), potassium (K^+) and magnesium (Mg^{++}) to the surface, the edges or within the lattice of the clay crystal. These cations, although attached (bonded) to the mineral, may nevertheless be exchanged with other cations in the solution. This is known as 'cation exchange' and is measured by the cation exchange capacity or the total negative charge on the surface (cec) expressed as the number of milliequivalents (meq$=1$ mg of hydrogen) absorbed by 100 g of oven-dry material (meq $100\,g^{-1}$). The exchangeable cations are those that are attracted to the surface and the total of the exchangeable base ions (non-hydrogen) is known as the 'base saturation', expressed as a percentage thus:

$$\frac{Ca+Mg+Na+K}{Ca+Mg+Na+K+H}\times 100 = \text{base saturation }\% \qquad (4.12)$$

If the clay colloids hold many H^+ ions, the clay is an acid clay; the reverse is true if the clay holds many ions capable of forming bases. Bases are generally released from silicate minerals in the order $Ca>Mg>Na>K$, and held in the order $H>Ca>Mg>K>Na$. Exchanges therefore tend to be dominated by H^+ and Ca^{++}, with Na^+ often being removed.

In addition to clay colloids, organic colloids, mainly derived from humus, also possess high cation exchange capacity. They are less stable than clays and can fluctuate rapidly in the weathering environment. In addition, they have a high capacity for water retention and thus greatly enhance the weathering process.

Fixation

Fixation occurs when inorganic material absorbs ions so that they cannot take part in further simple exchanges. A common feature of detritus derived from potassium-rich silicate rocks is that minerals such as muscovite and K-feldspars often remain as recognizable particles long after other minerals have broken down. Although potassium and sodium are often present in igneous rocks in similar proportions, in natural fresh water and sea water, sodium is ten times more abundant. This suggests that potassium is, in some way, often immobile. In particular it is believed that it is trapped in expandable lattice clay minerals such as montmorillonite. This process and the conditions that control it are imperfectly understood but some work indicates that, in environments that favour the production of layered lattice minerals, some cations may be preferentially fixed and their removal retarded.

Chelation

Chelation involves the formation of ring structures between a metallic ion and complexing agent. There is more than one bond between the agent and the ion so that the complex ring structures so formed are stable and therefore effectively remove ions from the weathering solution. The agents are mainly derived from plant debris and organic secretions, and by the direct removal of metallic ions from a mineral (Lehmann, 1963). This condition is one of the reasons why weathering can be very effective in organic-rich environments. It is even used as an explanation of increased weathering under lichen-covered surfaces, for lichens are known to excrete chelating agents (Schatz, 1963) and increase the concentration of iron, silicon and calcium in surface rinds. In addition some ions are made more soluble as chelates than they would otherwise be in solutions without chelating agents.

Redox potential (Eh), oxidation and reduction
Krauskopf (1967) defines redox potential as the ability of a system to bring about an oxidation or reduction reaction. The stability of any oxidation state, such as the ferrous or ferric compounds of iron, depends on the energy change needed in the addition or removal of electrons. Measured quantitatively this is expressed in terms of the standard 'oxidation' of hydrogen to hydrogen ions, the arbitrary value of 0·00 volt

$$H_2 = 2H^+ + 2e^-, \quad Eh = 0{\cdot}00 \text{ volt}$$

being given for unit activity of the hydrogen ion (i.e. $pH = 0$). In aqueous solution, the upper limit is given by

$$H_2O = \tfrac{1}{2}O_2 + 2H^+ + 2e^-, \quad Eh = 1{\cdot}23 \text{ volt} \tag{4.13}$$

For higher values, water is not stable and oxygen is released; at negative values hydrogen is released. This yields a useful diagram of natural environments when Eh is plotted against pH, since Eh varies with the concentration of the reacting substance (Figure 4.12). Where H^+ or OH^- is involved, the Eh varies with the pH of the solution, Eh becoming lower as pH increases. Eh is also dependent on climate, topography, the amount of organic matter and accessible oxygen.

The chemical definition of oxidation is the increase in the positive valence or the decrease in the negative valence of an element or the loss of an electron (e^-) from an element. In weathering this commonly means a combination of a mineral with oxygen to form oxides or hydroxides, and usually occurs with atmospheric or soil oxygen dissolved in the intermediary of water. Oxidation thus occurs where there is ready access to the atmosphere and oxygenated waters. Well-drained conditions, the destruction of organic matter and higher temperatures all favour this process and the minerals most commonly involved are iron, manganese, sulphur and titanium. Iron, for example, can exist in several valance states that rely for their relative solubilities on the redox potential, the various states being removed or precipitated at various Eh–pH values.

Oxidation can act as a distinct process of rock breakdown in which the electrical neutrality of a mineral is disturbed; ions must leave the lattice in order to achieve a balance and the lattice then either collapses or the vacancies are occupied by other ions, leading to further weathering. In other cases oxidation only takes place, even if the environment is an oxidizing one, if minerals such as iron are released by some other weathering process.

Reduction is the addition of electrons to an element. The process must be considered together with oxidation because both reactions are controlled by the redox potential. When minerals, which are often in an oxidized state, are placed in an environment where oxygen is excluded, reduction will take place. Such conditions are usually below the water table, in areas of stagnant water, waterlogged ground and gley soils. Cooler and wetter conditions, which often favour organic accumulation, encourage reduction. A typical example is the reduction of ferric to ferrous iron which is then soluble in the correct pH conditions and can be removed in solution.

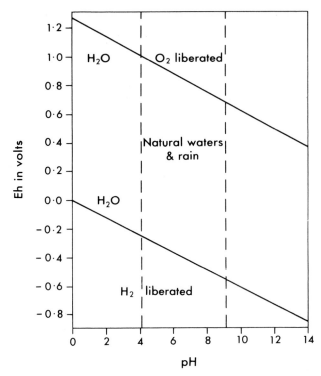

Figure 4.12 Relationship between EH and pH to define the limits of the natural weathering environment. The scale extends in both positive and negative directions, positive readings being oxidizing and negative readings reducing environments.

Physical weathering

The forces applied to a rock mass are discussed elsewhere in this volume (p. 134). Here, it is only necessary to comment that physical weathering involves the comminution of rock by mechanical processes alone, and hence all the processes depend on some applied force. These include gravitational forces such as overburden pressure, normal load and shearing stress; expansion forces due to temperature change, crystal growth or animal activity; and water pressures or tensions which are important in the study of processes controlled by wetting–drying cycles as well as in calculations of the 'effective' stresses applied to a rock mass. In addition, since many of these forces are applied both at the surface and within a rock mass or mantle, it is necessary to state the lateral and vertical confining conditions in order to understand responses such as linear expansion or rock fracture.

Pressure release
Pressure release or unloading was suggested by Gilbert (1904) as the mechanism responsible for producing shear planes, curved and horizontal joints and the subsequent distintegration of rock masses. It is now widely believed that wherever the removal of overlying rock has caused the vertical pressure to be relieved, fractures will develop approximately parallel to the ground surface. If the ground surface is curved, radial release of confining pressure will

tend to produce curved sheets of rock (exfoliation sheets), and if the rock face is vertical, vertical joints and tension cracks are likely to develop. Nearly all textbooks quote spectacular examples such as the exfoliation of inselbergs in some arid areas, immense rock sheets on the igneous domes of Yosemite, California (Plate VII), or vertical rock slabs such as Threatening Rock in New Mexico. The process is used to explain the joint patterns and slope failures of recently deglaciated cirques and glacial trough walls; the dome-like form of some Dartmoor tors, the heaving of mining tunnels and the sudden rock-bursting of new quarry faces and open-cast mines. It is even possible that expansion of freshly exposed clays and the cambering of escarpments may be due to this process.

Plate VII Dilatation joints (sheet structures) in granite, Yosemite, California. (C.E.)

Despite this interest, however, it has to be admitted that the process is poorly understood. A common explanation is as follows. The strength of a confined mass increases in proportion to its confining, overburden pressure. If overburden is removed, the strength of the buried rock decreases and the confining pressures are reduced. The internal stresses are then relieved by expansion. This produces cracks parallel to the surface which increase in spacing and eventually disappear at depth. The qualitative nature of this discussion emphasizes that there is a need for geomorphology to develop explanations based on principles of rock mechanics (Labasse, 1965) and the nature of rock strength.

As an example, Carson and Kirkby (1972) have summarized the work of Terzaghi (1943, 1962a) and Bjerrum and Jorstad (1968) with respect to the stresses involved in the unloading of rock walls. They show that in cases where lateral pressures are removed (e.g. by incision of a stream), it is possible to calculate the vertical pressure:

$$\sigma_z = \gamma \cdot z \tag{4.14}$$

the horizontal pressure:

$$\sigma_x = K \cdot \sigma_z \tag{4.15}$$

and the change in the horizontal pressure following the removal of lateral support:

$$\Delta\sigma_x = E \cdot \Delta\varepsilon_x \tag{4.16}$$

where, in the active state of stress, σ_z = vertical pressure, σ_x = horizontal pressure, γ = bulk unit weight of material, z = depth below the surface, K is the coefficient of earth pressure, E = modulus of elasticity and $\Delta\varepsilon_x$ = strain in the x direction (Figure 4.13).

The minimum horizontal pressure or the condition of limiting stability is given by

$$\sigma_x = \gamma z \tan^2(45 - \phi/2) - 2c \tan(45 - \phi/2) \tag{4.17}$$

and the maximum height of a cut before failure by

$$H'_c = \frac{4c}{\gamma}\tan(45 + \phi/2) - z \tag{4.18}$$

where c = cohesion, ϕ = angle of shearing resistance, H'_c is the maximum height of a rock face before failure and z is the depth of the tension crack.

Of most interest in terms of the unloading argument is that a zone of tension often develops in the upper part of the slope which may produce open cracks. The thickness of this zone is given by:

$$Z_0 = \frac{2c}{\gamma}\tan(45 + \phi/2) \tag{4.19}$$

Finally, it is worth noting that the response of the rock to pressure release may be very variable over time. In extreme cases (quarry walls), response can be very rapid. In schists exposed by the retreat of the Franz Josef Glacier, Gage (1966) reported sudden fracture a 'few years' after unloading. Elsewhere, in massive rocks, glaciated slopes cut by 'valley joints' have remained stable since the last glaciation (Bjerrum and Jorstad, 1968) but during that time they have been closely jointed.

Crystal growth: frost weathering

Frost weathering is defined as the mechanical breakdown of rock by the growth of ice within the pores and discontinuities of a rock subjected to repeated cycles of freezing and thawing.

There has been considerable discussion on the nature of the most effective freeze–thaw conditions (Cook and Raiche, 1963; Embleton and King, 1975b). The critical temperature points in defining freeze–thaw cycles have never been agreed since the temperatures of the air or soil are not in themselves accurate indices of the extent of freezing. Rocks have differing thermal conductivity, and other factors such as water content, rate of change of temperature and dissolved ions all influence the effectiveness of the process. Each weathering case has to be examined separately and it will probably never be possible to define a universal, 'effective', freeze–thaw condition.

Following Hewitt (1968), a single frost cycle can be defined as in Figure 4.14 and Table 4.18. The number of frost shifts (one crossing of the 0°C boundary per period) is a useful

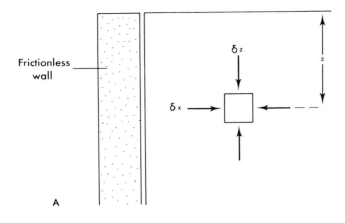

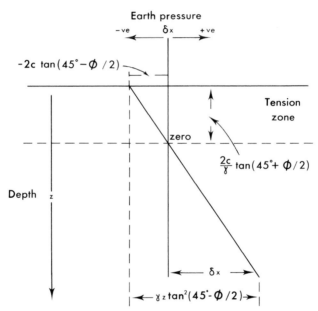

Figure 4.13 The stresses involved in analysing the unloading of rock walls (Carson and Kirkby, 1972).

measure of the variability of the environment. The wavelength of the cycle and the rate of temperature change are vital descriptions of the intensity of the process. One freeze–thaw cycle is composed of three crossings of 0°C, at the beginning, middle and end of each cycle. The freeze–thaw ratio Tn/Tx, where Tn is minimum temperature of the cycle and Tx is maximum temperature, yields a value of unity for a symmetrical curve about freezing point and <1 for a dominant melting phase. The mean monthly freeze–thaw ratio measures a freeze or thaw period. It has to be admitted, however, that there has been insufficient work relating these parameters to rock breakdown.

Perhaps the most interesting parameter is the rate of freezing. Rapid freezing favours breakdown by wedging since it results in the freezing of near-surface water. This seals the

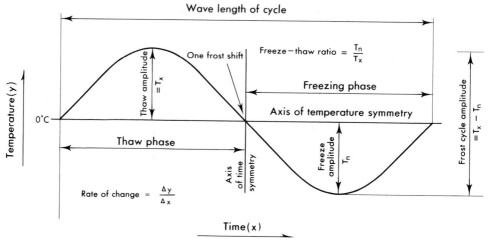

Figure 4.14 The terminology used to define a single, complete frost cycle (after Hewitt, 1968).

voids containing unfrozen water, creating a closed system in which high pressures can develop (Battle, 1960; Pissart, 1970). In fine-grained soils, slower freezing can assist breakdown since it permits the development of segregated ice (Chapter 6) and maximum growth of ice crystals.

The whole subject of the pressures exerted by ice crystals and ice formation, and the validity of freeze–thaw action as a weathering process is vigorously debated (White, 1976). Most textbooks note that water, on freezing, increases in volume by 9·05 per cent. The cryostatic

Table 4.18 The terminology of the freeze-thaw environment

Parameter	Definition
Frost shift	Any crossing of the 0°C boundary
Freeze–thaw cycle	Three crossings of the 0°C line at start, middle and end of cycle
Absolute no. of frost shifts	No. of crossings of 0°C in a time period
Mean no. of frost shifts	No. of crossings of 0°C/time period
Frequency	Wave length of cycle (daily, annual, etc.)
Intensity	Amplitude of temperature change and slope of temperature curve
Effective temperature shift	The temperature shift which is effective for a given process (e.g. soil freezing)
Thaw amplitude	Max. temp. (°C)
Freeze amplitude	Min. temp. (°C)
Freeze–thaw ratio	Freeze amplitude/thaw amplitude
Frost-cycle amplitude	Max. minus min. temperature
Rate of change	Δ Temp./Δ Time
Short frost cycle	Less than diurnal (at least 4 frost shifts in 24 hours)
Daily frost cycle	Approx. 24 hours; freezing at night, thaw by day
Several-day cycle	One melt, one freeze–thaw in >36 hours
Monthly	Deduced from mean monthly max. and min. temperatures
Annual	Defined from mean temperatures of warmest and coldest months (freezing in winter, thawing in summer)
Several-year cycle	Thawing only during warm summers

pressure will continue to increase in a closed system as temperature falls to $-22°C$, the point at which ice (Ice I) changes to Ice III with a lower volume. The maximum pressure (Bridgman, 1912, 1914; Grawe, 1936) of $2115\,kg\,cm^{-2}$ can only be reached if the process takes place in a closed system, with a one-component system of water at $-22°C$, and in a confined space able to resist the pressure. These conditions will never be found in nature since the water system usually includes air and the maximum tensile strength for rocks is about $250\,kg\,cm^{-2}$ (granite $70\,kg\,cm^{-2}$, marble 49–$63\,kg\,cm^{-2}$, limestone $35\,kg\,cm^{-2}$, sandstone 7–$14\,kg\,cm^{-2}$). Battle (1960) has shown that 'for maximum shatter to take place, freezing must be induced from the crack downward, forming a closed system in which pressure can build up'. A similar condition would have to occur with respect to water trapped in soil pores, the freezing of the outer layer preceding the freezing of inner water, thus allowing grains or flakes to be prized loose.

Recent work by Mellor (1970) has confirmed that the rate of freezing and the resulting freezing strain are important. Rapid freezing of pore water can cause its sudden expansion, initiating disruptive strains which approach and, theoretically, exceed the tensile failure strain of rocks. The most favourable conditions occur when water saturation exceeds 50 per cent. Other experiments (Washburn, 1973) have shown that the amount of water available partly determines the response, a fact that perhaps explains the intense weathering sometimes observed on shorelines and seepage lines.

Despite the uncertainty about the nature of the process there are many field observations of ground and rock being cracked open in cold conditions (Washburn, 1969) and several freeze–thaw laboratory experiments (Potts, 1970; Tricart, 1956; Wiman, 1963) that show the production of fine particles, loose mineral grains and flakes from experimental rock fragments. There are few observations and no laboratory experiments on the wedging of large rock fragments. Nevertheless, most authors suggest that frost weathering produces angular fragments over a wide range of sizes (10^{-5} to 10^{1}m). The ultimate reduction is probably to silt-sized debris (Hopkins and Sigafoos, 1951) and rarely to clay.

Other processes, such as heaving and thrusting, can result from frost action, and these are covered in Chapter 6.

Crystal growth: salt weathering

The general heading of salt weathering includes three major processes of rock breakdown (Cooke and Smalley, 1968):

1 The thermal expansion of salt crystals.
2 The hydration of salts.
3 The growth of salt crystals.

The efficiency of these processes has been widely recognized (Birot, 1968a; Wellman and Wilson, 1965) and has recently been the subject of considerable research (Goudie, Cooke and Evans, 1970; Evans, 1970; Kwaad, 1970; Goudie, 1974; Fookes and Collis, 1976).

Thermal expansion of salt crystals   The expansion of salts when they are heated depends on temperature and their thermal properties. Temperature ranges in deserts are often as high as $30°C$ (Hume, 1925) with surface diurnal ranges as extreme as $50°C$ and maximum surface temperatures of $80°C$ or more (Cloudsley–Thompson and Chadwick, 1964; Peel, 1974). Many salts including calcium, sodium, magnesium, potassium and barium have high coefficients of volumetric expansion (Figure 4.15). Their volumes increase approximately linearly with temperature up to $300°C$, and the expansion may be as much as three times that of granite under similar conditions. At the high temperature ranges quoted above, a salt such

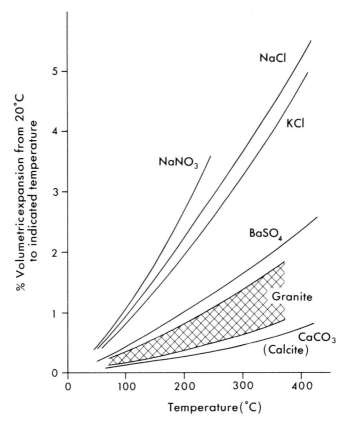

Figure 4.15 The volumetric expansion of some common salts found in desert regions and also, for comparison, granite (Cooke and Smalley, 1968).

as sodium chloride might expand by 1 per cent and thus salt crystals in near-surface pores might well cause splitting or granular disintegration.

Goudie (1974) describes experimental work on this process. 3 cm cubes of sandstone were immersed in saturated sodium nitrate solution for 24 hours, dried to a constant weight at 60°C and subjected to temperature cycles of 60°C for six hours and 30°C for 18 hours, not unlike a 'normal' desert surface régime. Unfortunately, no measurable change in sample weight occurred over 58 cycles. In a second test, Goudie subjected similar salt-saturated cubes to heating, under an infra-red heater, for 2160 cycles at 60°C for one hour and 20°C for one hour, over 180 days. Again there was no detectable loss of weight or cracking. The results were inconclusive, perhaps because the method used did not lead to salt growth throughout the rock pores so that any expansion merely filled up available pore space. Indeed, Goudie does not show that preliminary salt crystallization occurred at all. Possibly tests with other salts will prove more effective.

Hydration of salts   The hydration and dehydration of salts such as gypsum-anhydrite can take place in many environments. Cooke and Doornkamp (1974) suggest that the process may involve the following steps. During the day, in high temperatures, salt crystals that are low in the water of crystallization may be formed. At night the anhydrous salts absorb water from the atmosphere and higher hydrates are formed. Important to this process are relative humidity and temperature as well as the dissociation vapour pressures of salts.

The pressures involved have been calculated by Winkler and Wilhelm (1970):

$$P = \frac{(nRT)}{(V_h - V_a)} \; 2 \cdot 3 \log \left(\frac{P_w}{P'_w} \right) \tag{4.20}$$

where P=hydration pressure (atmospheres); n=no. of moles of water gained during hydration to the next higher hydrate; R=gas constant; T=absolute temperature (°K); V_h= volume of hydrate; V_a=volume of original salt; P_w=vapour pressure of water in atmosphere (mm mercury at given temperature); P'_w=vapour pressure of hydrated salt (mm mercury at given temperature). These calculated pressures can reach considerable levels as long as the hydration takes place in 12 hours in a sealed pore, and provided that they do not exceed the tensile strength of the rock.

Salt crystal growth   Salt crystallization has long been accepted as an effective weathering process. Soil scientists point to the presence of K horizons in soils, in which sub-surface horizons are so impregnated with carbonate that the soil morphology is determined by the carbonate habit. Usually all mineral grains are coated in carbonate and separated or pushed aside by crystal growth. In desert areas, especially in seasonally wetted gravels or sabkhas, beds can be forced apart and the surface heaved into polygonal form by gypsum crystals and sodium chloride. It is probably the most effective of all salt-weathering processes.

The actual pressures can be calculated (Cooke and Doornkamp, 1974);

$$P = \frac{RT}{V} \cdot \lg n \left(\frac{C}{C_s} \right) \tag{4.21}$$

where P=pressure on the crystal; R=gas constant; T=absolute temperature; V=molar volume of crystalline salt; $\lg n$=natural logarithm; C_s=concentration at saturation point without pressure effect; C=concentration of a saturated solution under external pressure P.

There have been many attempts to demonstrate the effectiveness of the process in the laboratory and engineers regularly use standard salt-weathering tests on building materials. Tests to simulate natural environments (for example, immersion in saturated salt solutions at 17–20°C for one hour, dried at 60°C for six hours, and at 30°C for remaining part of 24 hour cycle, repeated 40–50 times and measured by weight-loss) are used by geomorphologists (Goudie, Cooke and Evans, 1970; Cooke and Doornkamp, 1974; Goudie, 1974) who report that:

1 The most effective salts (Figure 4.16) are, in order, Na_2SO_4, $MgSO_4$, $CaCl_2$, Na_2CO_3, NaCl, $MgCl_2$ and $CaSO_4$.
2 Chalk breaks down most readily, followed by limestone, sandstone, shale, gneiss, granite, dolerite and diorite (Figure 4.17).
3 The rate of disintegration is closely related to water absorption capacity (porosity and permeability).
4 Surface texture and grain size control the rate of disintegration which diminishes with time for fine materials and *vice versa* for coarse.
5 The process may be a more effective weathering process than thermal expansion of salt,

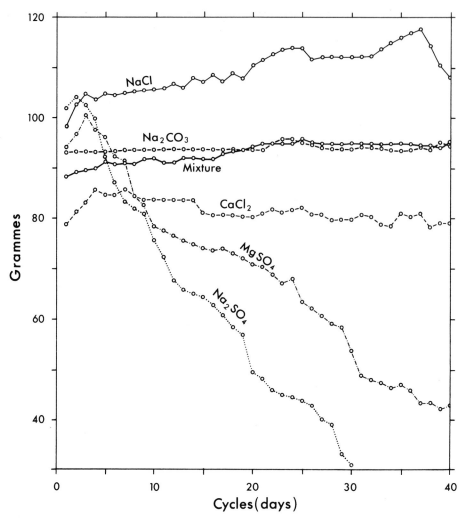

Figure 4.16 The effect of some common salts on the change of weight of Arden Sandstone when used in laboratory weathering texts. The efficiency of Na_2SO_4 and $MgSO_4$ should be noted (Goudie, Cooke and Evans, 1970).

insolation weathering, wetting and drying or even frost weathering. Goudie (1974) found that maximum disintegration was achieved, however, by a combination of salt crystallization and frost, confirming the view (Wellman and Wilson, 1965) that these processes are particularly effective in Antarctica.

Insolation weathering (thermoclastis)

The physical comminution of rocks by their expansion and shrinkage caused by diurnal temperature changes is one of the most keenly debated aspects of rock weathering. Early explorers reported hearing the sound of rocks splitting, and recorded the occurrence of grains, rock scales, broken pebbles, crazed rock surfaces, cleaved boulders (Kernsprung) and even very

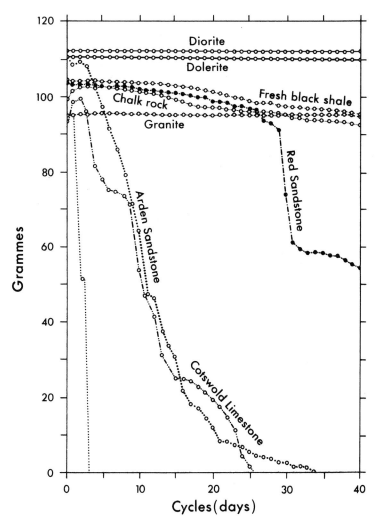

Figure 4.17 Examples of salt weathering of different rocks when treated with sodium sulphate in laboratory experiments (Goudie, Cooke and Evans, 1970).

large detached rock slabs. In 1892, von Streeruwitz noted that, in the Trans-Pecos region of Texas, where diurnal ranges reached 42°C:

> I frequently observed on the height of the Quitman Mountains a peculiar crackling noise and occasionally loud reports ... careful research revealed the fact that the crackling was caused by the gradual disintegrating and separation of scales from the surface of the rock, and the loud reports by crackling and splitting of huge boulders.

In summarizing the earlier literature, Merrill (1897) put forward the essence of the theory thus:

> Rocks ... are complex mineral aggregates of low conducting power, each individual constituent of which possesses its own ratio of expansion, or contraction, as the case may be

... As temperatures rise, each and every constituent expands and crowds against its neighbour, as temperatures fall, a corresponding contraction takes place. Since in but few regions are surface temperatures constant for any great period of time, it will be readily perceived that almost the world over there must be continuous movement within the superficial portions of the mass of a rock.

Subsequent work showed (Table 4.19) that these movements were very small but they were believed to be sufficient to cause 'a decided weakening', and in 1925 Hume, quoting Ball, suggested that the Eocene limestones of Egypt would expand 0·25 mm m^{-1} under a 25°C temperature change.

Table 4.19   Some early figures for the expansion of rocks by heat

Author	Average expansion of granite inch per foot/°F × 10^{-5}	Average expansion of marble inch per foot/°F × 10^{-5}
Bartlett (1832)	0·4825	0·5668
Adie (in Merrill, 1897)	0·438	0·613

This would exert pressures in *confined situations* of up to 25 kg cm^{-2}, a figure that exceeds the tensile strength of sandstones and some limestones but is less than the strength of marble and granites. Hume (1925) concluded that some surface heaving and extension of rock cracks might be produced by this process, an opinion supported by Brown (1924) who described fractured pebbles from the Peruvian desert.

This widely accepted hypothesis was challenged when Blackwelder (1933) and later Griggs (1936b) demonstrated in laboratory experiments that, even when the rocks were subjected to extreme temperatures, they remained uncracked. Blackwelder heated rock specimens to temperatures as high as 200–300°C and cooled them by the same amount. Griggs used cycles of heating and cooling with a range of 110°C, equivalent to 244 years of diurnal temperature change. In a recent experiment, Goudie (1974) reports no significant weathering under a normal desert temperature range of 60°C for six hours and 30°C for 18 hours. The hypothesis seems, therefore, to have been substantially disproved, except in the case of weak jointed rocks. On the other hand, many authors including Griggs (1936b) and Birot (1960, 1962) have shown that disintegration does occur if the rocks are cooled with water, a sensible conclusion since water has a much higher coefficient of thermal expansion than most rocks and may also promote other weathering processes.

A few geomorphologists, notably Ollier (1969), remained unconvinced, however, and in continuing to search for field examples of the process, point out that the laboratory experiments do not take account of the possibility that cracking will only occur if the rocks are confined, so that maximum pressure can be exerted. It is also necessary to carry out experiments in which the duration of stress is of diurnal length so that maximum fatigue can occur.

Despite this long history of study it seems strange that there has, in fact, been little work on the relevant physical properties of rock such as thermal conductivity, specific heat, volumetric heat capacity and thermal diffusivity. Neither is there readily available much accurate information on such aspects as the diurnal range of surface temperatures in deserts (Table 4.20A), the temperature gradients with depth (Table 4.20B) or the stress fields within the rock (Gray, 1965).

The figures available (Table 4.20) suggest that surface temperatures of rocks can reach 80°C, with ranges of approximately 50°C. Temperature gradients within the outer 0·5 m

Table 4.20 **A** Some data on the surface temperatures of rocks
B Measured and theoretical thermal gradients

The maximum theoretical value is calculated from the equation

$$\left(\frac{\delta T}{\delta Z}\right)_{max} = -\sqrt{2\frac{AT_o}{D}}$$

where T=temperature, Z=depth, AT_o=the temperature amplitude, D=the damping depth (after Ravina and Zaslavsky, 1974)

A

Author	Rock	Area	Excess T°C rock over air	Max T°C	Min T°C	Diurnal range
Hume 1925	Slate	Lower Egypt	4°C	—	—	—
	Flint	L. Egypt	18°C	—	—	—
	Sand and gravel	L. Egypt	20°C	—	—	—
	—	L. Egypt	—	—	—	30°
Cloudsley-Thompson & Chadwick 1964	—	—	—	>80	—	50°
Peel 1974	Basalt Varnished	Tibesti (Wour)	—	78·5	35·5	43·0
	Cambrian Sandstone	Tibesti (Wour)	—	79·3	45·1	37·0
	Unvarnished Sandstone	Tibesti (Wour)	—	78·8	41·8	34·2
	Dune	Tibesti (Wour)	—	—	—	38–39
	Sand-filled joint	Tibesti (Wour)	—	—	—	29·6

B

Author	Rock	Area	Temperature range in rock	Temperature gradient outer 30 cm
Roth 1965	Quartz Monzonite	S. California	24°C over 60 cm	0·5°C cm^{-1}
Peel 1974	Sandstone	Tibesti	—	0·85°C cm^{-1}
Ravina & Zaslavsky 1974	—	Theoretical (maximum)	50°C over >10 cm	7°C cm^{-1}

of the rock are of the order of $0 \cdot 7°C \, m^{-1}$. The temperature change is exponential with depth so that, not far beneath the surface ($<1 \, m$), diurnal changes have little effect and temperatures remain constant. This is, of course, related to the poor thermal conductivity of the rock (Joffe, 1949) but again there are few applications of the figures to weathering studies. The information available suggests that values of $4 \cdot 7 - 6 \cdot 6 \times 10^{-3} \, cal \, cm^{-1} s^{-1} °C$ for sandstone and $7 \cdot 4 \times 10^{-3} \, cal \, cm^{-1} s^{-1} °C$ for lava are typical (Birch, 1942). At these rates there is a time lag between heating at the surface and temperature response at depth, so that it is possible for cooling to be taking place in the interior while heating is commencing on the surface.

These rather unsatisfactory figures probably reveal that there is a complex series of temperature variations in the near-surface zone of rocks capable of creating differential expansive forces, but whether these are large enough to cause fatigue remains to be demonstrated by stress-field analysis and experiment.

Some information is available for temperature variations caused by fire (Blackwelder, 1926; Emery, 1944; Birkeland, 1974). Usually this produces spalls of rock up to 6·0 cm long with sharp edges. Records are particularly frequent in the western USA where forest fires are common. Merrill (1897) reports that heating rocks by fire is a common method of quarrying in India where wood fires are used to lift sheets up to 15 cm thick. In one instance, an area of 69 m², 13 cm thick, was loosened from the quarry floor by this method.

The search for an adequate explanation of diurnal rock cracking has led Ravina and Zaslavsky (1974) to suggest that when water condenses due to the increase in relative humidity at night in desert areas, there is an interaction of the electrical double layers at the rock–water interface in closed cracks. The consequent high electrical gradients create high pressures in the water layer which fluctuate with changing humidity and temperature. These, together with thermal pressures, are suggested as a cause of weathering and may find some confirmation from the laboratory observations that water application can cause rapid breakdown. The sudden cooling effect caused by desert rains may well accentuate these processes.

Hydration

Although the nature of hydration is more fully considered in the discussion of chemical processes (p. 111) and the hydration of salts (p. 112), it is worth noting here that considerable swelling and expansive stresses can accompany the hydration of many minerals. The forces of hydration can reach 2000 kg cm^{-2} (White, 1976), particularly at low temperatures when water is drawn in by ice nucleation. Differential expansion between mineral grains can also lead to granular separation.

Many authors have drawn attention to this process which Wilhelmy (1958) termed hydration shattering (*hydrationssprengung*). Although direct evidence and laboratory experiment are lacking, some indirect information suggests that it might be a process producing effects hitherto explained by frost action. Cook and Raiche (1962) have shown that diurnal frost rarely penetrates more than a few centimetres into the soil in arctic and alpine areas. This would limit the effective frost cycle to the annual cycle and therefore it is still necessary to explain rock shattering in areas dominated by the diurnal cycle. White (1976) recommends hydration as the most likely process.

Other processes

Slaking   Closely related to hydration is slaking, the alternate wetting and drying of rocks which may cause large-scale splitting or complete disintegration into flakes. It is possible that the linear expansion of rocks as they soak up water is sufficient to loosen fragments along cleavage planes but Ollier (1969) has pointed out that more complete disintegration could be due to ordered-water molecular pressure. Surface boundaries of minerals attract water molecules that are oriented by polar charges into an electric double layer. If, on wetting and drying, successive layers built up by joining positive and negative ends of the molecules, layers of 'ordered water' could develop which, as they grow, exert expansive forces. The idea is supported by the experiment (Anon., 1966) that rocks slaked in highly polar liquids broke down but those wetted by non-polar liquids did not. This process is similar to that suggested for desert areas by Ravina and Zaslavsky (1974).

Colloidal plucking   Reiche (1950) suggested that the drying-out of colloids adhering to mineral surfaces may cause a form of physical weathering in which minute flakes are plucked from a crystal surface. The process is suggested by the observation that gelatine, drying in a glass vessel, will detach glass flakes from the surface. There is, however, no field or laboratory evidence to support this idea.

Physical weathering by biotic activity   Rocks and minerals are regularly comminuted by plants and animals. As noted earlier, biotic activity can cause or aid chemical processes and is probably underestimated in this respect. Physical changes can also be effective. Burrowing by rodents, rabbits and other animals causes mixing and mineral transfer. Similarly ants, termites and earthworms move large quantities of material. Earthworms bring several tonnes of soil per hectare to the surface every year. Termites can add up to 3 cm to the surface and are even thought to be responsible for the production of stone lines within the weathered mantle. In some cases, such as molluscan fauna on coastal rocks, mechanical boring can cause high rates of erosion and the production of fine debris. Chemical changes from secretion and digestion may also be important.

Plant growth contributes significant stresses to rocks by growth in cracks, by the penetration of tap roots and by heaving of the surface. There are few measures of the magnitude of the process but the combined effect of physical stresses, provision of channels for water movement, shading and changes to the micro-climate, and transfers of nutrients and bases must be very large.

Conclusion

This chapter has concentrated on the nature of weathering processes. It is not the purpose of the book to discuss in detail the products and effects on landforms. These are well covered in existing texts such as Ollier (1969) or Birkeland (1974).

In summary it is only necessary to draw attention to two facets of the subject. First, study of Figure 4.1 shows that, after the rocks are broken down from their primary forms, they suffer a complex series of transformations and translocations that involve many other geomorphological processes. Residual minerals and products are moved through the landscape, deposited and precipitated. Some of these subjects are discussed in further chapters but it is worth noting that the weathering of rock is an essential preliminary to all of them.

Secondly, it will be clear from this chapter that our knowledge is still very meagre. Modern work is moving towards an exact specification of the physical and chemical weathering environments, rock properties, and the thermodynamics and kinetics of the processes. Until this is done a true understanding of the mechanisms involved will not be achieved.

5

Mass movements

D. Brunsden

The general term *mass movement* is applied to those processes that involve a transfer of slope-forming materials from higher to lower ground, under the influence of gravity, without the primary assistance of a fluid transporting agent. They merge imperceptibly with processes in which transporting media, such as air, water or ice are involved and which are generally called *mass-transport* processes. The movements may be slow or rapid, shallow or deep and include one or more of the mechanisms of creep, flow, slide or fall.

The safety factor

The stability of a slope against failure is assessed by the safety factor or the ratio of resistance to force. Mass movement will only occur when the disturbing forces become greater than the resistance of the slope-forming materials. This occurs when the ratio falls to unity.

Many methods are available for the quantitative analysis of slope stability. It is not possible to provide here a critical review of these methods since this would involve a more thorough discussion of the theories of soil mechanics than space permits. Instead, a brief discussion of the main problems involved is provided and some of the more common techniques are summarized in Figure 5.1. Readers who require more information are referred to standard books on soil mechanics such as Terzaghi and Peck (1967), the US Corps of Engineers Manual (1952), the Highway Research Board Special Report No. 29 (Eckel, 1958) or Skempton and Hutchinson (1969). Geomorphological texts providing introductions include Carson and Kirkby (1972) and Statham (1977).

Stability analyses are mainly confined to rotational or translational failures of single or compound form. There are few investigations of falls or flows (Hutchinson, 1970, 1971a). In consequence the analyses are based on an assumption of either a planar, circular, compound circular, logarithmic spiral or compound planar-circular form of the failure surface.

The methods also depend on the assumption of an average shearing resistance along the slip surface as estimated from a limited number of sample points and analyses, and allowance must be made for gross variations. It is probably true to say that the most critical decision in the procedure is the choice of methods to determine the shear strength of the materials involved in the slide.

It is important to realize that the calculations are based on the geometry of a simple two-dimensional section through the slide and detailed procedures are often based on analyses of separate slices of that section. Assumptions must be made that the conditions along each slice or cross-section are applicable to the rest of the slide area. It is possible for three-dimensional analyses to be made but it is generally accepted (see Kenney, 1956) that this yields only a small increase in accuracy (<10 per cent), and the considerable increase in calculation is therefore usually avoided.

When the position of the surface of failure is unknown, or where it is required to estimate

A

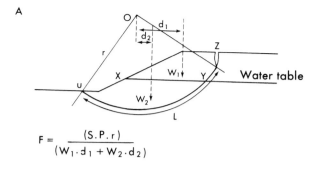

$$F = \frac{(S.P.r)}{(W_1 \cdot d_1 + W_2 \cdot d_2)}$$

B

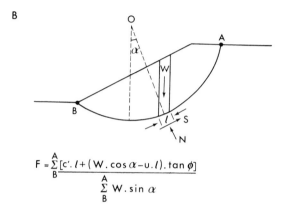

$$F = \frac{\sum\limits_{B}^{A} [c'.l + (W.\cos\alpha - u.l).\tan\phi]}{\sum\limits_{B}^{A} W.\sin\alpha}$$

C

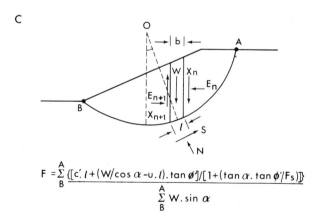

$$F = \frac{\sum\limits_{B}^{A} \{[c'.l + (W/\cos\alpha - u.l).\tan\phi]/[1 + (\tan\alpha.\tan\phi'/F_s)]\}}{\sum\limits_{B}^{A} W.\sin\alpha}$$

Figure 5.1 Examples of stability analyses to determine the safety factor of deep-seated slips (after Carson and Kirkby, 1972). W_1 and W_2=weight of different units; S=shear strength per unit area along the arc; W=weight of slice; N=total normal force at base of slice; u=pore pressure at base of slice; O=centre of circle that includes failure arc; E_n, E_{n+1}=resultants of the total horizontal forces between the slices; X_n, X_{n+1}=resultants of the total vertical forces between the slices.

the potential surface of failure, a necessary exercise is to determine the most critical case, such that the ratio between (a) the shear strength of the soil along the surface of sliding and (b) the shearing force tending to cause sliding, is at a minimum. This is usually achieved by testing a variety of failure-surface positions using graphical or computer-based repetitive calculation.

In addition, the position of the piezometric and groundwater surfaces must be determined by instrumentation or estimated from available groundwater data. This is essential if the calculations are to be achieved in terms of effective stress (see Chapter 2). Analysis is usually carried out in terms of effective stresses where the shear strength (S) mobilized under conditions of limiting equilibrium is:

$$S = \frac{c'}{F} + (\sigma_n - u)\frac{\tan \phi'}{F} \tag{5.1}$$

Where F is the safety factor
  c' effective cohesion
  ϕ' the effective angle of internal friction
  σ_n total normal stress
  u pore pressure at all points on the failure surface

Effective-stress calculations are the most useful methods for geomorphological purposes since they are relevant to the long-term condition of slopes. Total-stress analysis in which the shear strength is independent of the total normal stress on the slip surface, and where pore pressure does not need to be measured, is less useful except where the slopes consist of saturated clays in which drainage has not been able to occur since the time of slope formation.

Finally, in considering the methods summarized in Figure 5.1, it should be realized that only idealized cases have been presented. Few natural slopes are composed of simple, homogeneous materials. Discontinuities, variable strata and old failure surfaces will all control deviations from the simple planar or circular cases and invalidate the results if they are not discovered.

Temporal variations in the safety factor

At any one time most hillslopes will be in a stable state (safety factor > 1). We can envisage failure conditions as occupying one tail of a probability-distribution curve (Figure 5.2A), the exact shape of which will vary from region to region, rock type to rock type, process to process and time to time as the variables controlling the distributions of force and resistance change. In any one area, it is likely that more slopes will be subject to minor forms of mass movement, such as creep, than to large-scale displacements such as deep-seated failure (Figure 5.2B). Through time, slopes subject to failure by a given process will also be represented by a family of probability curves which vary as the resistance to failure changes. Changes in pore-water pressure may alter the shear strength of a material, for instance, (Chapter 2), so that the safety factor will also vary (Figure 5.2C).

If the safety factor is plotted against time, the resulting curve (Figure 5.3, A) represents

Figure 5.2 Hypothetical probability-distribution curves for the safety factor. **A:** For any set of slopes in a specified environment; **B:** For different processes for all slopes in a given environment; **C:** For changes in time on one slope. $t=1$, $t=2$: seasonal changes in pore pressure causing displacement of the safety-factor curve.

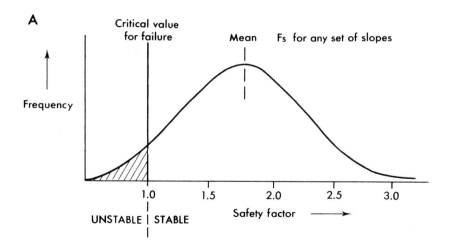

A

Critical value for failure

Mean   Fs for any set of slopes

Frequency

UNSTABLE | STABLE   Safety factor

1.0   1.5   2.0   2.5   3.0

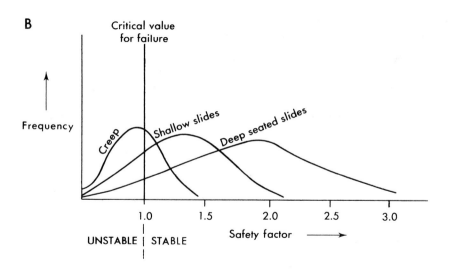

B

Critical value for failure

Frequency

Creep   Shallow slides   Deep seated slides

UNSTABLE | STABLE   Safety factor

1.0   1.5   2.0   2.5   3.0

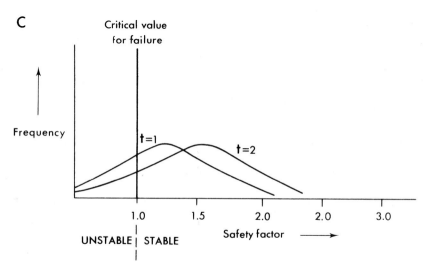

C

Critical value for failure

Frequency

t=1   t=2

UNSTABLE | STABLE   Safety factor

1.0   1.5   2.0   2.0   3.0

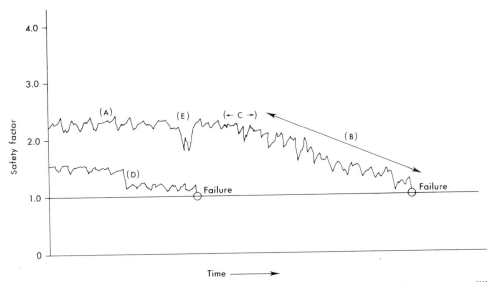

Figure 5.3 Hypothetical changes in the safety factor with time. **A**: Changes affecting an equilibrium slope; **B**: Seasonal changes due to pore-pressure variation; **C**: Long-term trend caused by progressive change in strength; **D**: Sudden changes due to unloading of slope; **E**: Effect of an exceptionally wet period.

the safety factor régime for a given hillslope and will reveal any long-term changes that may cause instability (Figure 5.3, B). Thus seasonal rainfall and evaporation changes will control groundwater levels, pore-water pressure changes and ultimately seasonal variations in shear strength, which will be reflected in seasonal variations in the safety factor (Figure 5.3, C). Should there be a long-term trend of change in groundwater levels, or changes in strength due to weathering, these will show as a trend imposed on the seasonal variation. Sudden changes in the time curve will be due to short-term variation in either the strength of the materials or the forces applied to the slope (Figure 5.3, D). The causes of variation in the safety factor can therefore be regarded as the fundamental causes of mass movement.

Causes of mass movement

All slope-forming materials have a tendency to move downward under the influence of gravity, and this tendency is counteracted by the mobilization of shearing resistance. If the force exceeds the resistance, failure will occur and the hillslope will move to a new position of equilibrium. This response will employ a variety of mechanisms (fall, slide, flow, creep), and will involve a change in slope geometry, a dissipation of pore-water pressures and a redistribution of the force and resistance parameters.

The causes of landslides can be divided into two types (Terzaghi, 1960):

1 External causes that produce an increase in shear stress but no change in the shearing resistance of the slope-forming materials;
2 Internal changes of shearing resistance without any change in the shear stresses.

It should be noted that the external changes are often closely associated with subsequent, autocatalytic, internal changes to shear strength.

External changes in stability conditions

Geometrical changes

The natural erosion of a hillslope by undercutting, the removal of rock by glaciation, stream incision, and artificial excavation of materials during construction all cause slope steepening or increase in slope height. The increase in gradient and height increases the total stresses in the rock. The shear stresses on the potential surface of sliding increase since more weight is progressively acting on that surface. These relationships are often expressed by engineers in the form of stability charts that express the critical cases for a slope in terms of height, angle, shear strength and unit weight (Figure 5.4) (Fellenius, 1927; Taylor, 1937; Bishop and Morgenstern, 1960).

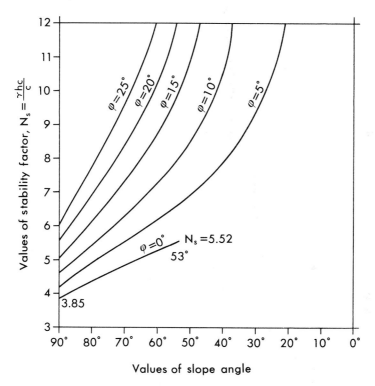

Figure 5.4 Slope-stability curves for a slope in material having values for cohesion and friction as calculated by Taylor (1937).

A useful example of the effects of changing geometry on slope stability and slope evolution was described by Skempton (1953) for hillslopes cut in till in County Durham. The external changes involved a sequence of stream incision, producing deep V-shaped valleys whose floors were later widened to form floodplains which protected the side-slopes from undercutting. The materials involved were described as sandy boulder clay for which values of $\phi'=30°$, $c'=0.84 \, \text{kg m}^{-2}$ were estimated as reasonable (Skempton and Hutchinson, 1969). In the earlier stages of stream incision, stable, straight slopes at 30–35° were established, followed by some shallow sliding as cohesion was decreased by weathering. The slopes remained at this angle despite creep and shallow failure until the slopes reached a critical height (Figure

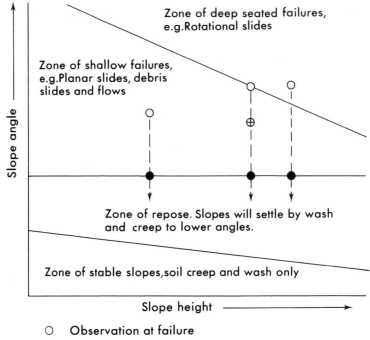

Figure 5.5 Theoretical slope-angle and slope-height relationships (after Skempton, 1953).

5.5) of approximately 45 m when deep-seated slides occurred. This field observation agreed with values which could be calculated and plotted on the slope height-angle graph (Figure 5.5) using a general formula of the following type:

$$h_c = \frac{4c'}{\gamma} \frac{\sin\theta \cos\phi'}{1-\cos(\theta-\phi')} \tag{5.2}$$

(after Culmann, 1866; see Carson, 1971a)

where h_c is critical slope height
c' is cohesion
ϕ' is the angle of internal friction
γ is unit weight of material and
θ is slope angle.

This situation persisted until incision ceased and floodplains developed to protect the bases of the slopes. At this point the slopes flattened through shallow sliding to an angle of limiting stability against landsliding at approximately 22°.

Unloading
The removal of material or load from an initial clay surface, by excavation or rapid valley incision, can lead to a progressive series of internal changes including lateral expansion of the newly-formed slope, opening of fissures and tension cracks (see Chapter 3), increase of

mass permeability, long-term softening and reduction of strength (Skempton and Hutchinson, 1969), a progressive change in the safety factor and, ultimately, slope failure (Bishop and Bjerrum, 1960).

The removal of material reduces overburden pressure and total normal stresses on the potential surface of sliding (Point P, Figure 5.6). This is accompanied by an increase in

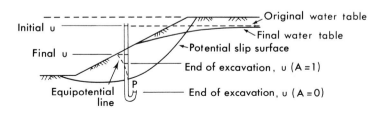

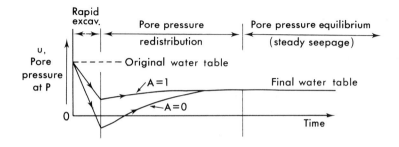

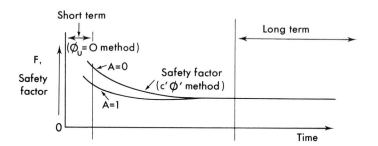

Figure 5.6 Changes in pore pressure (*u*) and safety factor (*F*) arising from the excavation and unloading of a slope cut in clay (after Bishop and Bjerrum, 1960, from Skempton and Hutchinson, 1969). *A*=pore-pressure coefficient (Skempton, 1954).

shear stresses as incision proceeds. The removal of load causes the clays to expand and, if these materials are relatively impermeable so that equilibrium pressures cannot be restored quickly, a subsequent fall in pore pressure occurs. With time, however, the clays can slowly transmit water and pore pressures recover equilibrium. Since there will be no further change in overburden pressure, shear strength will slowly diminish, as will the safety factor, until a steady position is reached. This will be in equilibrium with a steady seepage pattern suitable for the new slope profile. This is known as the *long-term condition* of the slope. The *short-term condition* is the period of rapid excavation. In sands and gravels the short-term condition of low pore pressures will be short-lived since drainage can occur rapidly. In clays, however, the condition may last for months or years after the initial incision. In the extreme case where there is no drainage during the period of slope formation, the change in overburden pressure is the same as the change in pore pressure and the shear strength therefore remains unchanged. Thus although the stresses at a point P on the potential surface of sliding have increased, the strength is the same and can be regarded as being independent of total stress. This is known as the $\phi=0$ case in which only the undrained cohesive strength of the soil may be considered. Generally, however, drainage does occur and shear strength will diminish for some time after slope formation has ceased. It is this fact that explains the long-term failure of many slopes cut in cohesive soils.

The processes of unloading (see also Chapter 4, p. 116) have been much neglected in geomorphology. Glacially eroded slopes, coastal cliffs, granite sheeting and clay failures are the most commonly quoted examples but it seems likely that the process should be considered wherever rapid erosion occurs.

Loading

The addition of material to a slope by artificial or natural means increases its height and the effective weight of the slope-forming materials. This, in turn, increases the shear stresses on the potential surface of sliding. Pore pressures also rise, which in cohesive strata lowers the shear strength and may cause movement (Figure 5.7).

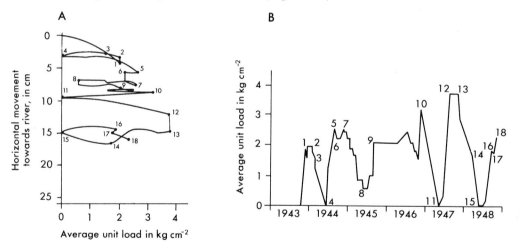

Figure 5.7A: Relationship between unit load at the base of an ore pile and the horizontal outward movement of a point. **B**: The variation in load over time. Numbers 1–18 refer to the positions of the point in **A**.

Very rapid loading can cause highly unstable conditions if the pore pressures rise so quickly that drainage cannot take place, leading to rapid failure of the slope. This may occur under natural conditions where the supply of debris from the upper slopes may load materials lower down to initiate a condition of *undrained loading* (Hutchinson, 1970; Hutchinson and Bhandari, 1971; Hutchinson *et al.*, 1975). The examples described by Hutchinson refer to mudslides on the Kent, Isle of Wight and Antrim coasts. High pore pressures were generated at the head of the mudslides by the discharge of debris from steeper upper slopes. This generated a forward thrust from the loaded area which enabled shearing movements to take place on slopes considerably flatter than those corresponding to limiting equilibrium.

Other cases where this mechanism can be applied include submarine mass movements on delta fronts, the overriding of solifluction lobes (Zhigarev and Kaplina, 1960), the toes of landslides and beneath man-made fills or embankments.

Shocks and vibrations

The occurrence of earthquakes, explosions or machinery vibrations frequently cause slope failure. They increase the horizontal forces acting on a slope and increase the moments tending to cause rotational movement (Figure 5.8). Slightly cemented materials, submerged loose sands and steep, jointed rock faces are particularly vulnerable, for the movements also tend to disturb intergranular bonds and cements, decreasing cohesion and internal friction. If the reduction of strength reduces the safety factor to unity, a landslide will occur. Sudden shocks or slope failures can be followed by several associated processes. These include liquefaction, remoulding, fluidization, air lubrication and cohesionless grain flow.

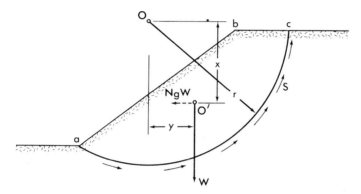

Figure 5.8 Diagram given by Terzaghi (1960) to illustrate the conventional method for computing the effect of an earthquake on the stability of a slope.

$$\text{Safety factor} = \frac{Slr}{Wy + NgWx}$$

where x and y are distances as defined in the Figure, O' is the centre of gravity, l is the distance ac, S is the shear strength along the potential slip surface and Ng is the horizontal acceleration.

Liquefaction Liquefaction is a process that causes a saturated mass of soil suddenly to lose its shear strength and to behave like a fluid. In certain cases a sudden disturbance to a slope (shock, slope failure, etc.) can affect the grains of saturated fine materials, upset the soil skeleton, and tend to force the particles to compact into a dense arrangement (Figure 5.9). This process is accompanied by a decrease in porosity and, if the water in the voids is unable to escape, the sand grains may become, temporarily, partly supported by the pore fluid. The external forces (load) are then transmitted to the fluid instead of to the grain contacts, thus yielding a transient increase in pore pressure. Should this increase to a point where pore pressure equals normal load, a 'quick' or fluid condition will exist, reconsolidation will be temporarily prevented and the material will spread laterally as a liquid, usually in the form of a flowslide (p. 163) or a spreading failure (Terzaghi and Peck, 1967, Fig. 49.11).

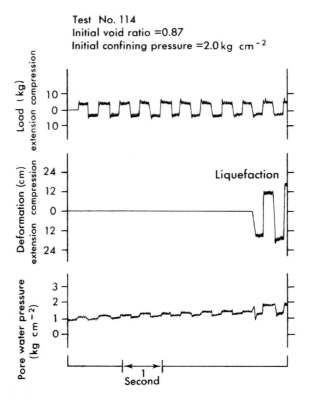

Figure 5.9 Record of a pulsating load test on loose sand causing liquefaction (after Bolton Seed, 1968).

Recently, Bishop (1973) has shown that, although flowslides are commonly associated with fine sands and coarse silts in a loose saturated state (Casagrande, 1965, 1971; Koppejan *et al.*, 1948; Terzaghi, 1957; Bjerrum, 1971; Hvorslev, 1960), they may also occur in coarser materials and where continuous particle degradation occurs with strain, even at relatively low stresses. Bishop (1967; 1973) proposes that the potential post-rupture behaviour of many flowslides is indicated by the brittleness index in both drained and undrained conditions:

$$\text{Drained } I_B = \frac{\tau_p - \tau_r}{\tau_p} \tag{5.3}$$

where τ = resistance to shear for a given value of effective normal stress.

Subscript p = peak state at failure
Subscript r = residual state

$$\text{Undrained } I_B = \frac{c'_p - c'_r}{c'_r} \tag{5.4}$$

where c' is apparent cohesion and subscripts p and r are the peak and residual states where $\phi = 0$ (Figure 5.10).

Bishop shows that materials that are strain softening or brittle are particularly vulnerable.

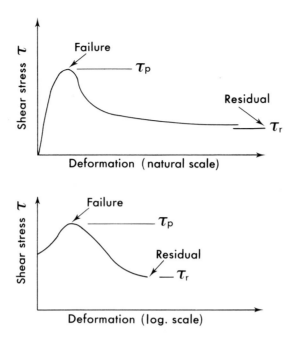

Figure 5.10 The definition of the brittleness index, I_B (after Bishop, 1973).

The highest index values were found in very loose, saturated sands and quick clays under undrained conditions, but may also be found in drained materials where the cohesion component of shear strength is large relative to the effective normal stresses. Some colliery and limestone waste materials that contain extensive fines, that degrade to a finer state during large displacements, or that may be in a loose state of compaction, were also shown to be subject to 'brittle failure'. For example, the coal-tip debris at Aberfan, Wales, showed a brittleness index of 50–60 per cent and a decline in the angle of shearing resistance from $\phi' = 39.5°$ to $\phi'_r = 18°$ as the fragments degraded to a fine paste along the failure surface.

Remoulding Earthquake shocks, vibrations, explosions or simply the collapse of a slope due to a deep-seated failure may also lead to remoulding. This is a process in which the kneading or working of a clay, at unaltered water content, makes that material softer. The softening is due to the breakdown of the initial structure of the clay by the movement as well as the rearrangement of the water molecules in the adsorbed layers. Clay samples that have been fully disturbed from their original state are called remoulded samples and the term 'sensitivity' is used to describe the effect of remoulding on the consistency of the clay.

$$\text{Sensitivity } (S_t) = \frac{\text{unconfined compressive strength (undisturbed)}}{\text{unconfined compressive strength (disturbed)}}$$

The value of S_t ranges from 2–4 for most common clays, to 4–8 for sensitive clays, and >8 for extra-sensitive or quick clays such as the late-glacial marine clays of Norway or the Leda clays of Canada ($S_t \sim 25$). Clays of low sensitivity usually exhibit only local deformation or bulging on failure. Extra-sensitive clays, however, may become liquid, carrying flakes of the more intact surface crust and typically leading to quick-clay slides of disastrous proportions. In these cases the failure of the clay in a deep-seated slip is often sufficient to work the clays into a remoulded state, by collapsing the clay structure, reducing the undisturbed strength by up to 90 per cent, and converting the initial slide into a flow.

Fluidization   Several authors (Kent, 1966; Casagrande, 1971; Bishop, 1973) have postulated that fluidization of rock grains may follow from the collapse of the structure of relatively dry slope-forming materials. Fluidization with air or gas is a well-known industrial process for the transport of solids. It has been defined in industrial texts as a 'process in which finely divided solid is kept in suspension by a rising current of air' (Thrush, 1968). In this process, air or gas is blown through a bed of solid particles so that friction between the particles is reduced and they become highly mobile. It is not suggested that natural failures are caused by gas-emission or blown air, but it is possible that a slide in motion could trap air *within* its mass.

In terms of natural slope failure it is noted that the flow of mixtures of particles and fluids (air) involves both a fluid-transmitted stress and a grain-transmitted stress; total stress is the sum of these (Bagnold, 1954a). When rapid failure of dry material occurs, it is possible for the normal load to be transferred from the grain-to-grain contacts to the pore fluid (air) and for there to be a sudden decrease in friction between the particles. The result is a catastrophic flow of the solid grains until the air can drain from the moving mass. It is likely that many catastrophic flow slides, rock slides and loess flows can be explained by this mechanism, which has also been suggested as a possible mechanism for *nuées ardentes* (Anderson and Flett, 1903; Perret, 1935; Williams, 1942; Fenner, 1950; MacGregor, 1952; Macdonald and Alcaraz, 1956; McTaggart, 1960).

Air lubrication   An alternative mechanism for such catastrophic failures is the air lubrication hypothesis of Shreve (1966, 1968a, 1968b), who compared the mechanism to that of a hovercraft sliding on an air cushion. According to this hypothesis, catastrophic rock falls override and trap a cushion of air that becomes compressed. This enables the material to slide along at high speed, with little frictional contact with the ground except at the edges. Such movements would continue until the air drained from the mass, bringing it to a sudden halt. Shreve also pointed out that escape of the air might itself lead to fluidization and continued movement.

Direct evidence of such entrapped air is, of course, lacking but Shreve argues that the circumstantial morphological and depositional evidence strongly supports a sliding rather than a flowing mechanism. This evidence includes the pattern of transverse fissures in the debris, lateral ridges on the margins caused by air drainage, and the piling-up of debris at the terminus. Shreve suggests that the mechanism might be used to explain the high velocities and excessive travel of many earthquake-generated rock falls and catastrophic disintegrating rock slides. The slides of Elm in 1881, Frank in 1903, and Sherman Glacier in 1964 are all quoted as examples to be explained by air-layer lubrication. The hypothesis may also apply to avalanches (see Chapter 9).

Cohesionless grain flow   The processes of fluidization and air-layer lubrication as possible mechanisms for catastrophic rock slides (Sturzstrom) have been challenged by Hsü (1975), who believes that the eyewitness descriptions and post-failure morphology of these events are better explained by a fluid motion imparted to the mass by cohesionless grain flow. The process has been fully described by Bagnold (1954a, 1956) and can be defined as a fluid motion in which stress is transmitted by grains as a result of grain collisions. Bagnold suggested that a mass of grains concentrated in any flowing medium should be regarded as a fluid. The grains are not transported, however, by the fluid stresses of the medium but by the tangential force on the grains generated by the effective weight of the grains themselves. The grains are dispersed by a stress, normal to the motion of the flowing grains, which originates by collisions of grains in high concentration. This reduces the effective normal pressure and thereby reduces friction on the bed.

When the grains are but a diameter apart or less (volume concentration >9 per cent for spheres) the probability of mutual encounter, always finite for concentration of a random grain array undergoing shear, approaches a certainty. The grains must knock or push each other out of the way according to whether or not the effects of their inertia outweigh those of the fluid's viscosity, and both kinds of encounter must involve displacements of grains normally to the planes of shear. So it seemed probable that the required normal dispersive stress between the sheared grains might arise from the influence of grain-on-grain (Bagnold, 1956, p. 240).

Bagnold (1954a) expressed the stresses and resistances involved as:

1 The gravitational stress (σ_g) propelling the movement:

$$\sigma_g = [\rho_f + (\rho_s - \rho_f) C] gd \sin \alpha \qquad (5.5)$$

2 The normal dispersive stress (P) (equal and opposite to the effective normal pressure on the bed):

$$P = (\rho_s - \rho_f) Cgd \cos \alpha \qquad (5.6)$$

3 The shearing resistance (τ) to grain flow caused by grain collisions:

$$\tau = P \tan \beta' \qquad (5.7)$$
$$= (\rho_s - \rho_f) Cgd \cos \alpha \tan \beta'$$

4 The initiation of grain flow requires that $\sigma_g \geqslant \tau$ \qquad (5.8)

5 The critical angle for grain flow (θ_f):

$$\tan \theta_f = \tan \beta' \left[\frac{(\rho_s - \rho_f) C}{\rho_f + (\rho_s - \rho_f) C} \right] \qquad (5.9)$$

where d is thickness of grain flow, C volume concentration, $\rho_s - \rho_f$ the difference in solid and fluid densities, α is bed-slope angle, and $\tan \beta'$ is the dynamic analogue of the coefficient of static friction between grains.

In the case of rapid mass movements, the grains would be the solid particles (of any size) of the failed rock mass. The inter-grain fluid might be saturated mud, air or dust created by the pulverized rock debris involved in movement.

The essential features of such movements are that the flow is a continuously arranged, relative movement in which stresses are transmitted from one grain to another so that finite forces are transmitted and internal friction expended. In other words, there is a continuous and irreversible deformation with time of a body under a finite force (Reiner, 1958). Hsü compared the motion to that of a relay race in which each runner slows down as he imparts his baton (stress) to the next participant. A fallen runner may cause the pile-up of ridges of bodies across the track. An important feature is the way in which stratigraphic order is preserved: the rearmost formation constitutes the rearmost deposit after it has transmitted its energy forward through the next rock mass.

These features are held by Heim (1882, 1932) and Hsü (1975) to be matched by the phenomena observed at actual Sturtzstrom events and they therefore argue for a cohesionless grain-flow mechanism rather than fluidization or air-layer motion.

Changes in water régime

Effect of rainfall   The fact that many landslide events occur during periods of high rainfall has led to the simplistic assumption by many authors that rainfall is directly responsible for failure, often by a process described as lubrication. Terzaghi (1925, 1960) has pointed out that the presence of water may actually increase the coefficient of static friction and thereby act as an anti-lubricant. In addition he has shown that, at least in humid regions, there is always sufficient water in the soil to provide the thin film necessary for the lubrication of grain contacts. An associated point is that it is not possible to lubricate a slide surface on a first-time slide because that surface does not exist until failure occurs.

The main effects of rainfall on the stability of a slope are:

1 The elimination of surface tension as air is driven out of the voids of fine-grained cohesion-less soils, and a reduction in apparent cohesion.
2 The removal of soluble cements.
3 The initiation of weathering changes such as softening, wetting and drying, hydration swelling and hydrolysis.
4 An increase in the unit weight of the soil.
5 A rise in the piezometric surface, pore pressure and a decrease in the shearing resistance of the soil.

The latter is by far the most important effect. It is worth recording that similar effects can be caused by changes in vegetation cover, which may lead to a decrease in evapo-transpiration losses and increased infiltration.

The apparently close relationship between rainfall and the frequency or rate of mass movement (Figure 5.11) has often been noted. There is, however, commonly a lag between the two events, due to the time needed for water to infiltrate, build up the piezometric surface and initiate movement. The principle that explains this mechanism is the principle of effective stress (σ') as defined by Terzaghi (1936):

$$\sigma' = \sigma - u \tag{5.10}$$

where σ is the total stress acting in a given direction in an element of saturated soil and u is the pore-water pressure of the element.

For the main long-term slope problems, partially saturated air pressures can be ignored and pore pressure can be defined as:

$$u = \gamma_w h \quad \text{(hydrostatic)} \tag{5.11}$$

where γ_w is the unit weight of water and h is the piezometric head. The Coulomb equation for shear strength is:

$$S = c + \sigma_n \tan \phi \tag{5.12}$$

where c is cohesion, σ_n is normal stress and ϕ the angle of shearing resistance.

This can be rewritten in effective stress terms as:

$$S = c' + (\sigma_n - u) \tan \phi' \tag{5.13}$$

Thus strength is seen to be dependent on changes in the piezometric surface. Pore pressure which imparts a lifting force at any point on a potential surface of failure can thus be regarded as a hydraulic jack. The more the piezometric surface rises, the greater the part of the total weight of the overburden that is carried by the water. The case can be reached where $\sigma_n = u$, and $c = 0$, when the overburden 'floats' (*cf*, the case of some glaciers, p. 64).

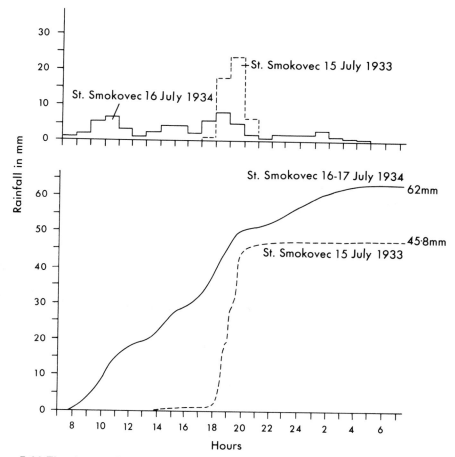

Figure 5.11 The characteristics of rainfall generating debris flows in the High Tatra mountains, 1933–34. *Upper*—hourly totals, *Lower*—cumulative curves (Zaruba and Mencl, 1969).

The importance of this principle to landsliding is that, if heavy or prolonged precipitation, thawing of ice, climatic change to a more pluvial period with higher groundwater levels, or similar change creates a condition where the resistance of the slope falls to a critical value, sliding will occur.

Drawdown A similar effect occurs in situations where the level of a body of water (lake, reservoir, river) is lowered. Normally this is accompanied by a slow outflow of water from the slopes or river banks so that the groundwater level and piezometric surface are lowered in sympathy. Should lowering occur rapidly, relative to the drainage rate, the fall in the groundwater levels will lag behind and there will be transient high pore pressures in the slope causing failure. Such conditions commonly occur in silts and other slow-draining soils.

These changes are listed as exceptional events or as a fundamental change in the long-term controls of the system. Pore pressure, however, also changes seasonally with variation in rainfall, evaporation and seepage, and so we would expect shearing resistance and the safety factor to show similar seasonal variation (Figure 5.3, C). Occasional obstruction of water exits (e.g. by freezing) will have similar effects in a rhythmic sequence over time. It should be noted that these changes do not necessarily cause slope failure, for this would have happened long before if the low values had coincided with a safety factor of unity. When failure does occur due to piezometric changes, and in the absence of sudden loading or shock or other external causes, we have to assume that it is caused by the first high-magnitude event (critical case) in the normal history of the slope, or that a change in some other variable has been progressively altering the stability conditions. Such progressive changes, which impart a trend to the graph of safety factor against time (Figure 5.3, B), are usually due to internal changes that alter the shearing resistance for no change in shear stress.

Internal changes in stability conditions

It has already been shown that a complicated series of internal changes to a rock mass can follow autocatalytically from an external change. Usually these changes, such as fluidization, remoulding or liquefaction, occur rapidly. In addition there are long-term changes that occur progressively through the lifetime of a slope and that cannot be related directly to any one external cause other than the processes that originally created the hillslope, perhaps thousands of years before. These changes involving a slow change in resistance include progressive failure by softening or stress concentration, weathering and seepage erosion.

Progressive failure

The mechanism of lateral expansion, fissuring and progressive strain softening of overconsolidated clays has already been discussed (p. 137) with respect to the stability of a new cut and unloading. This process may lead to 'delayed failures' that occur many years after a cutting has been excavated. Skempton (1970) has shown that first-time slips in overconsolidated clays occur at the fully softened strength, not the remoulded strength, at which the strength corresponds to zero cohesion but a critical ϕ' value for small strains (somewhat higher than the residual values of larger strains). This softening effect is related to the increase in water content through time on exposed clay surfaces in fissures. The clay is fully softened when there is no further increase in water content with time.

This process first came to light when Skempton (1948) attempted to explain why first-time slips on London Clay cuttings appeared to be time-dependent and occurred at less than the drained peak strength as measured in the laboratory. The results of this study showed that a vertical slope of 20 feet (6 m) in height would stand for several weeks, a 25° slope of similar height would fail after 10–20 years, an 18° slope after 50 years. Natural valley-side slopes were generally less than 10° (Figure 5.12).

This work was continued by Henkel and Skempton (1955), Skempton and DeLory (1957), and Hutchinson (1967) to show that the ultimate stability of many stiff fissured clays may be controlled in the long term by fully softened or by residual strength values and not by peak strength. In particular, the Jacksfield slide investigated by Henkel and Skempton (1955) demonstrated that the peak strength was reduced to a value approaching the residual without a visible failure occurring. The logical conclusion was that fissuring allowed water to percolate and thus to initiate softening at low strains until cohesion was reduced and a slide took place.

Progressive failure of jointed rock slopes has also been described by Terzaghi (1960, 1962a, b) who drew attention to the fact that, at any surface of potential failure in a hillslope, some parts will be represented by joints or other discontinuities along which strength is measured in terms of friction between the rock faces. Other parts will be joined masses of intact rock rather like 'dowels'. The total effective cohesion or resistance is then equal to the combined strength of these interfering blocks plus the friction between the joint planes. Slow yield will

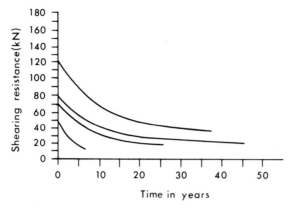

Figure 5.12 Diagram to show the decrease in shearing resistance of stiff, fissured London clay by progressive softening (after Skempton, 1948). Each curve represents a different locality.

cause the 'dowels' to snap one by one so that stress is increasingly concentrated on the remaining intact pieces and the process accelerates as a chain reaction. Eventually, when the reduction in strength has proceeded until the average shear resistance is equal to the average shear stress on the potential surface of sliding, the slope fails to a stable angle which will depend on the jointing pattern and the orientation of the discontinuities relative to the slope (Figure 5.13). This angle is likely to persist through time even if the relief tends to increase due to valley incision.

Terzaghi (1960) quotes the Turtle mountain slide of 1903 near Frank, Alberta, as being initiated by this mechanism, though here the progressive failure and stress concentration were aided by coal mining (Figure 5.13).

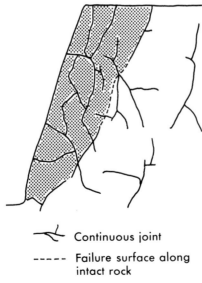

———⟨— Continuous joint

– – – – Failure surface along intact rock

Figure 5.13 The concept of effective joint area (after Terzaghi, 1962a, and Carson and Kirkby, 1972).

Weathering

In addition to the time-dependent processes of unloading, progressive failure and softening described above, which are in reality specialized forms of rock breakdown, there is a wide range of physical and chemical weathering processes that can affect the stability of rock slopes and residual weathering mantles (Chapter 4).

Perhaps the most often-quoted processes are those of freeze-thaw (p. 118) and desiccation (p. 128) which loosen and release joint blocks or mineral grains from rock faces under weathering-limited conditions. In all these processes weathering reduces the cohesion of the rock mass, lowers the friction between joint faces or opens joints and fissures, so that sections of the rock can be removed easily under the force of gravity. At the most dramatic, this can result in large toppling failures or rock falls moving large volumes of material. More commonly, isolated fragments are released from the immediate surface of the rock, which then accumulate as scree or talus slopes as the face above retreats. Where removal of rock waste is slow and slopes are more gentle, vegetated and soil-covered, the influence of weathering processes is more complex though no less profound.

From the point of view of slope stability, weathering processes lead to the following important changes:

1 The development of soil or regolith horizons that possess physical properties different both from each other and from those of the parent rock.
2 Increasing comminution of particles to smaller sizes and an increase in clay-sized particles.
3 An increase in water content.
4 A possible increase in pore-water pressures.
5 An increase in the number and size of voids and fissures. In coarse materials there may be an increase in void space and mass permeability. Where fines are produced, porosity and permeability may decline.
6 Collapse of original mineral structures and fabrics including a denser packing in some circumstances.
7 Changes in cohesion and in angles of shear resistance. Chandler (1969, 1972) has described some of these changes for the Lias Clay and Keuper Marl. He proposed four zones of decreasing weathering with depth, in which water content and grain-size changes were related to changes in cohesion from $28 \, \mathrm{kN \, m^{-2}}$ to $17 \, \mathrm{kN \, m^{-2}}$; the peak angle of shearing resistance declined from $32-23°$ to $24-18°$ with time.

The weathering of harder rocks leads to a different sequence of changes. The main effect is a change in grain size leading to either a continuous size distribution of particles or a multi-modal distribution depending on the nature of the parent material. Each distribution is related to a specific change in soil strength and subsequent soil behaviour.

Earlier (p. 146), it was shown that clay masses weather (soften) and fail to an ultimate angle of stability. Carson (1969) and Carson and Petley (1970), however, examining the reduction in strength of mainly cohesionless mixtures of rock rubble and sandy soils, showed that the weathering process was accompanied by abrupt changes in shear resistance. This was particularly noticeable where bimodal or trimodal distributions were produced. They noted that the slope tended to fail in three phases of instability: from a cliff slope to a scree slope, from a scree slope to a taluvium slope and from a taluvium slope to the ultimately stable soil slope. Carson suggested that these phases are represented in the landscape by groups of successively more gentle threshold angles. Each phase of instability is represented by break-points in the slope-angle distribution. The threshold hillslopes are states of temporary stability that depend on the mantle properties and the variation of pore pressure in the slope. When all rock types are considered, the slope-angle groups 45–43°, 38–33°,

28–25°, 21–19°, 11–8° are most common and are represented in Carson's theory by fractured and jointed cohesionless materials, loose cohesionless materials or talus, taluvium, sandy materials and clays respectively. The slope-angle groups are regarded as general classes or modes for these materials. The associated histogram range is related to variability in the soil mantle and groundwater distributions in the environment in question.

Where weathering produces a continuously graded soil-particle distribution, it is likely that phases of instability would be replaced by a process of continuous adjustment of the mantle to changing resistance, perhaps in the form of very small slips or creep. It might be expected, therefore, that parent materials that yielded such a mantle would be characterized by a continuous range of slope angles without distinct breaks between threshold angles. On the other hand, those slopes with multi-modal distributions of grain size are more likely to fail in shallow slides at successive critical stages in their history.

This exciting idea suggests that discontinuous form can result from continuously acting weathering processes, and goes a long way toward explaining the existence of abrupt breaks in slope-angle frequency, the periodicity of landslide events, and the complexity of response of hillslopes to conditions of slow change.

Finally, in discussing the relationship between weathering and mass movement, attention must be drawn to the effects of solution processes that can dramatically alter shear resistance by the removal of soluble cements or other binders. This is most dramatic in the case of the Norwegian quick clays. Bjerrum (1954) has shown that the mechanism responsible for the initial failures of quick clays is associated with the leaching of salts which reduces the bonding between particles. This increases sensitivity, and reduces the undisturbed strength over the geological time-scale so that failure occurs without an obvious immediate external cause. Bjerrum has also shown that failure seems to occur periodically as the streams are incised into the clay and leaching is enhanced. Thus 'waves of landslide aggression' act successively uphill. This should not be confused with Carson's 'phases of instability' but does represent a further case of discontinuous form produced by a continuous process.

Seepage erosion
The failure of a hillslope by seepage erosion or piping can be dramatic, and in artificial structures such as dams can be very serious since it often results in complete destruction and disastrous floods. Water percolating through a permeable soil may eliminate apparent cohesion by reducing the surface tension of drained soils; it may destroy cementing materials or it may physically entrain particles and cause failure by a form of retrogressive underground erosion. This may simply involve the washing-out of fine sands and the undermining of the slope, or, where the materials are slightly cohesive, it may produce a network of pipes. As the length of the pipes increases they progressively draw on a bigger groundwater catchment, their flow increases and erosion accelerates. Ultimately, the roofs of the pipes collapse, the sediment above breaks up and a slide begins.

On many slopes these pipes are shallow and generally develop into gullies (see p. 194). Should the pipe develop by the seepage of water through a dam, however, slope failure can be accompanied by the discharge of the water in the dam. Finally, on coastal cliffs, the washing-out of sand layers can cause failure of more coherent beds above, which generally collapse as slab slides or rotational slumps.

Modes of hillslope failure

The complex mechanisms described above result in many modes of hillslope failure. There are two important considerations: first, the nature of the initial rupture and secondly, the subsequent behaviour after movement has commenced and the material 'runs out' or is deformed away from its point of origin. In the account that follows, these two characteristics are not always separated since, together, they define some common types of mass movement. It is worth noting, however, that landslides that travel great distances are often placed under the general heading of 'flows' and that they require the occurrence of a secondary mechanism to generate high mobility.

Various criteria have been used to classify mass-movement phenomena ranging from parent material (Ladd, 1935; Zarubá and Mencl, 1969), the mechanics and causes of failure (Terzaghi, 1960; Yatsu, 1967); slope and landslide geometry (Ward, 1945; Skempton, 1953); type of material, shape of failure surface, post-failure debris distribution and soil moisture (Sharpe, 1938; Varnes, 1958). The most generally accepted method is the complex variable method of Varnes (1958) but it is often the case that particular movements cannot be fitted into the individual categories. The external controls of mass movement, namely, geology, climate, hydrology, slope geometry and vegetation are so complex that there is an almost infinite variety of mass-movement forms and rigorous classification may never be possible or desirable.

The approach adopted here is to discuss the criteria that may be used to *describe* landslides and to *define* the most commonly occurring types under the general headings of fall, slide, flow and creep. No rigorous classification is attempted; instead, dominant features will be used (whatever their nature) where these impart distinctiveness or a certain unity to the form and process.

Falls

Falls of rock or soil involve the free movement of material away from a steep slope. The size and shape of the detached mass is usually dependent on the nature of the discontinuities in the rock, the state of weathering of the rock mass and the slope geometry. Rock falls are always derived from the superficial layers of the rock face, a feature that distinguishes them from rock slides or rock avalanches which may be deep-seated.

The second characteristic is that the causes of failure are usually related either to climatic variables and surface weathering, or, in the case of larger rockfalls, to undercutting by erosion, unloading, lateral expansion, the development of tension cracks and stress-concentration processes. By far the most common causes of small falls are high rainfall, freeze–thaw and desiccation weathering. Since the process depends on weathering for the creation and enlargement of joints and the release of blocks, it may be described as a weathering-limited process.

This statement originates in the work of Terzaghi (1962a), Bjerrum and Jorstad (1968), Gardner (1970a) and Hutchinson (1971b) who show that the enlargement of joints and the loosening of blocks is accomplished by the freezing of water and frost-bursting (see p. 121). The highest frequency of rock falls coincides with the most severe winters, with the increase in freeze–thaw cycles in spring and autumn, with diurnal variations corresponding to the midday temperature maxima, and with spatial variations controlled by aspect, intensity of freezing and water availability.

Hutchinson's (1971b) work discusses rock falls in terms of simple laboratory tests and limiting-equilibrium methods of stability analysis. He shows that chalk falls on the Kent coast between 1810 and 1970 were closely related (Figure 5.14) to the average monthly effective rainfall as well as to the average number of days of air frost per month. The specific mechanisms were the creation of transient water pressures in pores and discontinuities. Frost action

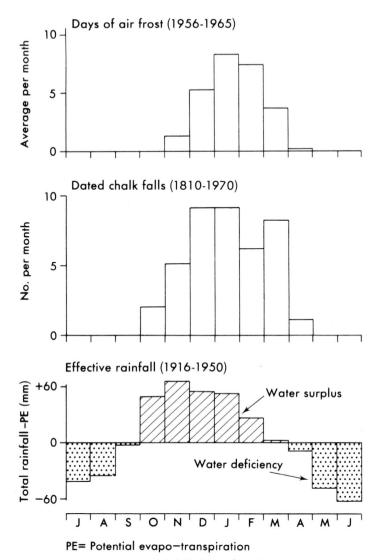

PE= Potential evapo−transpiration

Figure 5.14 The relationship between rock fall, days of air frost and effective rainfall for chalk cliffs on the Kent coast (after Hutchinson, 1971b).

loosened joint blocks and temporarily sealed water exits, thus allowing the development of positive pore pressures and the increase of rock weight by saturation.

Most studies have been made on hard, coherent rock faces and there is only limited information on other materials. Reference should be made, however, to falls of overconsolidated clays (Bazett, Adams and Matyas, 1961; Skempton and La Rochelle, 1965) since it is usually these materials that provide slopes steep enough to be subject to falls.

In addition to the detachment of small rock masses from a steep rock face by weathering, rock falls can adopt distinctive modes of initial failure when larger masses are involved. These are mainly characteristic of well jointed, bedded materials and take place along joints or deep tension cracks. Plane, wedge and slab modes of failure are the most common.

Small plane failures
These occur along highly ordered structures such as joints or bedding planes that are inclined towards an exposed face (Figure 5.15). The stability of the rock slab is given by the values of cohesion and friction along the discontinuity and by the pore pressures in the joint system. Thus the safety factor is given by Statham (1977) as:

$$F = c + \frac{(W \cos \alpha - u) \tan \phi_u}{W \sin \alpha} \tag{5.14}$$

where c=cohesion, ϕ_u=angle of plane-sliding friction between rock faces, W=weight of rock mass, α=dip of bedding or joint system, and u=pore pressure. The mechanism of these

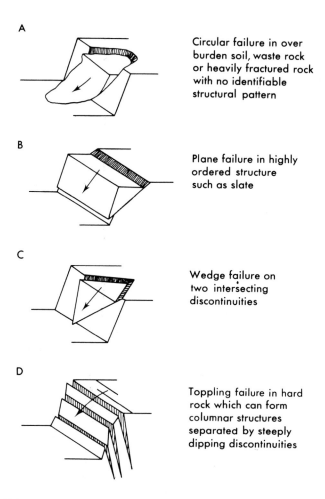

A
Circular failure in over burden soil, waste rock or heavily fractured rock with no identifiable structural pattern

B
Plane failure in highly ordered structure such as slate

C
Wedge failure on two intersecting discontinuities

D
Toppling failure in hard rock which can form columnar structures separated by steeply dipping discontinuities

Figure 5.15 Schematic diagrams of the major forms of rock-slope failure (after Hoek, 1973).

failures is similar (in the detachment phase of movement) to the much larger *rock slides* discussed below (p. 155). They are, however, included here as falls because by far the greater proportion of movement is by subsequent free fall, i.e. the failure takes place well above the slope foot and the resulting debris is indistinguishable from other masses supplied to a talus slope.

Wedge failures
A typical form of rock fall failure consists of a wedge of rock sliding on two intersecting discontinuities (Hoek, 1973). The geometry consists of the intersecting planes of the rock slope and the two discontinuities (Figure 5.15). Hoek assumes that water pressures will build up in the angle formed by the planes and that resistance to movement will be provided by the cohesive strengths and angles of friction on each of the planes to yield the stability equation:

$$F = \frac{3}{\gamma h}(c_A \cdot X + c_B \cdot Y) + \left(A - \frac{\gamma_w}{2\gamma} \cdot X\right)\tan\phi_A + \left(B - \frac{\gamma_w}{2\gamma} \cdot Y\right)\tan\phi_B \qquad (5.15)$$

where c_A and c_B are the cohesive strengths on planes A and B,
ϕ_A, ϕ_B are the angles of friction of planes A and B,
γ density of the rock,
γ_w density of water,
h is total height of wedge, and
X, Y, A, B, are dimensionless factors depending on the geometry of the wedge.

For failure to be possible in this configuration, the line of the angle of intersection of the planes bounding the wedge dips toward the rock face at a lower angle than that of the rock face itself. Hoek has shown that this is a most sensitive method of analysis which clearly isolates the effects of each variable on the safety factor.

Slab failure
This occurs whenever there is a strong development of vertical discontinuities parallel to the rock face or where unloading joints and tension cracks occur parallel to the ground surface as in the case of recently deglaciated valleys, rapidly eroding sea cliffs, or undercut river banks. The mechanism of unloading has been discussed in an earlier chapter (p. 116) and need not be repeated here. Well known examples are the failure of loess slopes in Iowa (Lohnes and Handy, 1968) and the 'Fall of Threatening Rock' from the sandstone edge of the Colorado Plateau (Schumm and Chorley, 1964). Although these falls are grouped under the heading of slab failures (after Carson and Kirkby, 1972) it is worth recording that the movement may occur in three forms. In clays and loess where the tension crack penetrates to the base of the slope, the accumulation of water may cause softening of a basal layer; alternatively, failure may occur on a weak stratum. The whole block then generally subsides rather like a strong man buckling at the knees. This process has been called 'sagging' by Ivanov (1899). The tell-tale signs are a distinct bulging at the base of the slope followed by rapid settlement and collapse of the column above. Where the rock flake is thin relative to the slope height, *toppling* failure may take place, especially if the slope base is undercut or buttresses of rock are isolated by faster erosion on either side. In other cases there is a transitional form where the growth of a tension crack does not reach to the full height of the face and the lower part fails on an inclined plane (Figure 5.15). The detached slab falls backward in failing and the tension crack closes during movement.

The maximum height h_c to which a slope can develop before this form of failure occurs is given by Terzaghi (1943) as:

$$h_c = \frac{4c}{\gamma} \cdot \tan\ (45 + \phi/2) - Z \qquad (5.16)$$

where c = cohesion,
  ϕ = angle of shearing resistance,
  γ = bulk unit weight of material,
and Z = depth of tension crack.

The worst case where the depth of the tension crack extends to the full height of the slope yields

$$h_c = \frac{2c}{\gamma} \cdot \tan\ (45 + \phi/2) \qquad (5.17)$$

which is the general statement for the occurrence of toppling, applying mainly to clay materials. In each case, however, provided the slope height does not exceed h_c, the slope can be regarded as stable against these forms of failure and will be subject mainly to the smaller varieties of rock fall, plane or wedge failure which tend to be weathering-limited rather than dependent on slope height or undercutting.

The landforms that result from rock falls, of whatever category, depend very much on the size of the mass, the height of fall and the geometry of the lower slopes. Where there is a gentler declivity beneath the free face, bouncing, rolling and dispersal of fragments will be common and the debris will fan out away from the original points of impact. If the descent is vertical onto a level surface a smaller degree of break-up will take place and the result will be a concentrated heap of debris which often adopts a lobate or circular plan form.

Where rock falls are common, scree or talus slopes will develop as long as the volume supplied is greater than the volume removed from the slope base by other processes (Plate VIII). The rates involved may be very slow. Rapp (1960) has shown that the retreat of schist faces in the Arctic near Kärkevagge may be as low as 0·1–0·15 mm a year or only one metre in 10 000 years. Rates are higher on more vulnerable rock types (e.g. chalk) but in all cases the rates diminish as the rock faces are consumed by talus growth. The highest rates may be expected wherever there is active removal of debris and erosion at the cliff base.

These points have some important implications for the development of slope form through time. Rock-fall events generally affect only the most superficial layers that are successively exposed to weathering. The processes are slow and probably act over the whole of a rock face if long enough time is available. The build-up of debris at the slope base depends on the balance between supply and removal, and whether or not the basal boundary conditions are stable or subject to lowering by stream incision. As a general rule these conditions dictate that the resulting slope forms will be rectilinear, consisting of steep free faces with small convex edges, talus slopes at or near the angle of repose of the debris, and gently concave or more gently sloping foot slopes. Where the slopes are undercut, the slope generally retreats parallel to itself at its stable angle. Where incision takes place the slope may, for a time, be convex, but as elevation increases to a critical point the angle will quickly be reduced by rock fall and sliding failures to a more stable state. These points form the origin of models of hillslope development under weathering-limited processes, of which those by Fisher (1866), Lehmann (1933), Wood (1942), Bakker and Le Heux (1946, 1947, 1950, 1952), Bakker and

Plate VIII Near-vertical rock face rising above Austerdalsbreen, Norway. The frequency of rock falls from the face, oversteepened by rapid glacial erosion, is shown by debris cones and boulders at the glacier margin, but debris is continually removed by the glacier. (C.E.)

Strahler (1956) are the most important. These are summarized and developed by Scheidegger (1961, 1970) and Carson and Kirkby (1972) to whom reference should be made for a full treatment of the subject.

Major rock slides

In high mountain areas, rock falls and slides sometimes occur on a catastrophic, high magnitude, low frequency scale. The rock falls are similar to those already described for smaller events except that when they are very large, the initial failure is often followed by a succession

Table 5.1   The geometry and other selected characteristics of some major rockslides. The term 'Fahrbös-chung' is the description given by Heim (1932) to the ratio of the height of the highest point on the scar to the horizontal projection of distance from this point to the tip of the slide, expressed as an angle, Shreve (1968) called this ratio the 'equivalent coefficient of friction'. The 'excessive travel distance' was calculated according to Hsü (1975) as the horizontal projection of travel distance beyond what

| Slide | Source | Material | Date | Surface of rupture | | | Angle, degrees |
				Width, km	Length, km	Scar height, km	
Beaver Flats N., Rocky Mts	Cruden, 1976	Limestone	—	0·40	0·14–0·79	—	26
Beaver Flats S., Rocky Mts	Cruden, 1976	Limestone	—	0·48	0·43	—	40
Jonas Creek N., Rocky Mts	Cruden, 1976	Quartzite	—	0·31–0·40	0·81	—	30
Jonas Creek S., Rocky Mts	Cruden, 1976	Quartzite	—	0·50	0·61	—	28
Maligne Lake, Rocky Mts	Cruden, 1976	Siltstone, chert, limestone	—	0·98	1·56–1·67	—	25–40
Medicine Lake, Rocky Mts	Cruden, 1976	Limestone dolomite	—	1·64	0·43	—	35–48
Mt Kitchener, Rocky Mts	Cruden, 1976	Limestone dolomite	—	1·93	0·16–0·40	—	30
Frank, Alberta	McConnell & Brock, 1904	Limestone	29.4.1903	0·97	—	0·87	45 av.
Saidmarreh, Iran	Harrison & Falcon, 1938	Carbonate	Prehistoric	1·45	—	1·49	20 av.
Flims, Switzerland	Heim, 1932	Carbonate	Prehistoric	—	—	1·98	18 av.
D'Ousoi, Pamir	Harrison & Falcon, 1938	—	18.2.1911	—	—	1·16	—
Sawtooth Ridge, Montana	Mudge, 1965	Carbonate	Prehistoric	9·66	—	0·06–0·37	50–80
Gohna, India	Holland, 1894	Carbonate, shale	–.9.1893	0·09	—	1·01	54 av.
Goldau, Switzerland	Heim, 1932	Carbonate	2.9.1806	—	—	0·97	20
Gros Ventre, Wyoming	Alden, 1928	Sandstone, claystone	23.6.1925	0·32	—	0·06	18 av.
Madison Canyon, Montana	Witkind *et al.*, 1962	Schist, gneiss, carbonate	17.8.1959	0·67	—	0·40	30
Elm, Switzerland	Heim, 1932	Schist	11.9.1881	—	—	1·05	42
Sherman, Alaska	Shreve, 1966	Sandstone, argillite	27.3.1964	0·3	0·45	—	40
Silver Reef, USA	Shreve, 1966	Limestone	Prehistoric	—	—	—	—
Blackhawk, USA	Shreve, 1966	Limestone	Prehistoric	1·83	—	—	—

of movements involving the rapid transport of rock debris as an 'avalanche'* or 'flow' over considerable distances. Major rock slides are defined by Sharpe (1938) as 'the downward and usually rapid movement of newly detached segments of the bedrock sliding on bedding, joint, or fault surfaces or any other plane or separation.' These may also be followed by extremely rapid and extensive movements of debris as the slide block disintegrates and moves away from its source. Whilst it may be helpful to distinguish between those *initial* failures which develop as rock falls and those which began as rock slides, the *subsequent* generation of fast-moving streams of debris exhibits close similarity and both types can be considered together.

The majority of major rock slides occur as planar failures along discontinuities that dip

* The term 'avalanche' originally referred to mass movements involving snow with or without other debris. Snow avalanches are dealt with in Chapter 9 (Nival processes) although clearly they are a form of mass movement and could logically be considered for inclusion in this chapter.

one expects of a rigid mass sliding down an inclined plane with a normal coefficient of friction based on $H = \tan \phi\, L$ where H=height of fall, ϕ=angle of friction and L=travel distance. These figures demonstrate the external mobility of the debris *after* it has slid from its source area and also show an inverse relationship of friction to volume.

| Travel of debris | | Area, km² | Volume, km³ | Fahrböschung, degrees | Equivalent coefficient of friction | Excessive travel distance, km | Height climbed, m | Minimum velocity, km hr⁻¹ |
Vertical drop, km	Horizontal travel, km							
0·35	1·10	0·30	0·004	—	0·32	—	—	—
0·30	1·22	0·30	0·005	—	0·25	—	—	—
0·88	3·25	1·08	0·005	—	0·27	—	—	—
0·92	2·49	2·06	0·01	—	0·37	—	—	—
0·92	5·47	4·66	0·50	—	0·16	—	—	—
0·32	1·22	1·55	0·086	—	0·26	—	—	—
0·66	3·22	3·63	0·039	—	0·21	—	—	—
0·83	3·00	2·67	0·037	14	0·25 (Shreve: 0·26)	2·1	120	175
0·91	14·52	165·8	20	4·5	0·08 (Shreve: 0·1)	16·5	450	330
1·22	16·15	41·4	12	7·5	0·13	12·3	—	—
0·73	—	10·4	2·09	—	—	—	—	—
0·06–0·37	4·02	11·6	0·65	—	0·09	—	—	—
1·49	1·61	>1·7	0·29	—	—	—	—	—
0·56	1·78	1·26	c. 0·04	12	0·21	4·0	—	—
0·64	0·09	0·81	0·038	10	0·17	2·5	—	—
0·40	0·49	0·53	0·029	15	0·27	0·9	—	—
0·61	2·01	0·63	0·01	17	0·31 (Shreve: 0·26)	1·15	100	160
0·60	5·0	4·5	0·03	12	0·21 (Shreve: 0·22)	4·1	140 (Shreve: 25)	185 (Shreve: 80)
—	—	—	0·23	7·5	0·13	—	45	105
1·22	9·14	0·75	0·28	7·5	0·13	7·6	60	120

at angles that are close to the angles of friction on those surfaces (Table 5.1 and Figure 5.16). Cruden (1976) following Terzaghi (1962a) has shown that a rock slope containing discontinuities dipping at α degrees out of the slope and undercut by a vertical slope of height h, will have the following stability relationships:

1 The force per unit area of discontinuity tending to produce failure along that discontinuity and through the slope base is

$$\gamma h \cos \alpha \sin \alpha \qquad (5.18)$$

where γ is rock density.

2 The force resisting the disturbing force, assuming no pore pressure along the discontinuity, is

$$c + \gamma h \cos^2 \alpha \tan \phi \qquad (5.19)$$

where c is the cohesion per unit area and ϕ is the angle of friction.

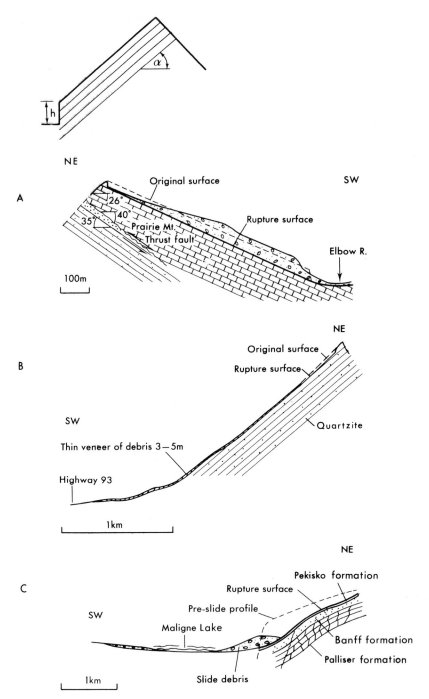

Figure 5.16 A model of a rock slide on a planar discontinuity and examples—**A**: the Beaver flats slide; **B**: the Jonas Creek south slide; **C**: the Maligne Lake slide (after Cruden, 1976).

3 The cliff will be stable as long as

$$h < c/\{\gamma \cos \alpha (\sin \alpha - \cos \alpha \tan \phi)\} \qquad (5.20)$$

Any increase in undercutting of the main slope to produce a vertical footslope of critical value h_c will lead to a rock slide.

It is clear from this statement that, if $c=0$, a planar failure slope will dip at an angle close to ϕ degrees. If cohesion exists, the amount of undercutting required for failure to occur will be greater, and the higher will be the required critical slope. Therefore the greater the value of c, the greater the possibility of a very large failure occurring. Similarly, for the same rock material, the volume of rock involved in a slide will be proportional to $h_c \cos \alpha$ and is therefore greater at gentler angles of failure.

Table 5.2 Some typical values of c and ϕ for rocks with discontinuities (from Statham 1977).

Rock type	Cohesion, bar (10^{-5} Pa)	ϕ degrees
Granite	1–3	30–50
Quartzite	1–3	30–50
Sandstone	0·5–1·5	30–45
Limestone	0·25–1·0	30–50
Shale	0·2–1·0	27–45

The cohesion value for rock material is difficult to estimate but is suggested by many authors (Jaeger, 1959) to approximate to twice the tensile strength of the rock. Some typical cohesion values are given in Table 5.2. The cohesion that can be mobilized in a rock mass depends on the proportion of the potential failure surface joined together by 'bridges' of intact rock. Terzaghi (1962) suggests that this approximates to

$$c = c_i \, A_g/A \qquad (5.21)$$

where c_i = cohesion of rock material and
A_g/A = proportion of surface where bridges are absent, to the proportion joined together.
We have already noted that many rock falls are created seasonally by freeze–thaw activity creating and widening discontinuities and detaching blocks. In the case of major rock slides, freeze–thaw is unlikely to have this effect due to the rock thickness involved. The formation of ice may, however, close drainage ways, increase pore-water pressures in the fissures and reduce the required critical value of h_c. An important process is undoubtedly the solutional effect of circulating water, which, over time, will reduce the 'joined-rock' area and concentrate the stresses on the remaining material. It is notable that, of the twenty cases shown in Table 5.1, at least fourteen occur in carbonate materials. It is reasonable to suggest, therefore, that the reduction of stability of a slope is controlled by such external time-dependent processes as river incision or slope undercutting and also by internal solutional weathering which destroys the cohesion of the mass.

The mobilization of rock-slide debris after the initial failure
Perhaps the most important and intriguing aspects of major rock slides are the form of the landslide deposits and the long distances that the material travels after the initial failure.

The plan form of the debris is similar to that of a lava flow or glacier. If unrestricted it extends as a wide lobe across a flat plain. If confined to a valley it extends down valley but still possesses a lobate margin. Where a slide descends into a narrow valley floor it commonly blocks the valley, forms lakes, and may run up the opposite side (e.g. Flims, Table 5.1). A typical feature is the ability of the slide debris to move over or around major obstacles. The area covered is always large and the volume is measured in millions of cubic metres.

The longitudinal profile of the debris lobe shows a gently sloping surface with very low relief, usually less than 10–15 m. There is sometimes a trough between the head of the slide and the base of the scar. The surface is hummocky, may include ponds and enclosed depressions, and is marked by both longitudinal and transverse ridges produced by shear between debris streams or as pressure ridges. The edges are sharp, marked by levées and with a pronounced distal rim.

The material is generally angular but remarkably unbruised, and heterogeneous in grain size with a tendency for fine material to concentrate in the lower layers, coarser material appearing on the surface. Very large clasts, often measuring several metres in equivalent diameter, may be transported. Some particles and surfaces may be striated but a common form is a jig-saw effect where blocks have been fractured but not segregated. The debris shows considerable ordering, the original stratigraphic succession being preserved, and is very porous, especially in recent debris.

All authors (Buss and Heim, 1881; Heim, 1882, 1932; McConnell and Brook, 1904; Alden, 1928; Harrison and Falcon, 1938; Witkind *et al.*, 1962; Mudge, 1965; Kent, 1966; Shreve, 1966, 1968a, 1968b; Hsü, 1975; Cruden, 1976), agree that a very characteristic feature is the efficient run-out of material away from its source. Heim concluded from empirical evidence that the travel distance depended on the initial height of fall, the regularity of the terrain and the size of the slide (volume). Certainly the morphological data (Table 5.1) show a tendency for large masses to travel long distances. The 'equivalent coefficients of friction' and 'excessive travel distances' reveal that the material travels many kilometres beyond what might be expected of a rigid mass sliding down an inclined plane with a normal coefficient of friction. The expected travel distance can be calculated from Coulomb's law of sliding friction

$$h = \tan \mu' L \qquad (5.22)$$

where h =height of rock fall,
μ' =coefficient of friction of pathway, generally
assumed to be 0·6,
and L =travel distance.

Small rock slides ($<0.5 \times 10^6 \, \text{m}^3$) generally have equivalent coefficients of friction close to the value predicted by this equation; for example, Heim quotes the Airola rock-fall coefficient as 0·64 and the Schächental slide as 0·58. Very large rock slides, however, have values between 0·1 and 0·3, and therefore obviously do not mobilize the same friction properties.

This fact has led to much discussion of the secondary mechanisms that could be responsible for the run-out phenomena. The methods proposed include fluidization (Kent, 1966) (see page 142 for explanation); air-layer lubrication (p. 142; Shreve, 1966, 1968a); cohesionless grain flow (p. 142; Heim, 1932; Bagnold, 1954; Hsü, 1975), or, as Heim expresses it, 'stress transmitted by colliding blocks'. Hsü also extended the latter hypothesis by proposing that the frictional resistance was reduced by the buoyant effect of dust suspensions acting as a medium between the 'flowing' blocks. Alternatively Heim suggested interstitial mud as a medium though this would not be the case in most of the dry rock slides. These mechanisms have been described earlier, but it must be admitted that they are not fully understood at present and we are still uncertain which, if any, is the true cause of this most dangerous form of mass movement.

Slides

Landslides are relatively rapid movements of slope-forming materials in which failure takes place on one or more discrete surfaces that limit and define the failed mass. In addition to rock slides, already described, sliding failures may occur in soils, as shallow planar slips, as debris slides, as rotational, deep-seated mass movements, as mud slides or in complex forms. The movement may be translational, rotating or retrogressive and in single, multiple or successive arrangements.

Shallow, planar, translational slides

In frictional, uncemented and weathered slope deposits, the most common mode of failure is rapid slipping on a relatively shallow plane, parallel to the ground surface. Typical forms include:

1 Narrow, long and shallow slips in debris mantles and weathered regoliths. The movements are translational and expose unweathered rock in the slide scar.
2 Slab slides in clays where there is a marked change in the state of the materials with depth and where movement lasts for only short distances downslope.
3 More complex forms that tend to disintegrate after the initial failure to form tongues of debris related to secondary mechanisms of debris flow or avalanche.

The conditions controlling such failures are given in Figure 5.17. There is a rapid increase in shear strength with depth which inhibits deep failure. The internal forces acting on any slice of material are regarded as equal and opposite and therefore cancel out. End-effects are ignored and the slope is assumed to be of infinite length. This is justifiable since the slide is much larger than it is deep (depth/length ratio $<0\cdot1$) so that end-effects become negligible. The failure surface is assumed to be a uniform inclined plane. A further assumption

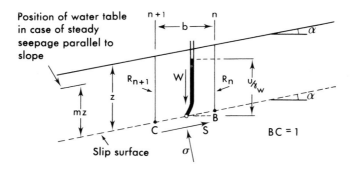

Figure 5.17 Conditions for the development of a shallow planar failure showing the forces for infinite slope analysis given by Haefeli (1948) and Skempton and DeLory (1957). For a slice of unit width *b*, the internal forces R_n and R_{n+1} are equal and opposite and therefore cancel each other.

α slope angle
σ normal stress
W   weight of the slice
z   overburden thickness
mz   height of water table above slip surface
γ_w unit weight of water
u   pore-water pressure

is that the materials are regarded as being uniform over the slope length and can be represented by one sample slope column. Therefore we need only consider the strength along the slide surface.

In terms of effective stresses $(c' \ \phi')$, the safety factor against landsliding is given as

$$F = \frac{c' + (\gamma z \cos^2 \alpha - u) \tan \phi'}{\gamma z \sin \alpha \cos \alpha} \qquad \text{(after Skempton and} \qquad (5.23)$$
$$\text{Hutchinson, 1969)}$$

(For notation, see p. 132 *et seq.*)

Modifications are made when variations in the flow of water in the slope are considered. For example, when flow is horizontal out of the slope, pore pressure $u = \gamma_w z$ (Figure 5.17) and:

$$F = \frac{c' + z(\gamma \cos^2 \alpha - \gamma_w) \tan \phi'}{\gamma z \sin \alpha \cos \alpha} \qquad \text{(after Haefeli, 1948)} \qquad (5.24)$$

When flow is parallel to the slope and ground water is at a height (mz) above the failure surface, pore pressure $u = \gamma_w mz \cos^2 \alpha$ and:

$$F = \frac{c' + z \cos^2 \alpha (\gamma - m\gamma_w) \tan \phi'}{\gamma z \sin \alpha \cos \alpha} \qquad \text{(after Skempton and DeLory, 1957)} \qquad (5.25)$$

Conditions for failure are at an optimum when the soil is in a weathered or a residual-strength state, where the water table is at the surface and where water flow is parallel to the slope. Slides occur when the ground is saturated after periods of heavy rainfall and in debris mantles that possess a marked decrease of permeability with depth. Failure surfaces are concentrated at cemented soil horizons, at the bedrock interface or, in the case of gelifluction lobes, at the permafrost boundary (Chandler, 1972). Many authors report that failure is associated with 'blow-outs' of water under pressure and with secondary disintegration of the failed mass, debris flow and gullying. These observations confirm that the initial failures occur wherever suitable water conditions can build up in the soil mantle.

Shallow planar slides often determine the course of slope evolution wherever cohesionless regolith develops on steep slopes. Sliding occurs when weathering has proceeded far enough to reduce strength toward a failure condition and where there is a sufficient thickness and water pressure. Generally short periods of slipping are separated by long periods of weathering for the slide usually removes all debris down to the bedrock surface. Erosion rates are therefore limited by the rate of weathering and by the frequency of occurrence of suitable pore pressures in the mantle. The spatial pattern seems to vary randomly across a smooth rectilinear slope, although where valley heads, percolines or other conditions cause water to concentrate on steeper slopes, there may be preferred areas of slippage.

Shallow slides do occur in clays (Henkel and Skempton, 1955; Skempton and DeLory, 1957; Chandler, 1970b), especially when the clay is softened (p. 146) by fissuring and water percolation toward a residual-strength condition so that a first-time failure takes place at a strength well below that of the peak strength. Since this process is most effective in the surface layers, a shallow failure takes place. A deep-seated failure is, however, prevented due to the peak-strength conditions found at depth.

The classic failures of this type were first described by Skempton (1948) for long-term failures in London Clay and by Henkel and Skempton (1955) for slides in fissured, over-consolidated, Coal Measure clay shales at Jacksfield, Shropshire, but there have been few studies reported in the strictly geomorphological literature.

Shallow slides that disintegrate on failure

Hillslopes with thick soil or regolith often fail in shallow, planar slides that subsequently disintegrate in the slide track and flow suddenly downslope (Kesseli, 1943). This category of mass movement has been described in the literature as debris slide, debris flow, debris avalanche, mud avalanche and flowside (Wentworth, 1943; Rapp, 1960; Prior *et al.*, 1970, Blong, 1973a).

They are characterized by single or multiple arcuate heads, cut into slopes of more than 20°, followed downslope by a parallel or widening track formed as the initial planar slide breaks up. The toe area consists of single or multiple lobes that form at the base of the failure slope. The morphometry is very variable. For example, the length–depth ratio ranges from *c.* 9 for the planar slide to *c.* 270 for the most mobile forms. They can be very efficient with a ratio between vertical drop and length of run-out of 0·36–1·0 and velocities of 1–14 m s⁻¹.

The disintegrating failures seem to be associated with fine, weathered materials (13–85 per cent sand size or finer); a solids content of 60–85 per cent by weight and moisture contents in the 15–40 per cent range. They may also be related to special groundwater conditions, for many cases have been recorded of 'rushes' of debris and water, soil piping in the scar region and a tendency to occur in shallow seepage hollows. It seems most likely, therefore, that the occurrence of high pore pressures or water seepage from nearby aquifers is necessary for their formation. It is possible that the initial slide failure may either release hillside water in a sudden gush or cause the transfer of load onto the pore fluids thus generating transient high pressures. This must, however, remain speculation in the absence of field measurement and accurate case histories.

A particularly distinctive form of disintegrating slide occurs in deposits with a high brittleness index (p. 140) such as fly ash, coal waste or fill. The general term for such failures is *flowslide* and there may be sub-aerial or sub-aqueous varieties. The initial failure of the slope, often rotational in form, causes a temporary transfer of part of the normal load onto the air or water in the voids, with a consequent sudden increase in fluidity, or liquefaction, of the debris (p. 139). The debris follows a straight and narrow track to form a bulbous lobe and levées along the margins. The velocities range from 2 to 44 m s⁻¹ at moisture contents of 15–58 per cent and can be enormously destructive as at Cilfynwydd in 1939, Aberfan in 1966 and Jupille, Belgium, in 1961 (HMSO, 1969; Bishop, 1973). Sub-aqueous forms occur on the Zeeland coast, Norwegian fiords and on the Mississippi delta (Figure 5.18) (Terzaghi, 1960; Casagrande, 1971). All these failures involved loose, saturated, medium and fine sands with loading of the upper portions by fill or deltaic deposition. The reported velocities range from 0·05 to 7·0 m s⁻¹ but they are capable of movement on slopes as low as 4·5°. Retrogressive failure at the head is common and bed scour may initiate further instability but the mechanism for run-out appears to be similar to that given for the sub-aerial variety.

Finally, an uncommon form of disintegrating failure occurs in parts of Canada and Scandinavia where sensitive (quick) clay has been deposited in recently deglaciated areas (Holmsen, 1953; Rosenqvist, 1953; Bjerrum, 1954). Quick clay is a sensitive marine clay that possesses an easily disturbed structure. The clay particles are bound by electro-chemical forces between the mineral particles and the recently saline ground water. This bonding has been reduced with time by leaching, so that if a slope is eroded to a failure condition, disturbance of the material as it slips away causes the sudden remoulding (p. 141) of the debris. The clay then spreads as a viscous fluid in a form known as a *sensitive-clay* or *quick-clay flow*. This further undermines the slope, initiates a new slide, which in turn causes remoulding and flow. Thus the failure takes place rapidly and retrogressively, perhaps involving several square kilometres of land and many million cubic metres of debris.

A

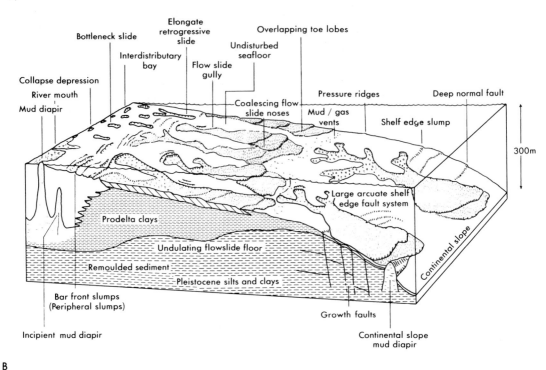

B

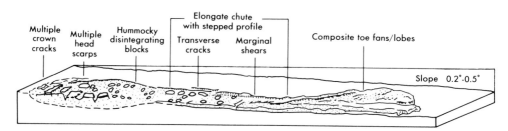

Figure 5.18 A: Schematic diagram of delta front instability features, Mississippi delta. **B**: Diagram of a typical elongate disintegrating slide (after Roberts *et al.*, 1978 in press; Prior and Coleman, 1978 in press and Suhayda and Prior, 1978).

As an alternative theory, it has been demonstrated by Cabrera and Smalley (1973) that glacially derived sensitive clays include both silt and clay-sized quartz and feldspar fragments. These form inactive cohesive bonds among the true clay particles, which significantly reduces the strength of the material. It is further suggested that such materials are capable of forming suspensions of particles in pore fluids when the slope collapses and this imparts great mobility to the debris. It is not yet confirmed which of these theories is correct but there is no doubt that the sensitive-clay slide is one of the truly dramatic hazards among the range of mass-movement processes.

Deep-seated slides

Deep-seated landslides occur in materials, particularly clays, where the rate of increase in shear stress with depth exceeds the rate of increase in shear strength, so that at a certain depth there is a critical surface where the mass is unstable. Thus if the height of a slope is increased at a given angle, slope failure will occur when the stresses reach the critical point (Figure 5.19).

Cohesionless soils that rely on frictional strength for stability usually fail in shallow slips; any increase in slope height or angle will lead to immediate adjustments in stability. This is so because both the shear strength, which is dependent on the normal load, and shear stress are, at least theoretically, zero at the surface and if these values are plotted against depth it can be shown that they will never meet at a critical point (Figure 5.19).

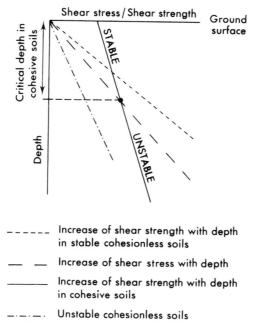

Figure 5.19 Diagram to show the increase of shear strength and shear stress with depth beneath a slope (Carson and Kirkby, 1972).

In clays and other materials possessing strength properties partly dependent on cohesion or cementation, the strength values do not begin at zero (the graph origin in Figure 5.19). If, therefore, the increase in strength with depth is less than the increase in stress, the strength line will cross the stress line and unstable conditions will be defined. Thus, deep-seated failures are most common in thick and relatively uniform deposits of clay or shale.

In homogeneous materials, the shape of the failure surface is usually a well defined, concave-upward curve, which imparts a backward-rotating movement to the slipping mass. The movement on this surface is deep (depth–length ratio=0·15 to 0·33), may be rapid and can take place without significant distortion of the mass except at the toe where overriding and bulging are common phenomena. The existence of the curved surface is explained by the tendency of an intact mass of soil to fail at an angle of $45° + (\phi/2)°$ to the major principal plane of the mass. Thus, in the case where streams cut vertically downward into a rock, the major principal axis of the rock mass is vertical with the major principal stresses acting on a horizontal plane surface. If there are no controlling discontinuities, failure will take place on a surface inclined at $\alpha = 45° + (\phi/2)°$ to the horizontal (Figure 5.20A).

In cohesive soils, however, vertical slopes rarely occur since valley downcutting is usually accompanied by some lowering of slope angle by weathering and small-scale degradation processes. In such slopes, the major principal axis varies from gently curving to vertical away from the slope face. On any potential slip line, the failure surface will still be oriented at $\alpha = 45° + (\phi/2)°$ to the principal plane and the surface of failure will therefore be curved (Figure 5.20C).

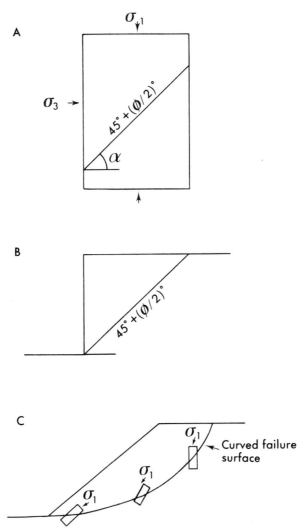

Figure 5.20 The relationship between the orientation of the major principal stress and potential slip surfaces for **A**: a triaxial test; **B**: a vertical slope and **C**: a curved failure surface on a gentle slope.

In some rock masses structural discontinuities or abrupt changes in strength occur at depth. In these materials failure will be controlled by the structures to produce a compound slide with both curved and planar elements. The usual topographic expression is that the failure is still deep-seated but is also more extensive in a horizontal direction into the slope (Figure 5.20C). Such failures vary in form from dominantly arcuate to planar or translational. The shallower the change in strength or discontinuity controlling movement, the more translational will be the slide until the form becomes a true shallow, planar slip.

Slides that include a planar element are usually described as non-circular failures. They exhibit severe distortion and fracturing of the mass and may consist of several distinct slide units. Usually they do not rotate backward on the curved surface but develop a 'graben' or subsidence form at their head caused by block settling as the front part of the slide moves away from the scar (Figure 5.20C). Classic cases are recorded for Gradot, Yugoslavia (Suklje and Vidmar, 1961) and Bekkelaget, Oslo (Eide and Bjerrum, 1954).

Slides that occur on curved failure surfaces are generally called rotational slides. There are three main modes of failure: single rotational slips, multiple rotational slips and successive rotational slips (after Hutchinson, 1967).

Single rotational slips   Single rotational slips (Figure 5.21A) possess a single concave failure surface and are most common in homogeneous clays. They are of simple plan form which may also be circular in outline. The radius of the scar in plan depends on the relief, the magnitude of the slide and any controlling fissures in the vertical plane. As a general rule the scar will tend to be a wide, flat curve on straight slopes but more constrained on concavities and deeper on convexities. In perfectly homogeneous rock, a straight slope could theoretically possess a three-dimensional failure form which approximates to a cylinder

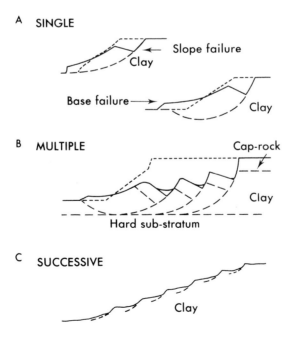

Figure 5.21 The main types of rotational landslide as defined by Hutchinson (1968).

rather than a portion of a sphere. Any fissures in the slope will impart angular controls to the plan form and cause departures from the curved outline.

The movement of a single slip is generally not rapid ($c.$ 5 mm s^{-1}) taking place in one dominant phase of movement, but is preceded and followed by creep movements ($c.$ 1 mm s^{-1} $\times 10^{-4}$) and can occur without distortion or fracturing of the slipped mass. The toe movements involve compressional forces and may bulge or ride over the lower ground surface. The main part of the failure surface is controlled by shear and the upper part, which is in tension, may crack open in the initial phases of movement. Later, however, tension cracks are closed as the mass back-tilts during rotation to form a depression between the back of the slipped mass and the slide scar. The depression may then be filled with water, sediment from the degrading scar or organic deposits. The lower slopes may show compression ridges and transverse cracks, especially if affected by the irregular topography downslope. Single slips occur frequently on cuttings, cliffs, river and canal banks in both normal and over-consolidated clays such as the fissured London Clay (e.g. Warden Point, Kent: Hutchinson, 1968), in tills (e.g. Selset, Yorkshire: Skempton and Brown, 1961), and in post-glacial clays (e.g. Lodalen, Oslo: Sevaldson, 1956). A wide range of failure records is available to demonstrate that this mode of failure can occur as first-time slides in the short or long term and in most types of clays. It is not surprising, therefore, to find that they are treated as a general model for which there are several methods of stability analysis (p. 131) and many published case studies.

Multiple rotational slips   These occur as a retrogressive series of slips, each on a curved surface that is linked tangentially to a common failure line (Figure 5.21B). The plan form consists of a sequence of arcuate blocks arranged one behind another in a staircase form. Each block is back-tilted with a depression between it and the front of the next unit. They occur on high, steep slopes that are being eroded at the base. A recurrent (diagnostic?) feature is the tendency for such slides to occur where a permeable but coherent material overlies a less permeable thick clay. This condition is found at several places on the British coasts and on the inland escarpments of the Weald (Gossling, 1935; Gossling and Bull, 1948), in West Dorset (Brunsden and Jones, 1971) (Plate IX) and on the Cotswolds. The best documented cases include the Folkestone Warren landslides (Hutchinson, 1969; Toms, 1953; Wood, 1955) and Fairy Dell, Dorset (Brunsden and Jones, 1974).

The importance of the cap-rock is that it probably retards degradation of the scar of the first slip. Thus the stresses on that slope are only slowly diminished with time, unless erosion at the base of the slope is able to destroy the original slide and steepen the scarp face. The cap-rock will slow down this process and maintain a steep upper slope so that, as the original slip slowly moves downslope, or is degraded, the stresses for failure can re-occur. It is therefore common to find that there are long intervals between the individual failures. Sufficient time has to elapse before the original failure has moved or degraded to re-create an unstable situation.

In Britain, most of the inland multiple landslides appear to be inactive, or to show movements of old blocks only. This is probably explained by the fact that the transport and erosion processes at the toe are now very slow, whereas in the late-glacial period when many of these slides are thought to have commenced, erosion rates and pore pressures would have been higher.

Successive rotational slips   Hutchinson (1967) has described how slopes in stiff fissured clays tend to fail in repeated, shallow, rotational slips as they approach their ultimate angle of stability. They are similar in appearance, and perhaps in mechanism, to successive terracettes

Plate IX Black Ven, Dorset. The cliff, about 200 m high, is the site of the most active mudslides in Britain. The photograph shows the slides which descended in 1958. Vegetated areas near the sea represent earlier slides. The upper cliff shows several rotational slips of multiple type. (N. Barrington)

formed by soil creep (Sharpe, 1938) and may represent the last stages in slope degradation by landsliding. As with terracettes, they may form regular or irregular patterns of stepped terraces or mosaics of small benches (Fig. 5.21C). As they tend toward stability they become increasingly subdued until all that remains is a faint rippling on the slope.

Quantitative information is lacking except for Hutchinson's observations (Hutchinson and Skempton, 1969) for the London Clay. They occur on slopes, between 8° and 13°, are of limited extent upslope (with terrace widths of 9–12 m and scarps some 1·5 m high) but they can be extensive along-slope. They have been observed at High Halstow, Kent, at Hadleigh Castle, Essex and elsewhere in the London Clay but they are not found below an angle of 8°. They are thought to originate at the foot of the slope and to progress upwards. In some respects they are similar to, and may be the clay-slope equivalent of, the successive shallow slips developed in frictional regoliths described by Rouse (1969) and Carson and Petley (1970). They are generally confined to steeper slopes and tend to be more elongated downslope rather than across the slope. Rouse, in particular, suggests a model of slope evolution by successive slipping which is so similar to the models suggested for the London Clay that it seems sensible to regard all the features as belonging to the same category.

Mudslides

Mudslides are 'relatively slow-moving lobate or elongate masses of softened, argillaceous debris which advance chiefly by sliding on discrete boundary shear surfaces' (Hutchinson

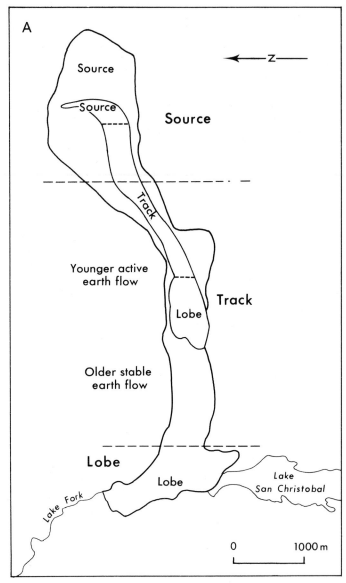

Figure 5.22 The forms of mudslides. **A**: Slumgullion, an elongate slide (after Crandell and Varnes, 1961); **B**: Beltinge, north Kent, a group of lobate slides (after Hutchinson, 1970).

and Bhandari, 1971). They occur in fissured, overconsolidated clays, glacial tills and, more rarely, in very fine sands and silts. They move on slopes as low as 3–4° but usually occur at angles of 5–15°.

There are two main types: first, an elongate variety (Figure 5.22A) such as the Slumgullion mudslide (Crandell and Varnes, 1961), the Stoss mudslide at St Gall, Switzerland (Von Moos, 1953) or the slide at Mt Chausu, Japan (Fukuoka, 1953). They can be very large (the Drynock slide in British Columbia is several thousand metres in length) and they have a length–breadth ratio of 10–17. They occur on slopes greater than the limiting angle for movement, with groundwater at, and flowing parallel to, the ground surface.

The second variety is a more lobate form (Figure 5.22B). These occur on flatter slopes, undergo erosion at the base and possess a length–breadth ratio of approximately 5. They may consist of single lobes or they may occur in groups as at Beltinge, North Kent (Hutchinson, 1970) and the West Dorset coast (Plate IX).

The morphology of mudslides is very distinctive. They consist of three main units, an upper supply area, a central neck or track and a lobate toe. The supply area generally consists of steep clay slopes from which material is derived by weathering, gullying or shallow, translational slides. This material often forms small feeder flows to the main track. The track, through which the mudslide material passes in the form of an accumulation lobe, is usually straight or gently curving. This is very sharply defined by discrete shear surfaces along the sides and base. The knife-cut nature of these shears is perhaps the most diagnostic feature of mudslides. The toe spreads in a bulbous form and may possess internal shears which rise sharply upward from the base to the surface. Both the main track and the toe area may be broken by open fissures which develop as Riedel shears at the edges, as radial cracks around the toe or as transverse cracks where the slide passes over steeper slopes. If the main track is intersected by a steep slope or cliff, the whole feature will be truncated so that the internal structure can be seen, and blocks of material can break away from the leading edge as the mudslide pushes forward. On the West Dorset coast a series of such truncated slides descends a staircase of terraces, with the material that breaks away from the upper slides rapidly supplying those beneath.

The material involved in movement develops a typical mudslide matrix consisting of hard, unweathered fragments of clay, shale or silty clay set in a matrix of softened and weathered

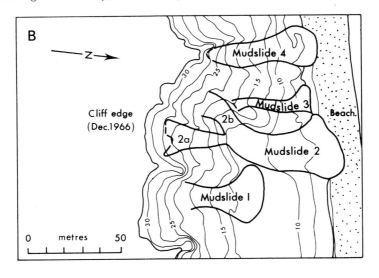

material. A dry crust may develop on the surface during the summer but this is rapidly softened during wet periods. There is a general tendency for the material to become stiffer toward the base and toward the better-drained portions of the toe, but the junction with the undisturbed material beneath the slide is always abrupt. Where fine sands or silts are supplied to the flow these tend to form lenses between successive clay layers.

The movements of the slide take place by sliding on the basal and side shear surfaces with little internal deformation except when the slide is very wet or where it is constrained by sudden changes of slope or width. Up to 90 per cent of surface displacement may take place by basal sliding.

The rates of movement are generally slow (Table 5.3) and occur seasonally. They may become stationary in the summer but during winter they show increasing movement as precipitation and groundwater levels rise. A typical feature is their tendency to surge forward at high rates and with little warning (Figure 5.23). These movements are generally related to the seasonal rainfall pattern but there may be a delay between the occurrences of any one precipitation event and a subsequent movement. It is quite possible for there to be movements apparently unrelated to a given rainfall event (cf. temperate glaciers, Chapter 8).

Table 5.3 Selected examples of the movement of mudslides
(Data mainly collected by Skempton and Hutchinson, 1969)

Site	Author	Slope	Period (years)	Average velocity $m\,yr^{-1}$	Max. daily velocity $mm\,s^{-1}$
Slumgullion	Crandell & Varnes, 1961	7·5°	13	5	—
Stoss	Von Moos, 1953	7·5°	1	7	—
Beltinge	Hutchinson, 1970	7·0°	3	8	0·03
Fairy Dell	Brunsden & Jones, 1974	9·0°	3	18	>0·06
Mt Chausu	Fukuoka, 1953	8·5°	20	25	0·03
Minnis North	Hutchinson *et al.*, 1974	29°–9°–35°	4	—	130

This behaviour can be explained by the mechanism of undrained loading (p. 138) first proposed by Hutchinson (1970). (See also Hutchinson and Bhandari, 1971; Hutchinson *et al.*, 1974). This work has shown that the movements are controlled by the pore pressures that build up within the mudslide itself. These are highest near the head of the mud slide and can be artesian, that is, water may rise above the mudslide surface. Pore pressures rise during winter and, on reasonably steep slopes, may cause movement without requiring additional loading. In many cases, however, the slopes are gentle and the previous year's accumulation of mud is well drained and stiff. In such cases, Hutchinson has shown that it is necessary to generate high pore pressures in order to overcome the resistance to passive failure in the drained areas and to enable movement to occur on slopes flatter than those required for sliding under conditions of limiting equilibrium.

Hutchinson suggests that the generation of high pore pressures results from the rapid, undrained loading of the head of the accumulation lobe by material descending the feeder slopes above. This leads to underconsolidation and provides the mechanism for sliding. It also explains the delays in movement caused when the toe area presents a temporary resistance that must be overcome by the development of a forward thrust from the loaded area. Sudden loading at the rear can also create the surging condition which is reported for all mudslides (Hutchinson *et al.*, 1974).

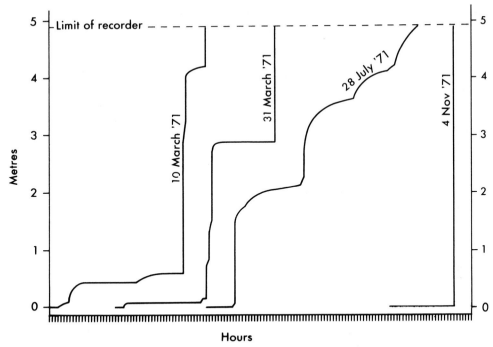

Figure 5.23 Cumulative traces of surge movement on a mudslide at Minnis North, Northern Ireland (Hutchinson *et al.*, 1974).

Debris flows

Debris flows represent a transitional set of processes lying between stream flow or mass transport and the drier forms of mass movement. The term is used here as a general designation for those processes that involve the mass movement of a wet mixture of granular solids, clay minerals, water and air under the influence of gravity, where intergranular shear is distributed more or less uniformly through the mass (Sharp and Nobles, 1953). In some areas of the world, notably the semi-arid USA, Caucasus, Alps, Rockies and Himalaya, they represent a hazard that annually causes severe damage and loss of life. Yet they are among the least studied and understood of all mass-movement phenomena. There is no satisfactory classification, there is much confusion in terminology and the mechanisms involved are poorly understood. The general characteristics are that they possess high fluidity during movement (from 1 to 93 m s^{-1}) and contain up to 50 per cent sand-sized or finer material, but show poor size sorting and lack of stratification, with numerous voids. A particular feature is their ability to carry very large clasts, often in their surface layers.

Debris flows (Figure 5.24) are common in areas with an abundant supply of loose debris, steep headwater slopes but low-angle run-out areas, and irregular but intense water supply across sparsely vegetated regolith. They are thus mostly found in mountainous, semi-arid and volcanic areas.

There appear to be three main categories: catastrophic mud flows, simple 'hillside' debris flows, and flows that originate in small mountain catchments (Hutchinson and Brunsden, 1974).

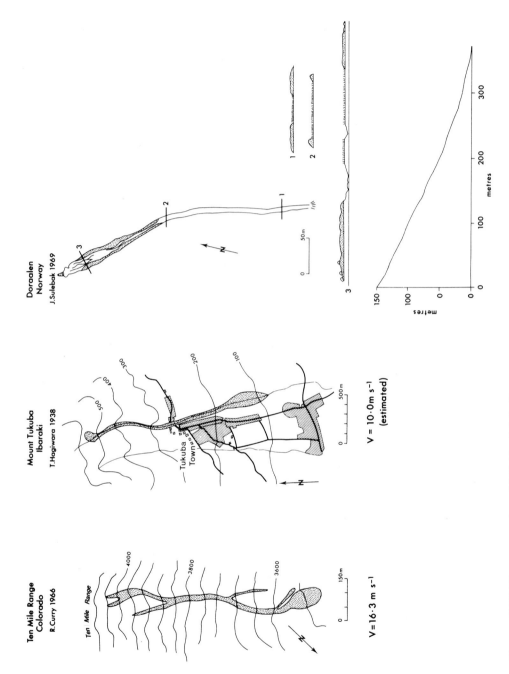

Figure 5.24 Three examples of hillside debris flows.

Catastrophic flows

These are characterized by a low water and clay content but possess an initial, high potential energy, rapid input of water and material, very high velocities, long run-out and efficient discharge. They may include phenomena previously described as lahars, rock-fall avalanches and debris avalanches but they all include the fundamental feature of originating in a large-scale event, such as a crater-lake breach, eruption or intense storm on a steep mountain slope. Typical cases occurred on the Gunong Kaloet (Java), Tokachi-daké (Japan), Irazu (Costa Rica), and Mount Rainier (USA) volcanoes.

The debris descends from its source at high velocities ranging from $13 \cdot 1 \text{ m s}^{-1}$ on Mount Pelée to 50 m a^{-1} on Tokachi-daké. It obliterates relief near its origin but later follows available depressions and valleys in one or two huge waves of debris of great thickness. The flows tend to die away fairly quickly on flat slopes but often release floods of muddy water which run on for many kilometres.

Hillslope flows (Hagiwara, 1938; Sharp, 1942; Curry, 1966; Sulebak, 1969)

In contrast, these are much smaller and less dramatic. The movement begins in a single or sometimes a multiple head, follows a long narrow track, down steep slopes of 30° or more, builds levées alongside the track and spreads out on the lower slopes into one or more elongated lobes (Figure 5.24). Such flows rarely erode their track down the hillside since the debris is derived from loose saturated accumulations on the upper slopes. The deposits are normally composed of 1–5 per cent clay, 40–60 per cent sand or finer material with abundant coarse clasts. Moisture contents are usually less than 10 per cent by weight.

The flow mechanisms involved are unknown but, from the few eye-witness reports, it seems likely that they originate in areas of loose regolith saturated either by intense rain or rapid snow melt, which then 'flows' downslope, with some violence, as a single or pulsating surge at velocities of 10–16 m s⁻¹. The lobe comes to rest when it drains both at its margins, providing sinuous levées of coarse material, and at the toe where only the finer material remains.

Valley-confined flows

Similar in properties to hillside flows, valley-confined flows originate in small ($<10 \text{ km}^2$) mountain catchments that have a dense, branching network of gullies cut in loose debris (Figure 5.25). The track may be an existing water course which is usually occupied by a stream.

The slope angles change rapidly at the base of the catchment slopes so that there is a sharp junction between the supply (15–20°) and run-out (6–9°) areas. The run-out can be as long as 30 km, even on gentle slopes, and it is again a diagnostic feature that the edges of the channel are marked by levées of coarse bouldery material. The deposition area is usually a fan of debris at the mountain foot made up of the deposits of successive flows. Apart from these generalizations, little is known about their morphological characteristics, but they seem to be efficient transporting agencies. The deposits are sub-angular, poorly sorted and weakly bedded, with abundant voids. Typically they include less than 15 per cent clay and more than 60 per cent sand or finer material. Moisture contents are 10–30 per cent by weight but there are few measured examples (Sharp and Nobles, 1953; Morton and Campbell, 1974). The main cause of movement again appears to be rapid supply of water-soaked debris onto steep slopes and the loading of material accumulating at the foot of the steep feeder slopes.

Wrightwood(California)

(Morton et al.,1974,Sharp and Nobles 1952)

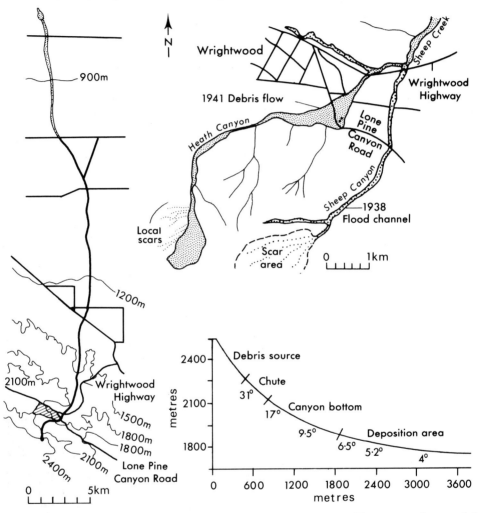

Figure 5.25 The Wrightwood mudflow: a valley-confined flow type. The map at the top right shows the upper parts of the mudflow track whose full extent is plotted on the left-hand map. At the bottom right is a long profile of the upper portion of the flow.

Soil creep

This term is applied to the slow downhill movement of slope debris by a variety of processes including true creep, flow, shear, sliding and heave. The material involved may be individual rock particles, loose scree, partially weathered regolith or well-developed soil. In most texts, the various processes are grouped rather erroneously under the general heading *soil creep* which, following Sharpe (1938), is defined as 'the slow downslope movement of superficial soil or rock debris, usually imperceptible except to observations of long duration'.

In this chapter the following mechanisms are distinguished: creep, expansion and contraction, heave and random forces. In a geomorphological sense these mechanisms, acting singly or together, are responsible for the processes known as rock creep, talus or scree movement, continuous creep, seasonal creep, and gelifluction.

Creep processes
Continuous creep All hillslopes are affected by the force of gravity which applies low stresses to the soil and, unaided by other agents, can cause the deformation of slope materials at a fairly constant rate (Plate X). This movement is usually known as *gravity, mass* or *continuous creep* and, in theory should be distinguished from soil movements caused by seasonally acting expansion or heave (Terzaghi and Peck, 1948; Terzaghi, 1960).

The loads involved are usually small and, if the shearing stresses do not exceed the fundamental or residual strength (Griggs, 1936a) of the materials, no movement will take place. If the stresses rise above this point the material will slowly deform or creep until a point is reached at which still higher values of stress cause a shear failure and rapid displacement. The phenomenon of mass creep is mainly restricted to clays or rock materials that can deform at low stress values and is similar to viscous flow (Johnson and Hampton, 1969; Ter Stephanian, 1965; Yen, 1969). Laboratory tests have shown that the rate of strain declines with time to a constant value unless stress is high enough to lead to progressive failure and sudden shear (Singh and Mitchell, 1968; Fleming and Johnson, 1971). Both the rate of creep and the time of attainment of a constant value are stress-dependent. The ideal condition for long-continued creep, therefore, is where the applied stresses are only marginally greater than the fundamental strength so that the upper boundary, for shear, is not reached.

There have been very few attempts to measure or test these ideas in the field. Terzaghi (1960) quoted examples of the creep of clay in test pits in Leningrad and also likened the process to the production of valley bulges and tectonic deformation. Ter Stephanian (1965) and Kojan (1967) used flexible tubes to show that the movements could occur at depths of up to 10 m with half the movement occurring in the lower layers. The surface rates were as large as 22 cm yr^{-1} with transport rates up to 2500 cm^3 cm^{-1} yr^{-1} (Table 5.4A). The shape of the velocity profile (Figure 5.26A) showed progressive retardation with depth and agrees with the theoretical results of Yen (1969). Table 5.4B gives data for a velocity profile recorded by Rudberg (1962) in Lappland. The only field measurements of the time dependency of creep are reported by Fleming and Johnson (1975) who were able to measure the rates generated by the excavation of a pit in a measured slope. This demonstrated that the rates of displacement do decline with time, as theory predicts, and that a constant rate is achieved quite quickly (35–40 days in silty clay).

The morphological effects of continuous creep are unknown but it is likely that slope settlement, valley bulging of clays and minor slump structures may owe their origin to long-continued, low stresses. Of greatest interest are the movements that approach the maximum rates for creep and that often precede or warn of the danger of shear failure and landslides. Such movements are characterized by acceleration (Table 5.4C) and often accompany progressive failure in stiff fissured clays. The magnitude of the displacement by creep is dependent on the thickness of the deforming layer and may therefore pass unnoticed in shallow slides. The movements are also larger in more consolidated materials.

Seasonal creep The surface materials on slopes are continuously affected by the expansion and contraction of the soil and by soil heave. Expansion can be caused by changes of temperature and moisture content, by the freezing of water, by the growth of crystals, concretions

Plate X Curvature of cleavage in shales, Abereiddy, Pembrokeshire, beneath a regolith of gelifluction debris. The curvature is due to downhill creep, probably under former cold conditions, but in some cases has also been attributed to drag by overriding ice. (C.E.)

Table 5.4 **A** Some measured rates of continuous creep

Author	Rate, cm³ cm⁻¹ yr⁻¹	Surface rate, mm yr⁻¹
Ter Stephanian, 1965	2500	130 max
Kojan, 1968	1321	10–223

B Velocity profile for creep measured in Lappland (Rudberg, 1962).

Depth (cm)	Rate, mm yr⁻¹
surface	50
10	38
20	28
30	17
40	8
50	3

C Pre-failure creep movements (from data gathered by Skempton and Hutchinson, 1969)

Author	Site	7 yrs	2 yrs	6 mths	8 days	1 day	Total movement before slip
Cooling, 1945	Kensal Green, wall	2 cm yr⁻¹		9 cm yr⁻¹			35 cm
Saito, 1965	Ooigawa, wall			0·4 cm day⁻¹ or 16 cm yr⁻¹; 0·1 cm day⁻¹, 5 cm yr⁻¹		1 cm day⁻¹; 10 cm day⁻¹	>20 cm
Suklje & Vidmar, 1961	Dorsan, slide				3 cm day⁻¹	30 cm day⁻¹	>40 cm
	Gradot, slide						>130 cm
Müller, 1964	Vajont, slide			70 cm yr⁻¹	0·3 cm day⁻¹, 110 cm yr⁻¹	6 cm day⁻¹; 20 cm day⁻¹	>250 cm

Note: The centimetre as a unit is retained in this Table in preference to expressing quantities in strict SI units (mm; mm s⁻¹, etc.) since the rates of movement are more easily comprehended.

A

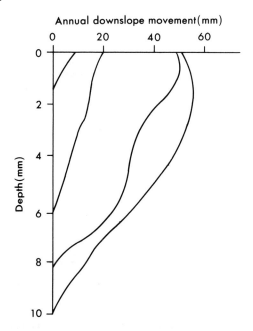

B

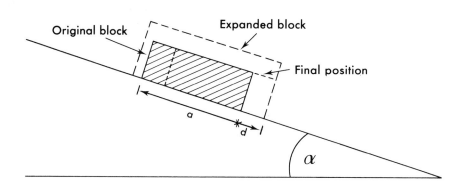

Figure 5.26A: An example of a continuous-creep profile (after Kojan, 1968); **B**: the effect of expansion and contraction on a soil particle after Davison (1889) and Kirkby (1967); **C**: the paths followed by the movement of wires inserted in a soil box during cycles of wetting and drying (Kirkby, 1963); **D**: soil creep actuated by freeze–thaw.

or roots and by chemical changes accompanying weathering. Together they are responsible for the phenomenon known as *seasonal creep* which is controlled by climatically induced changes and which affects loose rock particles on bedrock, scree, talus, humus and well developed soil.

C

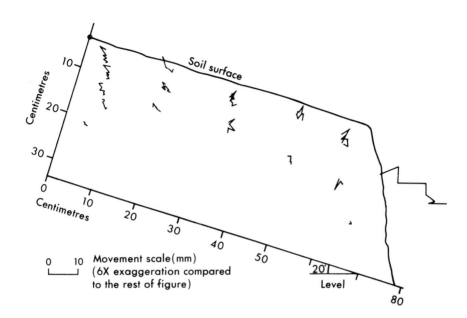

D

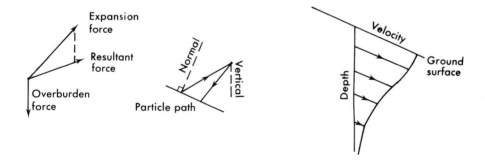

Expansion and contraction due to temperature change Temperature changes due to solar heating or fire were shown by Moseley (1869) to generate direct downslope movement of lead roofing of 23·5 cm yr[-1]. This idea was applied to soil by Davison (1888), Sharpe (1938) and Tamburi (1974) who have been able to demonstrate that insolation creep is effective on loose rock particles resting on bedrock surfaces.

Moseley suggested that the net movement of a body could be related to the following expression:

$$d = \frac{a\lambda \Delta T \tan \alpha}{u'} \qquad (5.26)$$

where d=net movement

a=particle length

λ=coefficient of linear expansion of the rock

ΔT=temperature range of one cycle

α=slope angle

u'=coefficient of sliding friction

and that the net movement occurred because both the expansion and contraction of particle length takes place under the influence of gravity which prevents a return to the original position (Figure 5.26B).

Controlled experiments by Davison and Tamburi confirm this view and show that stone blocks can move at rates of 13·17 mm yr^{-1} on a 17° slope. The controlling variables include the temperature range, the number of temperature shifts in a given period, the coefficient of friction, slope angle and particle size. Field measurements by Tamburi (1974) give lower values, 0·16–2·1 mm yr^{-1}, for single blocks placed on bedrock surfaces. There are, however, no records for insolation creep from screes, talus or soils and it seems doubtful whether such movements can overcome particle interlocking and the resistance provided by vegetation. Except in arid areas where loose particles rest on bedrock surfaces, the mechanism is probably slow and ineffective when compared to other mechanisms.

Expansion and contraction due to moisture changes   Expansion due to changes in moisture content takes place at the wetting front following rainfall or a rise in the groundwater level, but is more evenly distributed throughout the soil during the drying phase of a moisture cycle. This pattern, together with the development of open shrinkage cracks and the absorption of water by clay minerals during hydration, produces an almost continuous flexing of the soil with a constantly changing sequence of low stresses being applied to the soil particles. To some extent these will cancel out since both vertical and lateral movements can take place in any direction, into voids, but there will be a net creep downhill because of gravity.

The actual paths followed by individual soil particles can be very complex. Kirkby (1967) and Young (1960) both demonstrated in laboratory tests that the general downhill movement is accomplished along a zig-zag path corresponding to the actual expansion–contraction forces exerted on the particle Figure 5.26C). Fleming and Johnson (1975), showed by field experiment that, during the drying cycle, the direction of movement in silty clay was sometimes uphill or lateral and was determined by the position of shrinkage cracks. During the wetting cycle, the soil moved markedly downhill especially in the early part of the cycle.

The actual rates of soil movement generated by moisture changes are unknown since there have been few experiments designed to isolate each mechanism of soil creep. However, if it is assumed that, in temperate regions, moisture change will be a more important control than thermal or freeze–thaw processes, then the rough order of magnitude of such creep can be deduced from the published figures (Table 5.5).

Soil heave due to freeze–thaw processes   Many early authors believed that soil creep was mainly caused by the heave of soil particles due to the action of frost (Geikie, 1877; Kerr, 1881; Davison, 1889). Davison showed by experiment that individual particles could be lifted by 2 per cent of the depth of freezing and that the resulting downward creep after thawing was 0·6 per cent of the same depth. He proposed a model in which particles were lifted normal to the soil surface but contracted or settled in a direction that approximately bisected the angle between the vertical and normal (Figure 5.26D). As with creep caused by variations in soil-moisture content, individual particles will follow a zig-zag path downslope (Kirkby,

Table 5.5 Measured and estimated rates of seasonal soil creep

Author	Location	No. of years	Rate* $cm^3\ cm^{-1}\ yr^{-1}$	Surface rate $mm\ yr^{-1}$
Barr & Swanston, 1970	S. Alaska	—	15·0	6·0
Black, 1969	Wisconsin	—	—	5·0
Caine, 1963	Lake District	1	—	50
Capps, 1941	Idaho	30 000	—	4·5
Carson & Kirkby, 1972	Maryland	—	1·3	—
Chandler & Pook, 1971	Central England	10 000	0·3	0·04
Dedkov & Duglav, 1967	Tartar USSR	—	—	1–3
Everett, 1963	Ohio	2	6·0	0·25
Eyles & Ho, 1970	Kuala Lumpur	4	12·4	5·0
Finlayson, 1976	Western England	0·75	1·12–5·09	—
Fleming & Johnson, 1975	California	1	—	5–10
Iveronova, 1964	Tien Shan	8	>0<1	1·0
Jahn, 1960	Spitsbergen	—	—	120
Kirkby, 1963	Scotland	2	2·1	—
Lewis, 1975	Puerto Rico	4·4–5·4	1·53	2·13–10·4
Leopold et al., 1964	New Mexico	3	4·9	5·0
Leopold & Emmett, 1972	Maryland	—	—	0·5
Leopold & Emmett, 1972	Southern USA	2–9	—	0–8·6
Nieuwenhuis & Kleindorst, 1971	Holland	—	—	1
Owens, 1969	New Zealand	0·8	3·2	—
Rudberg, 1962	Lappland	—	—	75
Rapp, 1962	Lappland	—	—	300
Rudberg, 1964	Lappland	6	—	2·7–70·2
Schumm, 1956	S. Dakota	2	—	10–19
Schumm, 1964	Colorado	4	—	6–12
Slaymaker, 1972	Wales	2	2·7	—
Smith, 1960	South Georgia	—	—	50
Washburn, 1947	Arctic Canada	—	—	450
Williams, 1974	NSW Australia	2	1·9,3·2	—
Williams, 1974	NT Australia	3	7·3,4·4	—
Young, 1960	Derbyshire	2	0·5	1–2

* See note below Table 5.4.

1967) but the magnitude of movement will decrease with depth to the limit of freezing. The highest rates will be at the surface where all freeze–thaw cycles have some effect but will die away at depth by differing amounts according to the temperature range and effective depth of each cycle. Davison (1889) assumed that heave would decrease exponentially with depth to produce a concave velocity profile. Kirkby (1967) suggested that, in addition, overburden pressures and soil strength should be taken into consideration, and modified the Davison model to show that the resulting velocity profile would be convex–concave, the maximum zone of movement lying a little distance beneath the surface. The actual difference is small and neither model is fully confirmed by field measurement, perhaps because soils are variable and rarely meet the ideal requirements of theory.

The mechanisms described above have been shown to be quite effective in areas affected by frequent freeze–thaw cycles. Caine (1963) suggested that fine-grained scree materials in the Lake District could move at rates of 5–40 cm yr⁻¹ by frost creep, a figure that agrees with the rates of up to 22 cm yr⁻¹ recorded by Rapp (1960) in Kärkevagge and Owens (1969) in New Zealand. Schumm (1967) showed that the movement (d) was related to slope angle (β) in the form, $d \approx 100 \sin \beta$, though this has not been confirmed by other studies.

The rates increase in areas subject to pronounced gelifluction (see pp. 205–7) since the heaving mechanism will be aided by the growth of segregated ice lenses, wedges, needle ice and by wash, mudsliding (Chandler, 1972) or flow induced by the high pore pressures created during the summer melt. The rates (Table 5.5; data for Lappland) are often an order of magnitude greater than in temperate regions and the movements may be spatially separated into lobes and festoons of greater mobility rather than more generally spread across a hillslope.

Random movements   Culling (1963) has suggested that the continuous flexing of the soil by expansion, contraction and heave, combined with the production of root and animal passages, burrowing, the action of earthworms and termites, the removal of material by weathering processes and compaction due to treading, establishes a persistent set of molecular and macro-molecular stresses of low value which are so variable that they can lead to a random distribution of displacements. The actual movement of a particle depends on the presence of a suitable void into which the particle can move. This type of movement takes place at very low stresses and is quasi-viscous in nature. Plastic flow in which continuous alteration of form is produced by stresses exceeding a critical value (mass creep) only takes place when the voids occupy more than 50 per cent of the total volume.

This idea suggests that movement can take place in any direction as long as a suitable void is present and that material will tend to move from areas of low void concentration to areas of high void space. These movements are, however, balanced by consolidation processes that act under the influence of gravity and thus impose a net downhill transport proportional to the surface gradient. Culling's model, which employs stochastic processes in its formulation also shows that the smaller particles have a greater probability of movement and will tend to flow around coarser particles, the larger particles then rising to the surface as the finer aggregate moves away downslope to become more concentrated there.

There are no field or laboratory data to support this theory but the observation that fines are concentrated in this way, and that coarse particles do move to the surface of an aggregate when shaken in a box, perhaps provides useful circumstantial support. (See also Chapter 6 in connection with vertical sorting by frost, pp. 205–7.)

Gelifluction
Certain features of the periglacial environment are particularly favourable to mass movement of weathered debris: alternations of freezing and thawing, the accumulation of meltwater from snow or ground ice, the presence of frozen ground at depth which prevents downward percolation of moisture in thawed surface layers, and the fact that vegetation is often too sparse to restrain surface layer movements (Plate XI). Andersson (1906) proposed the term 'solifluction' for 'the slow flowing from higher to lower ground of masses of waste saturated with water' (p. 95). However, since the literal meaning of 'solifluction' is 'soil flow', it covers many different processes of mass movement occurring under a variety of climatic conditions, and also varying rates of movement, from slow downhill creep to relatively rapidly moving mud flows. Opinion is divided as to whether 'solifluction' should be used in such a wide sense. Cailleux and Tricart (1950), for instance, have used the term to include both slow creep and rapid mudslides under climatic conditions varying from periglacial to moist temperate. Rapp (1962), however, distinguished sharply between rapid sporadic movements,

such as mud flows and debris slides, the slow creep of coarse talus on steep slopes, and the equally slow creep of finer-grained material on less steep slopes. Only the latter of these movements he is willing to designate as solifluction. Dylik (1951) proposed the term 'congelifluction' to describe earth flow in the presence of permafrost, but excluding frost creep, sheet-wash and rapid mud flows, while Bryan (1946) had earlier suggested 'congeliturbation' to comprise all mass movements under periglacial conditions, including solifluction.

The term 'gelifluction' was used by Baulig in 1956 (a,b) to describe the flow of thawed material over frozen sub-surface layers. Washburn (1973) adopts this term to cover processes

Plate XI Small gelifluction lobes of coarse angular debris at the foot of a former shoreline in Devon Island, northern Canada. (C.E.)

of downslope movement, excluding frost creep, in the periglacial zone. It is unfortunate that gelifluction has been used both to denote the combined effect of periglacial slope processes in general and for a particular process (the flow of thawed debris) excluding frost creep. As has already been shown there are many practical problems of differentiating processes of downslope movement in the field and also in the laboratory. Although gelifluction (*sensu stricto*) may be more rapid than frost creep, the two processes may often operate simultaneously at certain times of the year.

Permanently or seasonally frozen sub-surface layers prevent downward percolation of moisture. The upper layer affected by seasonal thawing (the 'active layer') becomes soaked with water from melting snow, from melting of any temporary segregations of ground ice, or from rainfall. It is likely that snow melt is a more important source of water in some areas, and that in others, ground ice may be a major contributor. Differences of climate and of frost-susceptibility of the materials (see Chapter 6) must both be considered. The effect of excess water in the active layer is to reduce its shear strength. The shear strength of a material depends on internal friction and cohesion. The former depends in turn on the pressure at the contacts between individual grains and this varies with the pore-water pressure. The magnitude of cohesion in saturated debris depends on water content and therefore on the ratio of voids to solid. Significant gelifluction probably only occurs, according to Washburn (1973), at moisture values corresponding to or exceeding the Atterberg liquid limit, when soils have very little, if any, shear strength. Soaking of debris with water thus reduces both internal friction and cohesion and also increases the weight of the material. Snow patches resting on the active layer may also contribute to instability by their weight, increasing the shear stress; on the other hand, if the snow is not too thick, the ground may well be frozen up to the surface and easily capable of bearing the load.

Gelifluction may occur in areas of ground frozen permanently, seasonally, or diurnally: permafrost is not essential. The requirements in areas lacking permafrost are fairly deep and rapid frost penetration followed by thawing from the surface downward. In areas of seasonally frozen ground, summers must be relatively cool, as otherwise heat stored in the ground from the previous summer will promote thawing from below the level of winter frost penetration; this will be adverse to the preservation of a frozen layer in spring. Gelifluction on a large scale is therefore more characteristic of areas not experiencing warm summers and where the mean annual air temperature is not higher than 1°C (Williams, 1961).

Measurement of soil creep

It is clear from this discussion that the mechanisms of soil creep may involve many causes, and include shearing, flow, sliding, heaving or expansion of either individual particles or discrete blocks of soil. Measurement is, at present, unable to distinguish between these mechanisms and can only give average values for displacement as a whole, and over long periods. None of the methods employed is entirely satisfactory.

A full survey of methods would be out of place here and the reader is referred to a technical bulletin of the British Geomorphological Research Group for complete information (Anderson and Finlayson, 1975). The most common techniques involve the burial of markers or flexible tubes, or the precise survey of surface-marker displacements. Results are given as linear movements (mm yr^{-1}) of the surface, as volumetric measures of the soil moved across a plane parallel to the ground surface (cm^3 cm^{-1} yr^{-1}), or as angular measurements of surface tilt (rad yr^{-1}). Some available data are summarized in Table 5.5.

6

Sub-surface processes

C. Embleton and J. B. Thornes

In the sub-surface zone, the prevailing control of processes is the movement of water in the profile. Where water occurs as a solid (ice) it functions rather differently in relation to processes, serving more as a physical force than as a transporting agent or as a rate-regulator. Sub-surface processes relating to the occurrence and development of ice are dealt with in the second part of this chapter. The gaseous phase is also important in the sub-surface zone, notably because CO_2 is involved in chemical equilibria particularly of the bicarbonate group. The liquid and gaseous phases tend to replace each other. When water moves into the soil it displaces the gases found there and the reverse occurs on drainage.

Water in the sub-surface zone

The principal processes based on moving water are the transport of solutes of organic and inorganic materials and of suspended materials. Rarely does the sub-surface flow reach a velocity high enough for conventional fluvial processes to assume importance. The main processes are intimately related to weathering as indicated in Chapter 4.

Water movement

If we observe the variation in the moisture content of a soil with depth, using a neutron probe, we would eventually reach a point at which the soil is saturated, though in arid and semi-arid regions this could be quite deep. All the available pore space will be filled with water and this normally occurs when the volume of water is between about 30 and 50 per cent of the volume of the soil. Movement of water in this *saturated zone* is rather different in character from that in the *unsaturated zone*, though in both cases one is dealing with flow in a porous medium.

It is usual to treat the flow of water in such media on a fairly coarse scale, though for understanding chemical equilibria in the uppermost horizon it would help solve some difficult problems if the precise pathways of water through the soil could be established. For example, there is some disagreement about whether and where 'shunting' as opposed to 'by-pass' mechanisms are at work. 'Shunting' refers to the process whereby water moving through a soil replaces the water that is already there. In this way 'fresh' water continually replaces the 'stale' water in the pore spaces. Alternatively, if the soil already contains a lot of water, the infiltrating water coming from recent rain may flow only through the coarsest pores by-passing the finer pore spaces. Where the interface between soil and water in these fine pore spaces is by-passed, the opportunity for and amount of chemical exchange between soil and water is minimized. An extreme case is where the water is led through large voids called *pipes* which are open passageways from a few centimetres to several metres in diameter (p. 194). Evidently the river response, in the form of the lag between onset of rainfall and

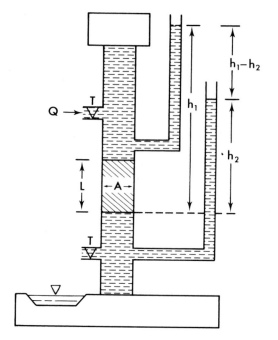

Figure 6.1 Diagram illustrating Darcy's experimental procedure.

onset of runoff, is quite different when pipes exist. However, only detailed tracing may solve the problem where pipes are absent. Some of the methods used are described in Atkinson (1978).

In the saturated zone, the velocity of flow in the porous medium (rock or soil) is related directly to the hydraulic gradient by a constant of proportionality which depends on the properties of the fluid and the medium. Darcy developed an experiment (Figure 6.1) for the measurement of this coefficient and used it in the expression known now as Darcy's Law:

$$Q = KA\left[\frac{h_1 - h_2}{L}\right] \tag{6.1}$$

where the terms are as shown in Figure 6.1. This expression is most frequently written as

$$\frac{Q}{A} = q = K \cdot i \tag{6.2}$$

where q is the discharge per unit area and i the hydraulic gradient. K is again the constant of proportionality and like q has the dimensions of velocity $(m\,s^{-1})$. The parameter K is dependent on the properties of the fluid, decreasing with viscosity and increasing with density. It is also related to the square of particle size. It may then be written as

$$K = BD^2 \frac{\rho_\omega}{\mu} \tag{6.3}$$

The term BD^2 is called the intrinsic permeability, where D is the particle size, B a coefficient involving shape and packing and ρ_ω and μ the specific weight and absolute viscosity respectively. K is then called the hydraulic conductivity. Only under conditions of constant density and viscosity (isothermal conditions) will conductivity reflect variations in the properties of the medium.

The other component of Darcy's Law is the potential gradient or *head*. This is the energy that drives the flow and comprises three parts, the gravitational head, the flow energy and the kinetic energy. The sum of these three sources is a constant, as given in the Bernouilli equation:

$$gZ + \frac{P}{\rho_\omega} + \frac{v^2}{2} = \text{const.} \tag{6.4}$$

where g is acceleration due to gravity, P is the pressure of the fluid and ρ_ω its density. This equation has the units of energy per unit mass $(m^2 s^{-2})$. Dividing through by gravity gives energy per unit weight and the result is in units of length (m). Z is the gravitational potential energy, $P/\rho_\omega \cdot g$ is the flow energy and $v^2/2g$ the kinetic energy. The first two terms are the potential energy (ϕ) and may be written as Darcy's Law by omitting the kinetic energy which is usually negligible at the velocities considered. Thus

$$q_x = -K \cdot \frac{d\phi}{dx} \tag{6.5}$$

where flow is now expressed in relation to a particular coordinate axis (x) and is shown negative since the flow is moving in the direction of decreasing values of potential. Maps showing the distribution of ϕ are called equipotential maps and can be used to indicate the magnitude and direction of flow in the saturated medium, as shown in Figure 6.2.

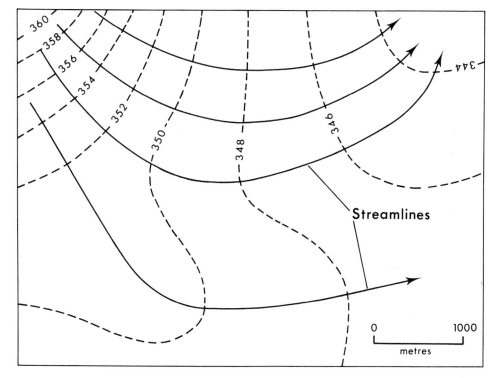

Figure 6.2 Water-table contour map showing the direction of groundwater flow. Contour interval—2 m (Ward, 1977, 2nd edition).

When the soil or rock is unsaturated the situation is more complicated. As the moisture content of the soil decreases, the water is held more tenaciously by the soil as a result of capillary forces. As the soil drains under gravity a point is first reached where the pressure is zero. Thereafter the pressure becomes negative as the soil applies a suction to the water. This negative pressure is called tension. In saturated conditions, the gravitational potential is usually greater than the pressure potential, but under unsaturated conditions the reverse may be the case. This is shown in Figure 6.3. Curve 'A' represents near-saturated conditions, tension is almost zero and the potential energy is almost a linear function of depth. As the

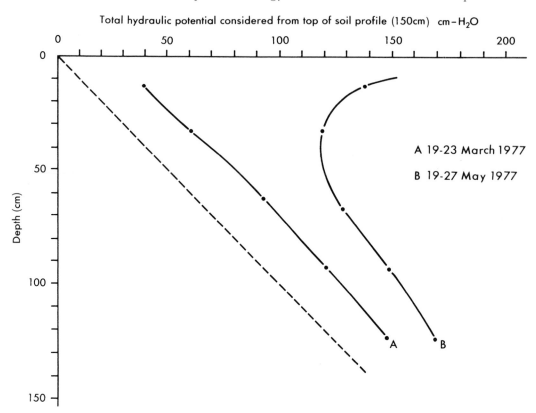

Figure 6.3 Distribution of total head in a tropical rain-forest environment (Manaus, Brazil) during **A**: a very wet season and **B**: a period of drying (after Nortcliff and Thornes, 1977).

soil dries out the distribution of head changes, reflecting the drainage of water from the uppermost horizons of the soil and the development of negative pressure (curve B, Figure 6.3). These effects may also be observed in an entire hillslope section (Weyman, 1973) and in three dimensions (Anderson and Burt, 1977).

Besides the shift from positive to negative pressures, the hydraulic conductivity is affected by unsaturated conditions and is again related to soil moisture content. The curve for a typical light clay soil is shown in Figure 6.4A. This is somewhat simplified because the moisture–tension curve normally exhibits a hysteretic effect, in which the tensions are higher for a given moisture content on the drying curve than on the wetting curve. These relationships, which are essentially controlling water movements, show quite significant variations with depth and laterally. Knapp (1973) gives the results of observations on the A horizon of a podzol in Wales all taken within a 1 m radius of a point on the surface. He showed that the tension–conductivity relationship may not provide a reliable measure of the characteristics of the soil. The scatter of sample results obtained was such as to make any interpretation from a 'characteristic curve' a hazardous procedure. Moreover, the rapid fluctuation of the relationships in very short distances makes the possibility of mapping such variations impractical.

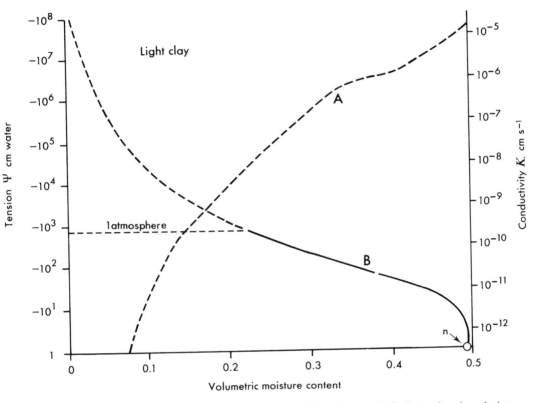

Figure 6.4A: Hydraulic conductivity in relation to soil moisture; **B**: Soil tension in relation to soil moisture.

The flow of infiltrated water is known as sub-surface flow. Sub-surface storm flow is defined as that part of the infiltrated water which moves laterally through the upper soil horizons towards the stream channels as unsaturated flow or shallow perched saturated flow above the main groundwater level. Baseflow (in the stream channel) is that part of the lateral inflow derived from deep percolation of infiltrated water that enters the permanent groundwater system and discharges into the stream channel.

Sub-surface flow is also regarded as important in generating both overland flow and stream flow. According to Freeze (1972), the prime requirement for sub-surface storm flow is a shallow horizon of high permeability at the surface. A downward reduction of permeability is commonly a result of horizonization of the soil and is particularly important at the base of the organic layer of an A horizon. If there is no infiltration from the surface, flow is essentially determined by the distribution of moisture and suction in the soil, and generally increases with depth and decreases through time. With evaporation (which decreases with depth), these effects will be emphasized. During infiltration, the surface layers become saturated, the suction at the surface drops to zero, and the locus of maximum through flow will occur near the surface and at depth. The maximum throughflow will occur when the soil is completely saturated and is equal to the saturated hydraulic conductivity, multiplied by slope and the soil-layer thickness.

There is no doubt that throughflow occurs but several workers have expressed doubt as to its efficacy for delivering sufficient water to contribute to storm runoff. Ragan (1967) observed sub-surface storm runoff in the hillside forest litter but found it to be quantitatively unimportant as a contributor to storm runoff. Dunne and Black (1970a) reached a similar conclusion. The issue is very complicated where extensive systems of pipes occur beneath the surface, which may carry large quantities of water during and after rainfall. They do not respond immediately after rainfall but only when storage deficit has been satisfied. Freeze's (1972) model simulation throws further doubt on the importance of sub-surface storm flow as a quantitatively significant generator of surface flow.

Upslope from a stream a unit of rainfall contributes less to sudden storm runoff in the channel and more to retention–detention storage. Hewlett and Hibbert (1967) theorize that only part of the water contributed to quickflow is actually new water, i.e. water from the current storm; the remainder is called 'translatory flow' or flow by displacement. This is the water already stored in the soil prior to the storm and will be released in large quantities only where the soil is extremely wet.

This translatory process has been studied in the laboratory by Horton and Hawkins (1965). They added 2·5 cm of tritiated water to a 1 m soil column drained to equilibrium. Then 2·5 cm of pure water was added each day and effluent was analysed daily for tritium. Only after 87 per cent of the original water flowed from the column did the tritiated water emerge. They concluded that in a deep soil profile rainwater likewise tends to displace water as it moves primarily through capillary pores. Martinec (1975) similarly used environmental tritium to trace the movement of snowmelt and concluded that infiltration of melt water causes a corresponding increase in the outflow of sub-surface water to streams.

Observations of forested catchments reveal a very rapid saturation of the zone adjacent to the channel. Outflow from this zone comprises two parts at any given time: that arising directly from runoff from falling rain, and that from upslope contribution. The latter is defined as the resultant flux (Harr, 1977) and can be written as

$$q_R = [(q_D + q_V \sin \alpha)^2 + (q_V \cos \alpha)^2]^{0.5} \qquad (6.6)$$

where q is the water flux $(\text{m}^3\text{s}^{-1})$; R, D and V denote resultant, downslope and vertical directions and α is the slope angle. The fluxes q_D and q_V are obtained essentially from the assumption that Darcy's Law can be used, in modified form, for unsaturated flow. The two components added together then yield an outflow (q) as:

$$q = I \cdot x_s + q_R \qquad (6.7)$$

where x_s is the area of the saturated zone and I the rainfall intensity.

Fluxes, calculated from field data, are shown in Figure 6.5 from the Amazonian experimental plot referred to earlier (p. 190). The data show the flux magnitudes and directions for a period of heavy rainstorms of high frequency, expressed in terms of unit cross-sectional area. The near-vertical resultant fluxes, with high values of 3–3·5 m day^{-1} indicate the freely draining nature of the soil and the lack of impervious strata and suggest that throughflow has an insignificant role to play at least in the wet season. In these soils, after a month of

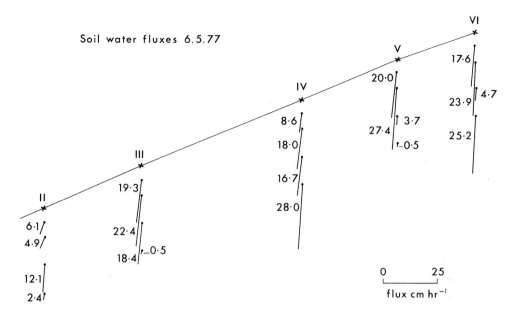

Figure 6.5 Fluxes in soil moisture in a tropical rain-forest environment.

drying, the fluxes remained nearly vertical but of much smaller magnitude (about one tenth). With vertical fluxes, as the saturated zone is approached, the potential gradient arising from capillary tension diminishes and movement is exclusively under saturated gravity flow. If the saturated zone is near the surface, as in the floodplains of small tropical basins in the wet season when heavy rain falls, the water table rises to the surface and water flows overland to the channel at rates that are much higher than sub-surface flow. In this way rivers may respond much more rapidly to rainfall. In temperate areas where inputs are smaller and less frequent and especially where permeability decreases rapidly with depth, the downslope component of flux may be equal to or of greater importance than the vertical component (Zaslavsky and Rogowski, 1969; Harr, 1977).

Mechanical effects

The movement of water through the soil produces important mechanical and mechanico-chemical effects. Sub-surface eluviation (suffosion) is the vertical and lateral transport of fine particles by flow. The process is probably volumetrically unimportant because of the low rates of movement and the self-destructive nature of the process, wherein the clogging of pore spaces occurs exponentially with time. Nevertheless it is probably important for its effect on the distribution of clay-size particles in the soil which determine the direction and magnitude of water flow, producing a positive feedback effect. The process is known to exist from formal experiments. Hallsworth (1963), for example, found that clay translocation in sandy soils diminished rapidly with the amount of added clay and that it was greater the larger the sand fraction in the structure. The latter reflects the need for a bimodal size distri-bution so that the particles of the smaller mode may move through the pore-spaces between those of the larger mode. Statham (1977) points out that with perfect spheres the ratio of large to small diameter needs to be at least 2·5:1 for the loosest possible regular packing and 7:1 for the tightest for such movement to occur.

Field evidence of translocation is more difficult to assess because the effects produced may be similar to those resulting from weathering. The evidence includes the development of clay-rich layers in the profile and the development of clay linings around roots. Occasionally, too, resurging water from slopes and river beds (Thornes, 1976) may carry with it significant amounts of suspended matter. The effects of throughflow variations (as measured by total flux) should be reflected by translocation of materials and Huggett (1977) has found evidence of this. He estimated the gains and losses of materials by comparing the amount of clay at any horizon with that of the lowest horizon in each profile. He found, for example, that the distances and volumes of translocation were greater in hollows than on spurs. Because convergent flow of soil moisture prevails in the hollows (whereas flow is divergent on the divides) the amount of throughflow per unit time is relatively larger there and compared with many parts of the landscape, less rainfall is required to initiate flow.

In addition to diffuse flow, sub-surface flow may be concentrated along discrete lines. Sometimes these form the heads of true channels and have no surface expression. They have been termed percolines and described by Bunting (1961). Downslope they may give rise to seepage lines, shallow depressions with waterlogged soil supporting water-loving plants. Pipes, also described from upland, well vegetated areas (Jones, 1971) have important effects on the hydrological régime, transmitting up to one-fifth of the runoff from a catchment. They are end-members of a system of sub-surface drainage channels which may be up to several metres across and are referred to as tunnels.

Pipes may extend from steep-sided channels where the water may be seen issuing after prolonged rain. They appear to be confined to zones having a steep hydraulic gradient and so are found on mountain slopes with impermeable bedrock or in conjunction with deeply incised gullies. Leopold, Wolman and Miller (1964) reported that where soil pipes were located up-valley from a gully headcut, the migration of the latter would be accelerated by the piping process. Many hypotheses have been put forward as to their origin of which rodent burrowing, root decay and soil fissuring are among the more common. Heede (1971) has developed and extended the argument that a dispersive agent in the soil may be responsible for some pipe development. He found that piping soils in Alkali Creek, Colorado, have a significantly higher exchangeable sodium percentage (ESP) than non-piping soils. The non-piping soils have an ESP of less than 1·0, whilst the piping soils have an ESP greater than 1·0 and an average of 12. The piping soils thus contain sufficient sodium to ensure dispersion when water moves through them and for swelling to occur. He suggested that the pipes are initiated along the lines of former soil cracks. In the soils with high ESP, dispersion of the soils in the cracks leads to widening of the crack at its base where the water runs out into the gully. In non-dispersive soils, erosion in the crack is so slow that the cracks close again during swelling of the soils.

Obviously in seasonally dry regions, where ESP tends to be high, this explanation may be satisfactory and the results of Stocking's work (1976) appear to confirm this. In temperate areas, the emphasis should probably be laid on the swelling properties of the clays, which is a related phenomenon. In both, mechanico-chemical effects associated with the solute–solid surface interaction are involved. The solid phase (soil) is penetrated by water molecules which weaken the binding force between the surface atoms and may lead to dissociation of the surface ions. As is pointed out in Chapter 4, clay molecules are particularly susceptible because of the substitution of Si^{4+} and Al^{3+} ions by cations of a lower valency such as Na, Ca, K and Mg. These dissociate from the surface in the presence of water but remain very close to it resulting in a (usually) negatively charged surface with an accumulation of posi-

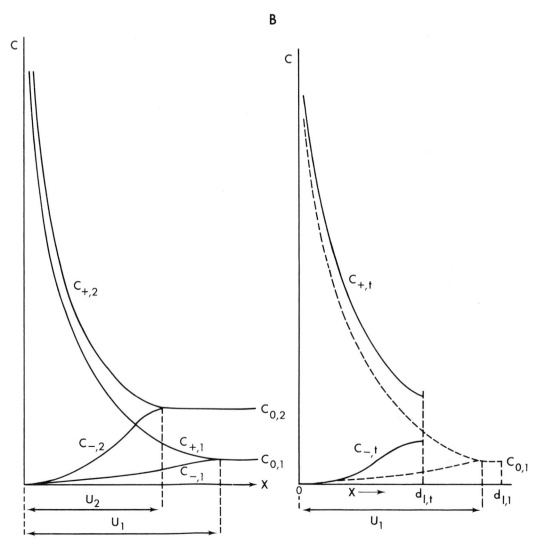

Figure 6.6 Illustration of the double electric layer.

tively charged ions close to it. Negatively charged ions will tend to be repelled by the surface charge and so increase away from the surface. The combined effects produce a modified exponential distribution away from the surface in which the concentration falls until at some distance it has the same concentration as the solute. This distance depends upon the concentration of the solute and the valency of the counter-ions, decreasing with increased valency (Figure 6.6).

If the moisture of the system falls, the width of the double layer cannot be greater than the thickness of the water layer and is truncated below its potential width. Such a truncated double layer reabsorbs water very forcibly until the potential width of the double layer is again satisfied. This produces a swelling pressure which may amount to several tens of bars.

The swelling pressure depends on the valency of the counter-ions and the salt concentration in the soil solution. The swelling results in the closure of inter-aggregate pores, a diminution of the infiltration rate and an increase in the tendency to generate overland flow.

This process is related to the ESP by the fact that the sodium ions constitute an important source of monovalent counter-ions in desert soils. The rapid swelling leads to breakdown of the aggregates and consequent dispersion. Repulsion of the anions also leads to a 'salt-sieving' effect in which the soil, usually clay with only very fine pores, acts as a leaky semi-permeable membrane, allowing water to pass more easily than salts. Differences in salt concentration may then cause bulk movement of water (Bolt and Groenevelt, 1972).

Chemical effects

Without doubt the most important sub-surface processes in geomorphological terms are those involving chemical solution, transport and precipitation for they are responsible for a large part of the total load removed from a basin in a given time period. Moreover, they control the relative balance of transport- and weathering-limited removal by surface processes and the character of the material supplied to the fluvial sub-system for transport. These sub-surface processes, like water movement, are intimately related to the vegetation cover and to the activity of the soil flora and fauna (see below).

Generally speaking, equilibration between the minerals of the soil and percolating water occurs according to the processes outlined in Chapter 4. At the simplest and most general level it may be assumed that this equilibration is dictated solely by the thermodynamic properties of the minerals and the solute. Assuming one is concerned only with inorganic soils and that the reactions and corresponding thermodynamic equilibrium constants for the chemical species involved are known, it is then possible to determine theoretically certain characteristics of the system. In particular, the solubilities of the minerals and their stabilities, one relative to another, as well as the concentrations of the dissolved minerals in the solute may be established. Thus, for example, in Figure 6.7 the solubility of some minerals in an aqueous solution at 25°C and 1 bar total pressure is given as a function of pH. Solubility here is expressed in terms of activities of the various mineral species. It will be seen that, with the exception of gypsum, quartz, gibbsite and kaolinite, the solubility of the minerals increases monotonically with a decrease in pH under these conditions. In this case, in the pH range typical for soils (4–9), the concentrations of dissolved aluminium are very low compared to those of the other solutes. Figure 6.7 also reveals that the simple process of equilibration according to the energy conservation principles need not necessarily correspond to the 'weatherability' of a mineral. For example, gypsum and muscovite have approximately equal solubilities at near-neutral pH. In fact, as van Breeman and Brinkmann (1977) point out, muscovite is very inert and weathers much more slowly than gypsum. Thus surface phenomena play a particularly important role (p. 194) though the simplification proves useful in modelling terms (Kirkby, 1977b).

The rate of equilibration is not given by this type of analysis but it is very important. The rate is usually expressed in terms of a first-order system in which reactions are rapid at first but become slower. The time taken for a 50 per cent change to occur in a given reaction is called the half-time, analogous to the half-life of a radioactive element. If the residence time of water in the soil is shorter than the equilibration time, only low concentrations would be transported away. If it is much longer, then although the eventual transport will be of a concentrated solution the total volume moved will be small. Equilibrium between cations in solution and exchangeable cations on the soil surfaces is usually very short, up to about half an hour according to the particle diameter and diffusion rate. The latter is

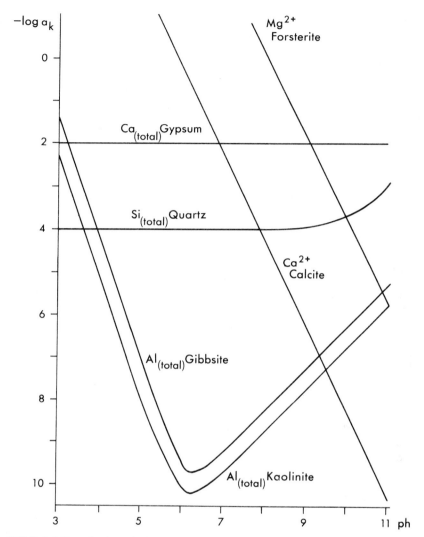

Figure 6.7 Solubility of minerals in a silicate solution at 25°C and 1 bar as a function of pH. Solubility increases from bottom to top of vertical axis.

usually more important and is related to pore space (see below). Cations showing half-times of reaction of the order of a day are less common though experiments show that potassium has this property, for example when being released from illitic clays. Smith and Dunne (1977) reported that both the release of cations and the uptake of silica in undisturbed soils may occur quite rapidly. Steady values were obtained in distilled water passed through soil columns, the time required for passage being only two hours, and the equilibrium state of the solute derived from thermodynamic considerations was close to that expected from the known characteristics of the soil.

Three processes are involved in the transport of solutes: diffusion, mass transfer and dispersion. The process of diffusion results from the random thermal motions of ions, atoms or molecules. For soils this is usually modelled by a modified version of Fick's first law which states that the rate of diffusion is a function of an impedance factor which accounts primarily for the tortuous pathway followed by the solute through the pores, the concentration gradient and a number of other complex effects. Naturally this impedance factor is related to soil moisture; for a dry soil it is, like hydraulic conductivity, very low. Moreover, the rate of diffusion depends on the characteristics of the molecules and this is revealed by the diffusion coefficients. On the whole, however, the distances moved are negligible compared with those due to mass transport. By comparison, chloride might move by diffusion $0.3 \, \text{cm day}^{-1}$ whereas water movement by throughflow could be $3 \, \text{m day}^{-1}$. A large pore 1 mm wide may require 1000s to achieve 95 per cent equilibriation by diffusion.

Mass flow not only transports the atoms but also disperses them. The latter occurs when a gradient of the solute occurs together with a flow-induced velocity pattern. The liquid phase travels faster in the wider pores than the narrower ones and part of the solutes dragged along in the liquid phase travels faster leading to mixing, as occurs in stream channels (Chapter 7). Porous aggregates tend to increase dispersion because there is not instantaneous equilibrium between the pore solution within the aggregates and that between the aggregates.

Finally, one must remember that movement of air in soils is also very important. Both temperature and the chemical composition of air (especially its CO_2 content) play important roles in the equilibriation processes. This movement of air occurs mainly by diffusion.

Biological activity

Biological activity influences virtually all the processes outlined above. At the simplest level fauna create voids through which water can move, and at the most complicated, the decomposition of plant materials leads to the production of complex organic acids that interact with the exchange processes already occurring in the soil.

The two most important effects, however, involve the role played by vegetation in the water and nutrient budgets of the soil. Water is used by plants to maintain the structure of the non-woody parts and to transport nutrients for photosynthesis. The plant draws water from the soil at a rate which is determined principally by the rate of energy supplied at the plant leaves, provided there is no limit to available soil water. As, for one reason or another, the available moisture in the soil diminishes, the plant can adjust by closing the leaf stomata and hence reducing the transpiration rate. Within the soil, the distribution of the uptake of water (and minerals in solution) is determined by the distribution of roots and the distribution of water. Generally speaking the distribution of root mass decreases very rapidly with depth. Gerwitz and Page (1974) found that plotting the logarithm of root quantity in a soil layer of unit thickness against depth tends to produce a straight line for some common rootcrops, and results for natural forest stands show a similar rapid decline with depth (Klinge and Herrera, 1977). Unfortunately most of the investigations have so far applied to commercial rather than natural vegetation cover, so that attempts to model the effect of root distribution on water uptake have met with only a modest degree of success (e.g. Gardner, 1964). Part of the difficulty is that plants appear capable of adjusting their root-water potential across the root system, so that there is differential water uptake in different soil horizons which may not be simply related to root mass or length.

The nutrient uptake by plant roots is even more complicated, for it occurs in relation to the concentration within the root relative to that in the soil environment around the root. This question is treated at length in textbooks such as that by Nye and Tinker (1977). The main point is that the volume of uptake of nutrients may be quite large even in a 'steady system' (Ovington, 1965; Rodin and Bazilevich, 1965; Jordan and Kline, 1973). The return input by leaf fall has to be appreciated if the weathering processes, as they vary across the different climatic environments, are to be completely understood. Nutrient transfer to the soil via leaf fall and nutrient release from decomposing litter may be much more important in the weathering process than is the supply of minerals by rainfall. Certainly this is true of the lowland tropical forests of Amazonia (Klinge, 1977).

Besides uptake of water and nutrients, the organic matter of the soil is directly involved in weathering through such processes as chelation and bacterial mineralization and most importantly by the production of carbon dioxide. Under aerobic conditions the CO_2 diffuses rapidly away from the root through the air-filled space, so that its effect is widely diffused over the whole soil where it plays an important role in the carbonate equilibrium system (Rich, 1973). A useful and up-to-date summary of these components is to be found in Trudgill (1977).

Sub-surface processes related to ground ice

The first part of this chapter has been concerned with sub-surface processes operating in the presence of moisture in its liquid phase. The change of phase from water to ice, and from ice to water, sets in motion a distinctive range of processes of great interest not only to the geomorphologist but also to the engineer and any others concerned with the practical utilization of areas liable to be affected. The processes associated with ground freezing include frost heaving, the sorting of sediments, thermal contraction cracking and various forms of differential displacement. When frozen ground starts to thaw, the sudden reduction in strength of the materials frequently leads to collapse and slumping, the phenomena of 'thermokarst'. These processes operate not only in the permafrost regions of the world (about one-fifth of the earth's land area at present) but also in neighbouring or high-altitude areas subject to seasonal or periodic ground freezing, so long as moisture is present in the ground. There are important differences between permafrost and non-permafrost areas in respect of ground freezing. In the former case, thawing only occurs from the surface downwards (giving the 'active' layer of the summer season up to 12 m in depth towards the permafrost margins), but freezing of the active layer will occur both from the surface down and from the bottom up because of the underlying permafrost. In non-permafrost areas, soils freeze only from the top down, but will thaw from the top and from below (owing to geothermal heat).

The process of soil freezing, which will be examined in the next section, is complex. A great deal of experimental work has shown that freezing does not necessarily occur at 0°C (because of the varying sizes of pore spaces and varying pore pressures, and because of impurities), that different soils freeze or thaw at different rates (because of variations in conductivity and moisture content), and that the duration and rate of freezing affect the changes taking place in the soil. If no moisture is present in the soil, lowering of ground temperature below 0°C will have no significant effect apart from slight contraction, but the presence of moisture before freezing will lead to ground-ice formation (Plate XII). Ground ice is defined as 'ice, of whatever origin or age, found below the surface of the ground, especially a lens, sheet, wedge, seam, or irregular mass of clear non-glacial ice enclosed in permanently or seasonally frozen ground often at considerable depth' (American Geological Institute, 1972). Table 6.1 shows a classification suggested by Mackay (1972). The term 'excess ice' is used to refer to the common condition when sediments are super-saturated with ground ice. This means that, on thawing, there is more water liberated than can be retained in the soil voids. The existence of visible masses of ground ice (for example, as lenses or layers) does not necessarily

Plate XII Ground ice in permafrost, Devon Island, Canada. Segregated ice is exposed in the pit to the right of the blade of the shovel. (C.E.)

indicate excess ice, for the sediment between the icy layers may not be saturated; nevertheless, except for pore ice, most ground ice exists in the form of excess ice, and therefore has very important geomorphological effects, both when forming and when thawing. Pore ice, which simply occupies available pore spaces in sediments, is only important geomorphologically as a slope-stabilizing agent.

Soil freezing

Williams (1967) has shown that the properties and behaviour of freezing soils are largely explicable in terms of a capillary model, in which the effects of interfacial energy (or surface tension) at small ice–water interfaces confined in the pore spaces of the soil are of fundamental importance. Different pressures of ice and water are associated with the curvature of these interfaces:

$$p_i - p_w = \frac{2\sigma_{iw}}{r_{iw}} \tag{6.8}$$

where p_i is the pressure of ice, p_w that of water, σ_{iw} is the surface tension between the ice and water, and r_{iw} is the radius of the interface. Thus the interfacial pressure difference is related to the pore size. The pressure relationship influences the freezing point which falls as $p_i - p_w$ increases or as the pore size decreases:

$$T - T_o = \frac{V 2\sigma_{iw} T_o}{r_{iw} H} \tag{6.9}$$

Table 6.1   Classification of ground ice based on origin (J. R. Mackay, 1972)

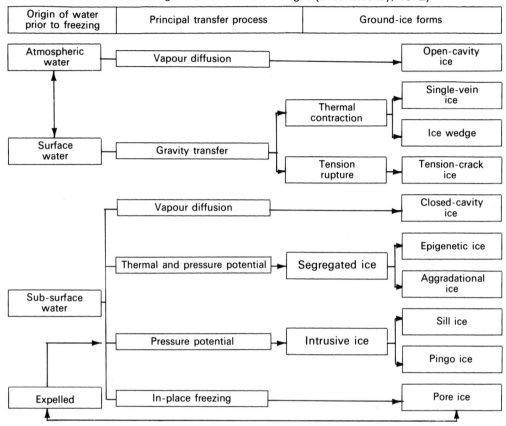

Origin of water prior to freezing	Principal transfer process	Ground-ice forms
Atmospheric water	Vapour diffusion	Open-cavity ice
Surface water	Gravity transfer → Thermal contraction	Single-vein ice / Ice wedge
	Gravity transfer → Tension rupture	Tension-crack ice
Sub-surface water	Vapour diffusion	Closed-cavity ice
	Thermal and pressure potential → Segregated ice	Epigenetic ice / Aggradational ice
	Pressure potential → Intrusive ice	Sill ice / Pingo ice
Expelled	In-place freezing	Pore ice

where T = freezing point °K, T_o = normal freezing point (273·15°K), V = specific volume of water, and H = latent heat of fusion. Thus, the temperature at which freezing occurs depends partly on the interfacial curvature and, therefore, on pore size (grain size), so that freezing and thawing of a mixed soil occurs over a range of temperature. Note that T_o = 273·15°K (0°C) only if the pressure on the ice phase is atmospheric, and that its value must be reduced by 0·0073°C bar^{-1} if there is overburden pressure. Equations 6.1 and 6.2 combine to relate the interfacial pressure difference directly to temperature:

$$p_i - p_w = -\frac{(T - T_o)H}{T_o V} \qquad (6.10)$$

Following equation 6.2, it will be seen that ice will form first in the larger pore spaces, and that, as temperature falls, it will be able to penetrate smaller openings. Many pores may remain water-filled because they are too small to allow the interfaces to grow through them, so that the freezing process is a gradual one. The proportion of water remaining unfrozen varies inversely with grain-size at a given temperature, so that clays and silts solidify at lower temperatures than coarser materials (Beskow, 1935) and moisture remains mobile for longer during the freezing process. Half of the water in a clay may still be unfrozen at −2°C.

Ice segregation

Early experimental work by Taber (1929, 1930) showed that the segregation of ice in the soil, leading to formation of lenses and layers, results from the migration of pore water to the freezing plane where it solidifies. As the ice lenses grow, ground-surface heaving results. This is not heaving due simply to expansion of water *in situ* as it freezes, which only amounts to approximately 10 per cent increase in volume: heaving needing an expansion of over 100 per cent has been frequently observed, so that there must be a considerable gain in water content during freezing. Examination of samples frozen in the laboratory revealed that the segregated ice layers consisted of crystals growing within the material normal to the layers, that is, in the direction in which heat was being most rapidly conducted away. Taber showed that grain size was a critical factor for segregation: in materials of grain size 0·01 mm or less, segregation could be induced without difficulty, though in clays it might be restricted by the fine pores reducing the rate of water migration to the growing ice crystals.

Subsequent work has broadly confirmed and amplified Taber's conclusions. The radius of the ice–water interfaces is critical in the behaviour of the freezing plane. The latter can only advance into the soil during freezing if the critical radius of the interface, r_c, determined by the size of continuous pore openings, is sufficiently small:

$$p_i - p_w \geq \frac{2\sigma_{iw}}{r_c} \tag{6.11}$$

Values of the right-hand term in equation 6.11 for a given soil (Table 6.2) can be determined from air-intrusion tests (Williams, 1967) which indicate the pressure of air necessary to begin to force water suddenly out of a saturated sample – the air-intrusion value depends on the radius of the largest continuous openings through the pore-structure of the soil. p_i at the freezing plane is approximately equal to the overburden pressure, and p_w, the pore-water pressure, can be estimated for a variety of field conditions.

Table 6.2   Values of $\dfrac{2\sigma_{iw}}{r_c}$ for various soil types as determined by air-intrusion tests (Williams, 1967)

General soil type	$\dfrac{2\sigma_{iw}}{r_c}$kg cm^{-2}
Coarse sands	0
Medium and fine sands	0–0·075
Medium silts	0·075–0·15
Fine silts	0·15–0·5
Silty clays	0·5–2·0
Clays	>2

If equation 6.11 is not satisfied, because of high pore-water pressure or small values of r_c, or both, then the freezing plane cannot penetrate through the soil pores and an ice lens will start to form, fed by capillary water from below. But as water migrates to the ice lens, pore-water pressure will tend to fall. If the fall of pressure is sufficient, the conditions of equation 6.11 will be approached and, if achieved, ice-lens growth will cease, the freezing plane then being able to advance through the pores. This may in turn allow pore-water pressure to rise and a new ice lens will form. Thus soils susceptible to the formation of segregated ice often contain alternations of ice lenses and frozen ground with pore ice, concentrated in the upper layer of the permafrost where overburden pressure is least. Thick lenses are therefore associated with a high groundwater level (and high pore-water pressure), high permeability, shallow depth (low overburden pressure), small values of r_c (fine grain-size) and slow rates of freezing. Each lens marks a temporary position of the freezing plane in the soil and a point of equilibrium between supply of pore water and rate of frost penetration.

In general, silts are more favourable to ice segregation than clays or sands. Finer-grained materials have a greater suction potential (i.e. can support a greater difference in pressure between the ice and water due to the sharper curvature of the ice–water interfaces), but the loss of permeability means a reduction of water movement. It has often been observed that ice at a penetrating freezing-plane in clays does not advance so uniformly but tends to follow cracks or discontinuities. Ice lenses may form in places, but are often surrounded by layers of unfrozen plastic clay until the temperature falls sufficiently. Clay mineralogy also affects ice segregation. Linell and Kaplar (1959) show that kaolinite is more favourable to segregation than illite or montmorillonite.

Some general rules of frost-susceptibility have long been employed in engineering practice. Casagrande (1932) stated that most frost-susceptible soils (i.e. those liable to ice-lensing and heaving) will have at least 3–10 per cent of particles finer than 0·02 mm, and Terzaghi (1952) suggested likewise that several per cent of the particles must be below 0·07 mm. These, how-ever, are very generalized criteria; some soils that satisfy neither of these requirements have shown lensing and heaving, yet others falling within the limits have remained unaffected.

In sands and coarse silts, segregation is relatively uncommon, as will be apparent from the values set out in Table 6.2. Lensing can then only occur when the pressure difference $p_i - p_w$ is extremely small. Coarse sediments can, however, provide an important source of excess pore water on freezing, when about 10 per cent of the pore water will be expelled due to expansion. For a sand of 30 per cent porosity, the excess water would therefore amount to about 3 per cent of the total volume of sand. It is possible that if this excess water were unable to escape freely, values of p_w could rise (perhaps to as much as 90 per cent of over-burden pressure: Penner, 1967) and ice segregation might then begin. Mackay (1971) has discussed the development of massive beds of ground ice, near the western Arctic coast of Canada, whose ice content may exceed 500 per cent (weight of ice to dry soil) for several metres in thickness. They tend to overlie thick sandy beds, occur beneath frozen clays, and from analyses of low ice-content layers (sufficient residue cannot be obtained from high ice-content specimens) the ice appears to have formed in silty clays favourable for lensing. Mackay suggests that freezing of the saturated sands beneath has provided the excess pore water for massive growth of ice lenses and that high pore-water pressures have built up be-neath a confining layer of aggrading permafrost.

Frost heaving and thrusting

Frost heaving in soils is principally due to an increase in moisture content following migration of pore water to the freezing layer. This additional moisture appears mostly in the form of ice lenses as explained in the previous section. Ice lenses are normally oriented roughly parallel to the ground surface (i.e. perpendicular to the direction of heat flow), but they may also be affected by soil stratification (to which they also frequently lie parallel) and by other soil structures such as cracks. Thus irregular-shaped masses of segregated ice may form. Heaving of the surface is the main response to the increase in volume caused by segrega-tion, but there may also be lateral displacements which Washburn (1973) distinguishes as *frost thrusting*. *Differential heaving* of the surface occurs because of spatial variations in the processes of ground freezing resulting in cracking and the formation of hummocks. The amount of heave, h, in time Δt, is given by

$$h = ki\,1{\cdot}09\,\Delta t \qquad\qquad (6.12)$$

where k is the coefficient of permeability, i is the hydraulic gradient to the freezing plane, and $1{\cdot}09$ is a factor allowing for the volumetric expansion of water on freezing. The latter factor accounts for only a small part of ground heaving: by far the greater part is due to ice segregation. As well as affecting the micro-relief of the ground surface, ice segregation can weaken or destroy the texture of a soil, producing in place distinctive cryogenic textures consisting of patterns of veins and lenses, breaking the soil into polyhedral blocks (Demek, 1978). Desiccation of the lower soil, as pore water is drawn up, may also produce desiccation cracks in it (Pissart, 1964). The forces involved in frost heaving are considerable, as engineers and others concerned with utilization of areas susceptible to frost heaving have long been aware: growing ice crystals can exert a pressure of over $140\mathrm{t\,m^{-2}}$.

Studies of frost heaving have been made in many areas, both with and without permafrost. At Hornsund in Spitsbergen, Czeppe (1960) has collected data on vertical movements due to the annual cycle of ground freezing and thawing. There is a steady and sometimes rapid uplift during the main freeze-up period of early winter, followed by a period of little movement till the spring when subsidence begins. The annual amplitude may exceed 15 cm. Washburn (1969) has over many years studied frost heaving in the Mesters Vig district of northeast Greenland. Cone targets and rods were inserted to various depths and their displacements recorded. Heaving, up to 8 cm in one year, was found to be greatest in wet sites and those with little or no protective vegetative cover. Most heave occurs in the initial freeze-up period. The more deeply buried targets (up to $0{\cdot}3\,\mathrm{m}$) moved later, showing the arrival of the freezing plane at depth, and these also showed the greatest amounts of heaving. Most heaving appears to occur in the upper part of the active layer, where most ice lensing takes place. Some negative movements of unfrozen lower layers have been measured, due to cryostatic pressure compressing the unfrozen material beneath.

Up-freezing and tilting of stones

Movement of stones in freezing soils has long been known. Taber (1943) cites rates of uplift amounting to 25 cm a year: the stones move up to the surface, often standing on end, and then fall over. Many hypotheses have been suggested to account for the phenomenon, but the two termed by Washburn (1973) the 'frost-pull' and 'frost-push' hypotheses seem adequate to explain most features. In the first, ground heaving by ice segregation lifts the stones, especially as ice lenses grip suitably shaped stones. Voids beneath the stones then tend to fill with slumped material, or become narrowed by frost thrusting, so that, on thawing, the stones cannot sink back to their original positions. Kaplar (1970) has demonstrated the process in the laboratory using time-lapse photography. The frost-push hypothesis states that, because of the greater conductivity of stones, ice forms around and beneath them, raising them. Again, on thawing, the stones never re-settle to their original position. Washburn suggests that both mechanisms may operate depending on local site conditions. Note that the second hypothesis cannot be applied to heaving of poor-conducting materials, such as the wooden piles of bridges.

During the frost-pull action, long thin or slab-like stones tend to become tilted more and more towards a vertical position. Pissart (1969) has shown in experiments how the top of a tilted stone may be lifted first while the lower end is still held by unfrozen material. Retaining its new inclination during thaw, the stone is gradually rotated to stand on end. Much depends, however, on stone shape and original inclination.

Frost sorting

Size sorting of sediments is one of the most distinctive features of many periglacial areas, helping to produce various forms of patterned ground such as polygons and stripes (Plate XIII). Although details of the processes of sorting remain unclear, laboratory experiments, especially those carried out by Corte (1966), have shown that repeated freezing and thawing is the main cause of sorting, and have identified the conditions that are most favourable to sorting. Corte distinguishes between 'vertical sorting', resulting in the concentration of finer particles in the lower layers of the soil and coarser particles at the top, and 'lateral sorting' when lateral components of movement are involved, giving complex patterns of sorting. It is important to emphasize that sorting of either type is not restricted to frost-susceptible soils; however, as vertical sorting leads to a progressive increase in fines in the lower soil layers, these layers can become frost-susceptible in time and ice-lensing can commence. In early experiments, Corte (1961) examined the vertical migration of particles in samples of sandy gravel, all of the non-frost-susceptible type, during freeze–thaw cycles ranging from $+20°$ to $-20°C$. After nineteen cycles, vertical sorting had occurred, the finer particles descending and the coarser moving up. Glass marbles placed in some samples were uplifted by up to 0·22 mm per freeze–thaw cycle. All the samples were saturated with water. As a result of many more experiments conducted under a range of conditions and on a variety of materials, Corte showed that the freezing and thawing of an unsorted soil from the top

Plate XIII Patterned ground, Arctic Canada, showing sorting of originally mixed sediment into areas of finer and coarser particles. (C.E.)

downwards will produce vertical sorting so long as moisture is present. Large fragments in the soil move up by frost-heaving, as already described, and may accumulate in the top layer to form a blockfield. The fines move away from the source of cooling, and for this migration to occur, a thin layer of water must separate each migrating particle from the advancing freezing plane. Sorting occurs because finer particles can migrate under a wider range of freezing rates than coarser ones, each particle size being related to a critical freezing rate. If the freezing rate is too rapid, the particle will be overtaken by the freezing plane and incorporated in the ground ice. Thus, three major factors in frost sorting are:

1 Moisture content. Sorting decreases as the moisture supply decreases.
2 Rate of freezing. A larger range of particle sizes can be sorted at slower rates of freezing.
3 Particle size. Finer particles can be sorted under a wider range of freezing rates than coarser ones.

Vertical sorting alters the 'packing' or arrangement of particles in the matrix, and frequently leads to increase in volume. Corte showed that the larger the uniformity coefficient (D_{60}/D_{10}), the greater the volume changes. In one sample studied, there was an increase in volume of more than 10 per cent after only five freeze–thaw cycles. Surface heaving can, therefore, be due to sorting as well as to ice-lensing, though in frost-susceptible materials the latter is quantitatively much more important.

In the case of vertical sorting, fine particles migrate away from the advancing freezing plane. The same applies to lateral sorting, when freezing and thawing operate from the side, for example on the edges of banks or terraces, or in relation to vertically placed stones or

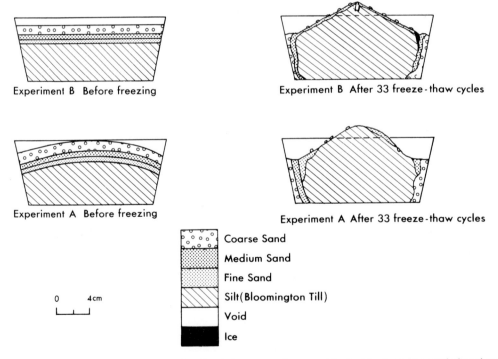

Experiment B Before freezing

Experiment B After 33 freeze-thaw cycles

Experiment A Before freezing

Experiment A After 33 freeze-thaw cycles

Coarse Sand
Medium Sand
Fine Sand
Silt (Bloomington Till)
Void
Ice

0   4cm

Figure 6.8 Sorting of sediments in aluminium containers subjected to freezing and thawing from the sides and bottom (Corte, 1966).

stony borders which are good heat conductors. The movement paths of the particles are more complex than in the case of vertical sorting because of gravity, and tend to be parabolic. The larger particles are left near the cooling sources. The differential movements of particles along curved paths can give rise to an appearance of contorted structure. Such structures, and the complex patterns of sorting associated with patterned ground, were at one time explained by convection currents (Gripp, 1926) or mass displacements (see p. 210), but many may be due to the combined operation of lateral and vertical frost sorting.

A further process that plays a part in sorting at or near the surface is movement by gravity of particles on the slopes of mounds. The mounds may form by differential expansion of the matrix during frost sorting, or by some other mechanism such as ice lensing in frost-susceptible materials. Figure 6.8 illustrates Corte's experiments using aluminium pans so that freezing and thawing proceed easily from the sides and bottom. Lateral sorting caused migration of fines away from the margins, which sank, and the formation of a central mound on whose slopes gravitational sorting occurred.

The phenomenon of frost sorting is complex and much work still remains to be done on the range of processes involved. Although it can operate in non-frost-susceptible materials, in other cases it combines with other frost-related processes associated with frost heaving and ice segregation; moreover, simple vertical sorting can turn a heterogeneous non-frost-susceptible matrix into one whose lower layers accumulate sufficient fines to become frost-susceptible. Depending on the processes operating, the nature of the material and its water content, a great variety of patterned ground results (see Washburn, 1973).

Thermal contraction

Cracking of frozen ground in arctic areas has long been attributed to contraction during periods of intense winter cooling (Leffingwell, 1915). The cracks form a polygonal pattern, the diameters of the polygons usually ranging between 5 and 30 m (the so-called 'tundra polygons'). In most arctic areas some melting occurs in the spring or summer so that the cracks fill with water which later freezes to give ice veins. Once initiated, the cracks provide planes of weakness which reopen under stress in each succeeding winter, the ice veins growing as further films of ice are added annually, developing into ice wedges (Figure 6.9). The latter may attain widths of many metres, representing hundreds of years of growth. In Victoria Land, Antarctica, Black (1973) reported that, for a total of 500 ice wedges at fourteen sites, growth rates ranged from 0·34 to 1·58 mm year^{-1}, with a mean of 0·84 mm year^{-1}. In arid areas, fine sediment rather than water fills the cracks giving sand wedges (Péwé, 1959), and composite ice and sand wedges have been described by Berg (1969), the gradation from one type to another representing varying degrees of availability of water and fine sediment.

Contraction cracking develops most spectacularly in areas of frozen ground rich in ice. Some permafrost contains as much as 80 per cent ice by volume and the rates of expansion and contraction will differ little from those of clear ice. The coefficient of linear expansion of clear ice is $52·7 \times 10^{-6}$ at 0°C and $50·5 \times 10^{-6}$ at -30°C, about five times greater than the corresponding values for silicate minerals. With a 30°C fall in temperature, cracks about 1·5 mm wide would develop around a prospective polygon 1 m in diameter (Lachenbruch, 1962). In ice-rich frozen ground, a drop in temperature of 4°C may be adequate to cause cracking but in the case of materials akin to rock, 10°C may be required (Black, 1963). Péwé (1966) has shown that ice wedges only form when the mean annual temperature is -6°C or colder over many years, and the southern limit of present active ice wedges in Alaska roughly corresponds to the mean annual isotherm of -6°C to -8°C. Contraction cracking, however, is not confined to permafrost areas: it has been reported from non-arctic

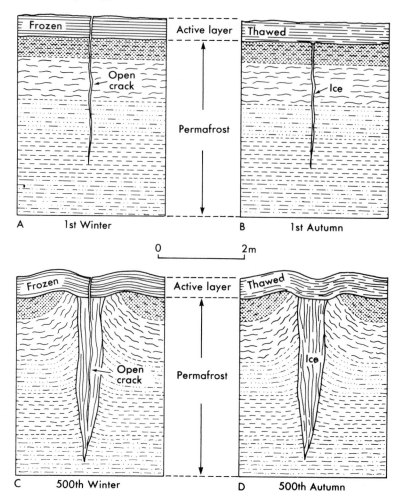

Figure 6.9 Evolution of an ice wedge according to the contraction-crack theory (Lachenbruch, 1962).

areas (e.g. Iceland: Friedmann *et al.*, 1971; New England: Washburn *et al.*, 1963) during occasional severe winters, but any ice in the cracks melts annually and ice wedges are not formed.

Benedict (1970) has drawn attention to the dangers of concluding that all superficial cracking of the ground following winter cold may result from cooling contraction. In part of Colorado, study of annual cracking of turf-banked solifluction lobes led to the conclusion that this was a tension phenomenon induced by differential frost heaving.

Cryostatic pressure

The increase in volume as water is converted to ice is 9·08 per cent, and in a closed system the pressures involved can exceed $2000 \, kg \, cm^{-2}$ at $-22°C$ (Bridgman, 1912). Such pressures are not remotely approached in nature due to weakness of the confining structures, but sufficient pressure can frequently be generated to cause deformation of sediments (involutions), up-arching of the ground and injection structures. In permafrost areas, cryostatic pressure can be set up between a downward-freezing active layer and the permafrost beneath. A particularly favourable circumstance will be when the active layer becomes frozen to the permafrost in some places before others, thus giving enclosed pockets of unfrozen material. Differential freezing of this sort may be related to variations in water content and pore-water pressure in the active layer, or to patchiness of any thermal insulation such as snow or vegetation at the ground surface. The pressures attained will usually be only quite low as there will often be some air entrapped or the thickness of the frozen crust above may be small. Pissart (1970) reports laboratory experiments confirming that the greatest pressures develop inside thick frozen layers composed of fine-grained sediments with high capillary water content. Figure 6.10 shows the changing cryostatic pressure for one freeze–thaw cycle lasting 4 days. After the period of the 'zero curtain' when, at 0°C, latent heat is liberated

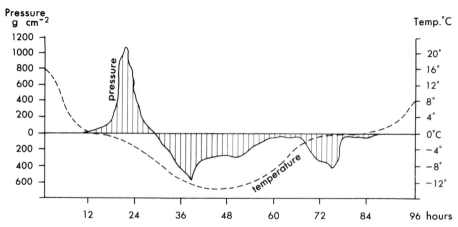

Figure 6.10 Graph showing pressure variation during one freeze–thaw cycle, measured at a depth of 5 cm in a layer of silt subjected to freezing from the surface downward (Pissart, 1970).

in the transition from water to ice, the increase in pressure can be very sudden. The maximum pressure registered in Pissart's experiments was a little over 4 kg cm⁻². Higashi and Corte (1972), however, contend that frost-heaving may be a more important process of surface uplift or sedimentary deformation than cryostatic pressure.

The formation of some types of pingo has also been attributed to cryostatic pressure. Essentially, it is a question of the freezing of a closed talik at shallow depth. As the freezing plane advances on all sides, water is expelled into the talik, where its eventual freezing and expansion lead to uplift and doming of the ground above. Mackay (1962) suggested that most of the pingos in the Mackenzie Delta area of northern Canada developed in the sites of former shallow lakes, beneath which taliks existed. When the lakes were drained, filled with sediment or invaded by vegetation, permafrost was able to form at the surface above the talik, thus sealing it at the top and forming a closed system within which cryostatic pressure increased.

It should be mentioned that another type of pingo has been recognized which does not form by cryostatic pressure. This is the so-called open-system type, attributed to artesian pressure (Müller, 1959). Here, groundwater flowing under artesian pressure in shallow taliks may force its way near to the surface where it freezes, the injected ice bulging up the surface.

In extreme cases, the ground may be ruptured, giving a temporary outburst of water and debris.

Other forms of differential displacement

In previous sections, several processes causing disturbance of sediments have been described, including frost heaving and thrusting, up-freezing of stones and frost sorting, and the effects of cryostatic pressure. The general term 'cryoturbation' (Edelman *et al.*, 1936) has been applied to disturbances of sediments by frost processes, and the term 'involutions' (Denny, 1936) applied to the deformed structures. It is also recognized, however, that apparently similar disturbances can be found in sediments far from any possible frost action. Kostyaev (1969) points out that some pre-Quaternary sediments are characterized by interpenetrating beds, ball-like deformations, folds and wedge-like structures (clastic dikes) such as are also seen in sediments disturbed by freezing and thawing. Among the non-frost processes that have been suggested to account for such deformations are the following:

1 Movements within the soil due to density differences. Heavier materials overlying lighter materials will be potentially unstable; depending on the viscosity of the lower layers, deformation or collapse can occur.
2 Effects of high pore-water pressure. By separating rock particles, pore-water pressures can reduce or eliminate shear resistance caused by friction, and squeezing or displacement of the liquefied sediments can occur. This can happen in areas with ground ice, whose melting can radically alter the amount of pore water. Plugs of fine liquefied sediment can be injected into coarser materials, whereas larger stones may sink.
3 Gravitational disturbance downslope, including gelifluction and sub-aqueous slumping.
4 Consolidation of loose sediments.
5 Dilatation of sediments due to disturbance.
6 Desiccation.

Kostyaev and others point out that although these processes can operate in frost-free regions, they can also occur in areas subject to freezing and thawing, and frost disturbance may also act as a trigger mechanism to initiate these processes.

Thermokarst processes

'Thermokarst' is a term introduced by Yermolayev (1932) to describe features and processes associated with the thawing of ground ice. The high ice content of some permafrost has already been mentioned; if there is 'excess ice' (see p. 199), thawing produces more water than can be retained in the soil voids. Further, the mineral content of ice-rich permafrost can be quite small in volume in comparison with the total volume of the permafrost. If the ice content is 300 per cent (weight of ice to dry soil), thawing of a 10 m layer will release about 8·5 m of water and 1·5 m of saturated sediment. Thawing thus inevitably leads to collapse features—pits, basins, caves and valleys—and slumping is frequent as high pore-water pressures reduce the shear strength of the thawing ground.

Czudek and Demek (1973) distinguish the following thermokarst processes:

1 Thermo-abrasion. This is associated with the shorelines of seas and lakes where the combined mechanical and thermal effect of wave-action on permafrost results in rapid undercutting of the land (Figure 6.11). Mackay (1971) describes how thermo-abrasion attacks massive icy beds in northwest Canada, producing large crescentic slump scars and causing recession of up to 10 m a year.

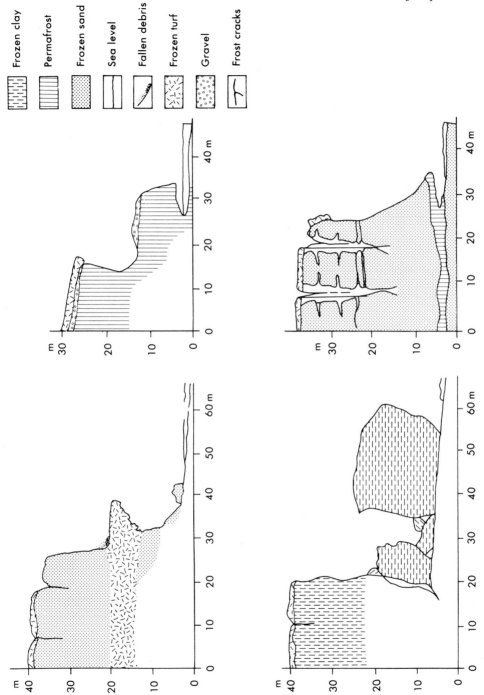

Figure 6.11 Effects of thermo-abrasion on Soviet Arctic coast profiles (Czudek and Demek, 1973).

2 Thermo-erosion. Flowing river water, because of its relative warmth, undercuts frozen banks at the waterline in a similar way.
3 Thermo-suffosion. This term refers to the mechanical washing-out of fines in the ground-water draining away from an area of thawing ground ice, leading to surface collapse.
4 Thermo-planation. Lateral degradation from river or lake banks, often proceeding most rapidly along ice wedges, can open out flat-floored recesses ('thermo-cirques') which grow in size and intersect. Alternatively, downward degradation can operate over a generally flat permafrost surface. Pools of water (alas lakes) frequently collect and, once formed, provide further heat stores for thawing the permafrost. Thermokarst lakes can attain depths of 40 m and widths of 15 km (Czudek and Demek, 1970); the depths are indicative of the minimum amount of ground ice that has melted.

The importance of thermokarst processes is only just being realized. Czudek and Demek (1970) estimate that more than half of central Yakutia is affected by thermokarst. Not all thermokarst is a natural occurrence: disturbance by man of the delicate thermal equilibrium of much permafrost, by clearing vegetation, cutting roads, levelling airstrips, removing peat or gravel, and so on, can initiate thawing and thermokarst.

7

Fluvial processes

J. B. Thornes

It is now almost a century since Gilbert carried out his famous investigation of the geology of the Henry Mountains. His chapter on 'Land sculpture' contained in the report of the United States Geographical and Geological Survey of the Rocky Mountain Region (1880) is a landmark in the investigation of fluvial processes because it was the first, and for a long time one of the last, attempts to combine the principles of fluvial mechanics with those of fluvial morphology. Engineers followed their pragmatic paths while geomorphologists 'dallied with description and classification' (Schumm, 1972). His work is a treasury of ideas and problems, and although his discussion of sediment transport and open-channel hydraulics is qualitative, Gilbert showed in this report how landform studies might have developed by an integration of the engineering and geomorphological approaches.

With some notable exceptions, such as Sundborg's (1956) study of the River Klarälven in Sweden, the drift away from mechanics has continued until quite recently. Even Sundborg was only partially successful in achieving explanation of *form* in terms of process. From the other side, and again with notable exceptions such as Ackers's and Charlton's (1970) continuing investigation of meanders and the work by Callendar (1969) on channel patterns in relation to dynamics, the hydraulicists have paid only relatively modest attention to geomorphological problems. Even geologists and sedimentologists, whose work lies at the nexus of the other two studies, have found difficulties in marrying together sedimentation, morphology and fluid mechanics.

What are the causes of this difficulty? Some of it almost certainly stems from methodology. The hydraulic engineers have adopted one of two approaches. In applied work they have sought and applied basic rules stemming from fundamental laws and parameterized by practical experiments and field experience. This methodology is amply demonstrated in Ven Te Chow's (1959) *Open channel hydraulics* or Raudkivi's (1967) *Loose boundary hydraulics*. In research work, engineers have sought deterministic or probabilistic mathematical models to describe and especially to analyse the behaviour of flow, of particles in the flow, or of conditions on the boundaries of the channel. The continuing creativity of the theoretical worker Bagnold (1966, 1973) has exemplified this school in recent years.

Geomorphological work on rivers, on the other hand, has largely been empirical. The approach is exemplified by the observation of temporal and spatial variations in channel characteristics or the qualitative description of process. In the early 1960s the debate between Mackin and Leopold and Langbein in Albritton's book *The fabric of geology* (1963) revealed the deep-rooted cause of this difference in methodology. On the one hand Mackin advocated 'rational' explanations in which full deterministic reasoning would be used to argue about the variations in channel slope. On the other hand Leopold and Langbein stressed the essentially indeterminate character of channel behaviour arising from the multiplicity of possible responses to a single change. This theme, emphasizing the probabilistic behaviour, has been sustained, developed and on many occasions reiterated in the intervening years. Maddock

(1970) wrote 'The behaviour of alluvial channels is a matter of tendencies—shifts in width, depth, velocity and slope *tend* to follow certain patterns. The patterns are not fixed, and they vary with the constraints placed on the variability of the factors governing stream-flow.'

The geomorphologist's view of the river is further conditioned by the time and space scales over which he operates. By and large these have tended to be long and small respectively. Major adjustments of channel morphology are responses to variables which, in the main, change over periods of tens or hundreds of years (Kennedy and Brooks, 1963). Those studied in hydraulics are responding to changes of seconds, minutes or hours. Moreover, geomorphologists have operated to a considerable extent in recent years in small river channels, whereas hydraulic models have concentrated on large rivers and estuaries.

In small channels, transient (non-equilibrium) behaviour is more in evidence; for example flood flow becomes relatively more important than base flow in small channels. This may be why geomorphologists have been less prepared to accept propositions based on steady-state operations of channel systems and have emphasized the complexities induced by change. Schumm has done most in recognizing the inherent instability of channels. Channels that change in a very unsteady manner, perhaps as the result of a single flood, pose much greater problems of analysis. The popularity of hydraulic geometry, which is the investigation of statistical regularities in river forms, can be largely attributable to the apparently uniform character of the observed river responses in width, depth and velocity to temporal and spatial variations in discharge (Thornes, 1978).

Possibly the third reason for the failure of geomorphologists to pursue deterministic explanations of channel morphology has been the relatively unsuccessful attempt to obtain an adequate model for sediment transport. The situation was such in 1972 that Yang was still able to show that, using the major equations currently adopted by practising engineers, a wide variety of results could be obtained, many of which compared poorly with the observed amounts or sizes of sediment observed in true channels. This failure is in part linked to the parameterization of the relevant models in laboratory channels that have little or no resemblance to true rivers. Bagnold (1966) wrote that:

> During the present century innumerable flume experiments have been done and a multitude of theories have been published in attempts to relate the rate of sediment transport by a stream of water to the strength of water flow. Nevertheless, as is clear from the literature, no agreement has yet been reached upon the flow quantity—discharge, mean velocity, tractive force, or rate of energy dissipation—to which transport rate should be related.

The basic problem is that no established branch of physics has interested itself in two-phase (fluid–solid) flow, so the hydraulic engineer solves the problems by empirical reasoning from past experience of like conditions. Bagnold himself went on to attempt to provide a suitable basis for such theory from general physics, and this forms the core of much current work in sedimentary and fluvial hydraulics.

Some time has been spent discussing the differences between the approaches of fluvial hydraulics and fluvial geomorphology so that the reader may appreciate the difficulties. The time when we have a deterministic, spatially distributed erosion–deposition model that operates over the time-scales of interest to geomorphologists is probably still a long way ahead, but that should be our ultimate goal.

Water on hillslopes and in channels

Rainfall

Rainfall may be regarded as an input to the basin system, and runoff on the slopes and in the channels the basic driving force of fluvial processes. Rainfall is highly variable both spatially and temporally, and the cellular pattern of storms means that in all but the smallest catchments, total precipitation, its intensity and time-flux will be difficult if not impossible to predict. Fortunately, because of the capacity for storage in the system, the pattern of runoff is much less variable.

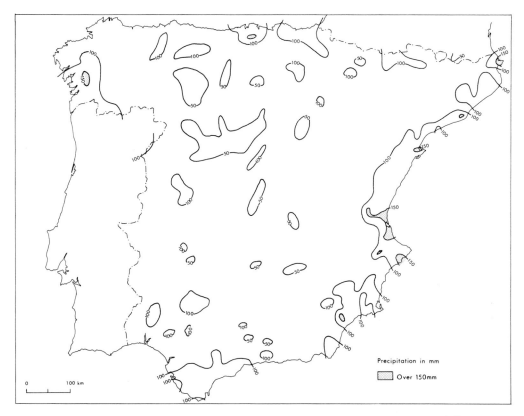

Figure 7.1 Rainfall-intensity map of Spain. The map shows the maximum rainfall amounts in a 24-hour period that may be expected to occur once every 50 years (based on data from Elias, 1963).

Rainfall intensity is an important control on the relative amounts of water that run off the surface and percolate into the soil. It is measured as the amount of rain falling in a given period. For Spain, Figure 7.1 shows the rainfall amounts that may be expected on average once every 50 years for a 24-hour period. The map may be taken as a crude indicator of spatial variations of rainfall as a geomorphological agent in that country. As the rainfall supply is exhaustible, higher intensities may be expected for shorter periods of time. Daily rainfall data give a limited indication of intensity because of the varying lengths of storms,

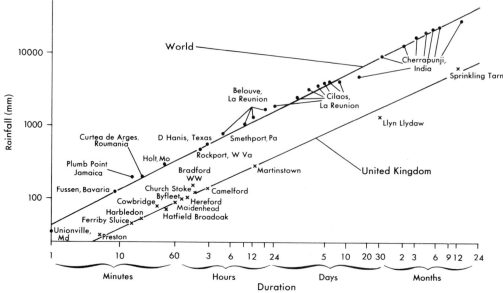

Figure 7.2 Rainfall magnitude and duration relationships for the world and the United Kingdom largest falls (Rodda, 1967).

and rainfall intensities vary seasonally. Some idea of the frequency of different intensities on a world scale is given in Figure 7.2 and at a particular location (Almeria, Spain) in Figure 7.3. Total energy in a storm is obtained by summing the instantaneous rate of energy production, e, over the storm and may be related to intensity by the regression equation:

$$e = a + b \log_{10} I \qquad (7.1)$$

where a and b are regression constants and I is the rainfall intensity (Carson and Kirkby, 1972), or by the type of formula used by Stocking and Elwell (1976):

$$e = a + b/I \qquad (7.2)$$

Some characteristic energy-intensity relationships are shown in Figure 7.4 (after Stocking and Elwell, 1976).

Higher rainfall intensities occur in tropical areas. It does not follow, however, that the highest soil-surface intensities also occur here because of the effects of interception by vegetation or artificial surfaces. The volume of interception depends on interception capacity (i.e. the ability of vegetation surfaces to collect and retain the falling precipitation), the evaporative losses, the duration of rainfall and its amount, and rainfall frequency. The last is important since if rain has just fallen most of the interception capacity is filled. Ward (1975) outlines the severe difficulties involved in the measurement of interception. Table 7.1 gives some values reported in the literature for interception losses on various kinds of cover. However, it may not be assumed that interception is necessarily very high where vegetation cover is itself dense. Nortcliff and Thornes (1977), for example, found that 83 per cent of free-falling rain reached the ground under heavy tropical rainforest. This figure, much higher than those usually quoted, reflects the high relative humidities in the canopy. Stem-flow also offsets the interception losses to some extent, especially under tropical conditions, but evidence suggests that the total volumes are not significant.

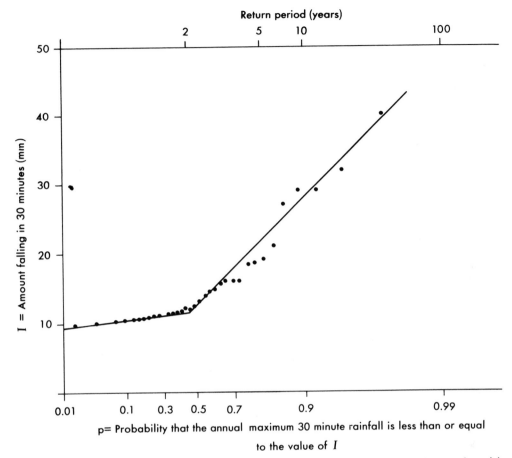

Figure 7.3 Rainfall intensity at Almeria over a 25-year period according to data gathered by Elias (1963).

Intensity and interception, as well as runoff, have an important role to play in erosive processes. Seasonal and inter-annual variability of total amounts is also important because of the effects on soil-moisture storage. Seasonality in runoff characteristics usually reflects seasonality in rainfall, except where snowmelt is important.

Basically the relationship between rainfall and runoff is determined by the capacity of the soil for taking in water; the process is called infiltration. Surface ponding and overland flow occur when the rainfall intensity exceeds the infiltration rate.

Infiltration

The infiltration process involves the movement of water into the soil. Horton (1933) originally described the process as comprising two parts, a period in which the available storage was being filled and one in which a limiting value was reached, with a constant rate equal to the transmission of the water through the filled soil. Some characteristic infiltration rates

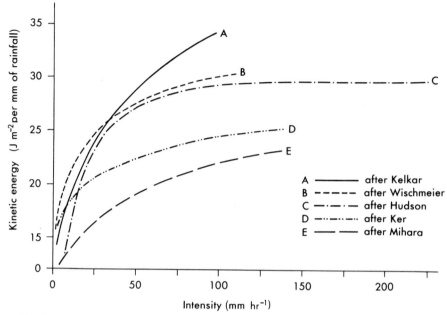

Figure 7.4 Energy–intensity relationships according to various authors (Stocking and Elwell, 1976).

through time from a standing fixed head of water, as measured in a ponded infiltrometer, are shown in Figure 7.5.

During flooding with the ponded infiltrometer the rate is controlled by profile properties. At first the rate is extremely rapid but it quickly falls to a steady rate called the infiltration capacity. These two components are represented by the Philip (1957) equation:

$$f = A + B \cdot t^{-\frac{1}{2}} \tag{7.3}$$

where f is the instantaneous rate of infiltration; t is the elapsed time since the beginning of rainfall; A the 'transmission constant' of the soil and B the storage component.

When the supply of water is varied, as with rainfall, the capacity of the soil to receive water may not be exceeded, with the result that the amount entering depends on the rainfall rate. This is called *flux control*. With *profile-controlled* infiltration, recognizable zones of soil

Table 7.1   The effects of interception by various kinds of cover vegetation

	mm	Per cent average annual precipitation
Conifers	0·1–7·7	25–35
Deciduous	0·2–2·0	15–25
Heather		35–66
Mature chaparral	2–4·0	12·8
Spring wheat		11–19

Source: Ward (1975); Rodda, Downing and Law (1976)

moisture develop. Except for a few millimetres of saturated soil near the surface, the soil remains unsaturated, and so it can exert a pull on the water above it. Within the unsaturated soil, three zones can be recognized: (a) a transmission zone, in which moisture content changes slowly with depth and time but which lengthens as the infiltration process proceeds; (b) a wetting zone, which exhibits fairly rapid changes in soil moisture with depth and time; and (c) a wetting front, which represents the visible limit of moisture penetration. The pattern of change for a simulated wetting front (A) and an actual field example (B) are shown in Figure 7.6.

During flux-controlled infiltration, three modes of rain infiltration may be observed on the basis of relative rainfall intensity (Rubin, 1966). These are non-pounding, pre-ponding and ponded forms of infiltration. When the rainfall is less than the infiltration capacity, non-ponding infiltration results. The surface soil does not become saturated but approaches a

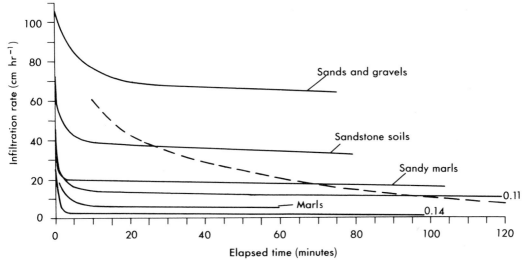

Figure 7.5 Typical infiltration curves for ponded infiltration of a small, fixed head of water on different materials in a semi-arid environment (Thornes, 1976).

limiting moisture content throughout the wetted zone; as the intensity increases, the limiting moisture content increases. When the intensity exceeds capacity, a pre-ponding mode occurs in which the surface-moisture content is approaching saturation but soil-water pressure remains negative. Finally, when the tension drops to zero, ponding actually occurs. The amount of pre-ponding infiltration or initial abstraction is therefore a function of the rainfall intensity and the initial soil-water content. If the rainfall intensity continues to exceed the saturated conductivity, ponded (profile-controlled) infiltration follows incipient ponding, and the time curves followed will be similar to those shown in Figure 7.5.

Smith (1972) introduced the idea of an 'infiltration envelope' (Figure 7.7): for a given soil the time taken to reach ponding will depend on the rainfall intensity. With a lower intensity more time will be needed to reach ponding. For a given intensity this time to ponding will vary with soil type. The final infiltration for a given soil type should have a limiting value which is the saturated conductivity. The dotted line tracing the start of ponded infiltration for different intensities is the infiltration envelope. Thornes (1976) has estimated the time to ponding from the final infiltrability for some Spanish soils, which are compared in

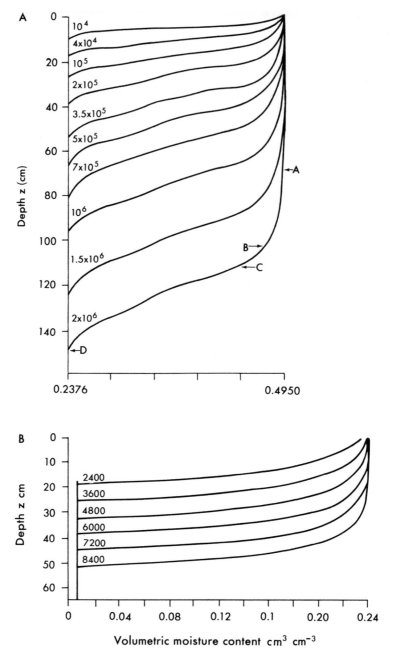

Figure 7.6 A: Simulated wetting front in a theoretical case described by Philip (1963). AB—transmission zone, BC—wetting zone, CD—wetting front.
B: Flux-controlled development; surface saturation is not reached (Rubin and Steinhard, 1963). In both cases times shown on different stages of penetration are in seconds. Diagram based on Dunin (1976).

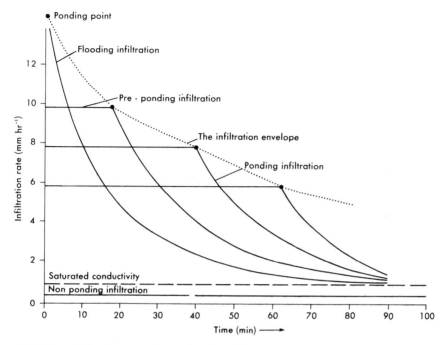

Figure 7.7 The infiltration envelope showing the time to surface ponding for rainfalls of different intensities (after Smith, 1972).

Table 7.2 with values provided by other workers. The times are seen to be remarkably short, even though the soils are assumed to be initially dry. The best fits for the Kostiakov and Horton types of empirical equation are also given. The data show that t_0 and final infiltrability are highly variable.

The basic causes of variability of the infiltration rate are variations in the storage capacity in the soil, the nature of the soil gradient and the extent to which pre-existing capacity is actually filled by moisture. These have been discussed in Chapter 6. It is usual to characterize a soil by a curve showing the relationship between soil moisture and hydraulic conductivity (Fig. 6.4). Hydraulic conductivity is the coefficient in Darcy's Law (1856) which relates the mean discharge flowing through the saturated soil per unit cross-section (q) to the total gravitational and potential force (ϕ). Darcy's Law is expressed as

$$q = K \cdot \text{gradient} \, (\phi) \tag{7.4}$$

K is the hydraulic conductivity and is, as we have just said, related to the soil moisture. Soil moisture and hydraulic conductivity reflect the basic non-capillary porosity of the soil and the structure and structural stability. Because these characteristics change seasonally, the infiltration rates are also seasonally variable. Thus the development of crusting, surface erosion, water repellency, accessions of salt in the profile at the end of a summer season and especially freezing of water in the sub-surface, all cause wide temporal variations in the rate. Vegetation cover also causes a considerable amount of seasonal variability. The presence of vegetation tends actually to encourage infiltration by inducing high soil–water deficits. In an experiment on semi-arid soils, Thornes (1976) found vegetation cover to be the most significant control on time to ponding. Rogowski (1972) shows that soil-water

Table 7.2A   Time to surface ponding and final infiltrabilities for some field and laboratory experiments on different materials and for different intensities. Reference 1 is for semi-arid conditions

Time (s)	Intensity (mm hr^{-1})	Final infiltrability (mm hr^{-1})	Conditions	Ref
46	45	520 (average)	Sand and gravel soils	1
26	45	154 (average)	Limestone soils	1
25	45	96 (average)	Marl soils	1
10	45	8 (average)	Decalcified marls	1
2260	62·2	52·1	Loam soils	2
640·8	62·2	40·6	Loam soils	2
21·9	2154·0	478·8	Lab, sand	3
81·7	1436·4	478·8	Lab, sand	3
165·7	1077·3	478·8	Lab, sand	3

Sources: (1) Thornes (1976); (2) Swartzendruber and Hillel (1975); (3) Rubin (1966)

Table 7.2B   Coefficients in the infiltration equations $f=At^{-B}$ (1) and $f=A+Bt^{-1}$ (2) for some field ponding experiments of infiltration rate (f) against time (t) where t is elapsed time in minutes

Condition	Equation	Mean A	Mean B
Sands and gravel soils	1	1·89	0·165
Limestone soils	1	0·81	0·204
Limestone soils	2	0·19	0·788
Marl soils	2	0·19	0·67
Decalcified	2	0·05	0·51

characteristics for retention and transmission are frequently highly variable with coefficients of variation approaching 70 per cent. As Dunin (1976) points out, this raises severe sampling problems and the complete hydraulic specification of the soil characteristics of a catchment is usually impossible. Apparently, the errors in infiltration in 'lumped' models involving spatial averaging over considerable areas could be as high as 75 per cent.

Overland flow and interflow

Surface runoff is the part of the runoff that travels over the surface of the ground to reach a stream channel and through the channel to reach the basin outlet. Sub-surface runoff is that part of the runoff that travels through the ground to reach a stream channel and through the channel to reach the basin outlet (Freeze, 1972). Overland flow is that part of the lateral inflow that flows over the land surface towards a stream channel.

Horton (1945) proposed a mechanism in which rainfall intensity exceeded infiltration rate to produce overland flow. The overland flow discharge per unit contour length (q) is then given by

$$q = (I-f) \cdot a \tag{7.5}$$

in which I is the rainfall intensity after interception, f is the infiltration rate and a is the area drained per unit contour length. For this to be the main generator of overland flow, most rainfall events must exceed infiltration capacity. This is only likely to be the case where infiltration rates are extremely low, for example on clay badlands, on frozen soil or on soil surfaces heavily compacted by agriculture. Several workers (e.g. Kirkby, 1969; Dunin, 1976) have shown that rainfall intensities are insufficient to sustain runoff of this type in most cases where data are available. More often, it seems, storm runoff comes from a small but fairly consistent portion of the upstream source areas that usually covers no more than 10 per cent and usually only 1–3 per cent of the basin area. According to Freeze (1972), even on these restricted areas only 10–30 per cent of the rainfall causes overland flow.

There is another way in which overland flow may be generated according to Hewlett (1961). If most of the rain infiltrates and the soils become saturated, the water comes to the surface in small areas. Ragan (1967) and Dunne (1970) argue that such areas are wetlands whose location is controlled by the topographic and hydrogeological configuration of the basin. Inevitably they occur in topographic lows near stream channels, stream heads or at seeps that feed small intermittent tributaries. Evidently the width is transient, the areas may be connected by running surface water and will fluctuate as the groundwater rises and falls. So the partially contributing area is a dynamic phenomenon, expanding in time and space subject to soil infiltration capacity as well as storm intensity and duration.

There have, as yet, been few rigorous field measurements of varying source areas during storm events, and attempts to estimate varying contributing areas have involved subtracting channel precipitation from total runoff volume and assuming a runoff coefficient of 100 per cent (i.e. an infiltration capacity of zero) for the contributing area. As Ward (1975) points out, this raises the problem of defining channel precipitation; moreover, the channel system itself may expand many times, as shown by Gregory and Walling (1968). The best known of these investigations is that of Hewlett and Hibbert (1967). Kirkby and Weyman (1973) used a dense network of hillside instruments at East Twin Brook in Somerset to record the presence of surface water and sub-surface saturation. The resulting maps showing area of surface water can be related to storm events. Both basins showed similar patterns of change, in which low-angle slopes near the channel and on the divides tended to saturate easily.

Another rather different type of observation of overland flow is that of Emmett (1970), which involved a higher spatial resolution than the catchment level or earlier plot investigations which were concerned only with the bulk quantities of rainfall, infiltration, runoff and the shape of the runoff hydrograph. Emmett used simulated rainfall on ground plots 2 m by 15 m and traced flow paths using dyes. Surface runoff within the plot areas tended to accumulate in several lateral concentrations. On nearby low-angle slopes, micro-relief features of the order of only 3 cm appeared to dictate the paths of flow concentrations. However, on steeper slopes micro-relief features did not appreciably alter the downslope gradients and their influence on concentrations of flow was negligible. Although flow rarely occurred as a uniform sheet of water, the concentrations were not regarded as sheet flow. In particular, vegetation played an important role in determining the flow pattern on some sites. The general appearance of runoff at most field sites was one of omnipresent surface detention, easily detected by the glistening of the sheet of water in the sunlight. Dye tracing showed that this sheet of water moved slowly downslope and often moved laterally to join the concentrated areas of flow. With ground litter, surface detention occurred in a series of puddles formed by barrier dams of organic debris, and surface runoff occurs in part by successive failures of these barriers. With increasing downslope discharge one would expect resistance to flow to decrease as relative roughness decreased. Other influences, such as ponding, played an overriding role and the roughness of field plots was about ten times that of laboratory plots.

An alternative approach to the Horton model of overland flow, derived from the work of Wooding (1965), is to regard the overland flow hydrograph as an unsteady wave-form. This describes many important aspects of overland flow (Woolhiser, 1975), and particularly the effect of different surface forms (shape of the overland flow surface) on the hydrograph. This is important because the characteristics of the unsteady flow are those that determine ultimately the temporal and spatial patterns of water erosion. Usually the water flow is modelled as a kinematic wave, in which the inertia is regarded as relatively unimportant. Brakensiek (1967) thought of a catchment as comprising many surfaces, one feeding into another, across which these waves of overland flow can move; he called the system so determined a kinematic cascade. Lane, Woolhiser and Yevjevich (1976) generated flows of this type on artificial surfaces to evaluate the effects of morphology on the runoff pattern. It was found, for example, that if the downstream profile is concave, the time to peak flow increased and the peak discharge decreased when compared with uniform slopes. If the hydrograph characteristics are important in determining erosion (see p. 246), there then exist interesting possibilities for geomorphological feedback.

Runoff from snow melt

Finally, in this section we must note that snow-melt runoff plays an important geomorphological role in slope and channel processes in many environments throughout the world, from the high Arctic over frozen soils, to desert regions where snow melt from high altitudes may be the only source of runoff.

The participation of snow melt in runoff has been evaluated by radioactive substances, which suggest (Martinec, 1976) that infiltration rates from snow melt are very important, about half the total available water being lost to infiltration. This water may stay in the catchment in the form of sub-surface storage long after the snowfall producing it has disappeared.

Dunne and Black (1971) examined the runoff from snow melt under limited infiltration capacities and found that about half the meltwater left the plots as overland flow. Discharge rates, total volumes and timing of this runoff were basically related to shortwave radiation. The contribution of snow melt sub-surface water to runoff was of course much more attenuated in time. The direct runoff was strongly influenced by the non-uniform characteristics of snow accumulation, the distribution of frost, the soil characteristics and the type and distribution of vegetation cover. This suggests that the patterns of ground saturation resulting from snow melt may be different from those for rainfall. There appear to be very few investigations of the effects of snowfall on hillslope hydrology.

Runoff in channels

We have already suggested (Chapter 3) that the critical properties of flowing water are its velocity, its weight (which is a function of depth) and its turbulence. These react with the materials of which the bed is formed in a way which will be discussed in this part of the chapter. For many years the channel morphology has been expressed in terms of discharge through the idea of a dominant channel-forming discharge. Now that greater attention is being paid to unsteady flow and as better methods are becoming available for the solution of the complex equations involved, a better understanding of the relation between channel form and discharge will become available. Notwithstanding these advances, Leopold and Maddock (1953) and many workers since have shown the relationships of width, depth and

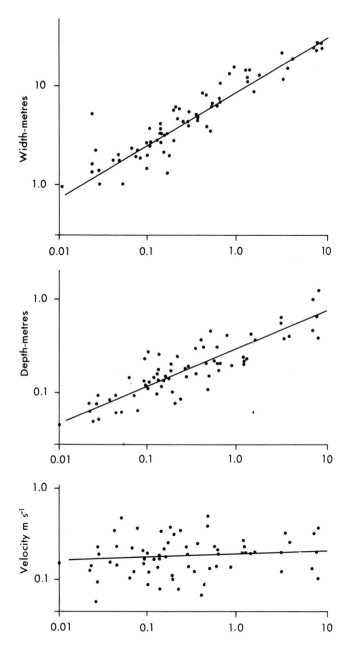

Figure 7.8 The hydraulic geometry relationships for a small channel, Brandywine Creek (Wolman, 1955).

velocity to discharge both at a particular location (at-a-station) and in a downstream direction. These graphs are shown in Figure 7.8. They are important because they illustrate (i) at-a-station: depth and velocity increase as discharge increases, whereas (ii) downstream: the response to increase in discharge is mainly through changes in width and depth. It is hardly surprising, given the consistently good correlations between these variables, that discharge has proved a useful proxy variable for depth and velocity relationships and so has figured large in geomorphological work on river channels. In dealing with glacial mechanics, on the other hand, it is more usual to deal directly with the forces involved.

Evidently, since there is a good correlation between flow volume and the forces shaping the stream bed, the distribution of flows will approximate to the distribution of forces, both through time and in space. If a bigger flow means greater force, then the product of the size of the flow and its frequency of occurrence will indicate the role played by this type of flow. This is, of course, the concept of effective work referred to in Chapter 2. The evidence suggests that, in rivers, a large proportion of the work is performed by relatively frequent events of moderate magnitude (Leopold, Wolman and Miller, 1964). This idea is further discussed in Chapter 12.

Patterns of runoff over time

Since most flows are measured at a gauging station, either by some kind of meter or against a staff, total discharge for a unit of time (hour, day, year) is usually recorded. These can also be manipulated statistically to give daily, monthly and annual mean flows. These relationships are expressed by a flow-duration curve (Figure 7.9), which shows the duration

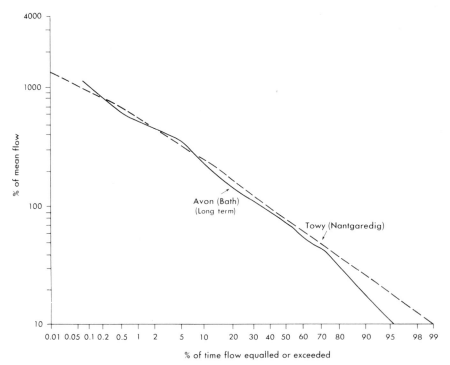

Figure 7.9 Dimensionless flow-duration curves (Rodda, Downing and Law, 1976).

or frequency with which various magnitudes of flow are equalled or exceeded. Data referring to the maximum floods or minimum drought flows are collected either as an annual series or partial series, and the return period of each event is computed. The return period t is usually obtained from

$$t=(n+1)/N \qquad (7.6)$$

where n is the number of years of record, and N the rank of a particular event. It turns out that the return period of the mean of the annual flood series is about 2·33 years and this is called the mean annual flood. Another discharge frequency which is often regarded as important is the bank-full discharge, which has a reported recurrence interval of about 1·5 years (Leopold, Wolman and Miller, 1964) or 0·5 years (Nixon, 1959).

   Although these kinds of data are useful for macroscopic comparisons of flow characteristics on a regional scale and even for pairs of basins, they tend to hide the important variations of flow which should correlate directly with scour and fill, i.e. at-a-station short-term changes. These changes are described by the flood-hydrograph, which is a plot of flow against time of the order of hours or days. The terminology associated with such hydrographs is shown

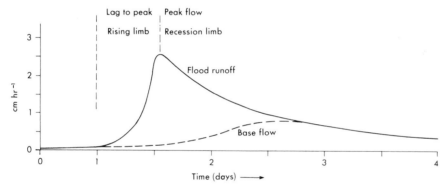

Figure 7.10 Hydrograph terminology.

in Figure 7.10, and lengthy descriptions may be found of the measurements and analysis of hydrographs in Ward (1975) and Gregory and Walling (1973). An important characteristic of the hydrograph is the peak flow, and it is this peak flow which is largely responsible for at-a-station variations in velocity and depth of water. Spatial variations in this peak flow are essentially a result of the propagation of flood waves over the network system and the way in which increments to the flow occur.

   Once flow is in the channel, theoretical methods exist to define its passage along a channel in time and space, using the equations of motion. A simplified approach was developed by Wooding (1965) and is known as the kinematic-wave solution to the flow equation. In this approach, which is well suited to small channels and overland flow, as mentioned earlier, the hydrograph is a solution to the continuity equation,

$$\frac{\delta q}{\delta d}+\frac{\delta h}{\delta t}=x \qquad (7.7)$$

where q is flow, d is horizontal distance, h is flow depth, t is time and x is rainfall excess, and the rating equation

$$q=\alpha h^n \qquad (7.8)$$

where α is a constant and n is the rating parameter. There are several different methods of solving this system of equations. The kinematic approach is a quasi-steady approach; it assumes that dynamic effects (inertia and momentum) can appear only as geometrical changes in the flow (i.e. depth). Changes in the water surface can result only from changes in the local flow rate and can only be transmitted in the direction in which the kinematic wave propagates, i.e. downstream. However, Woolhiser and Liggett (1967) show that the kinematic wave should provide a good approximation to a wide range of unsteady free-surface problems. Some simple 'basin' geometries and corresponding hydrographs using this technique are shown in Figure 7.11.

One of the problems of dealing with the kinematic routing procedure is that it is complex and expensive, and the assumptions are sometimes rather poorly satisfied. In particular, it

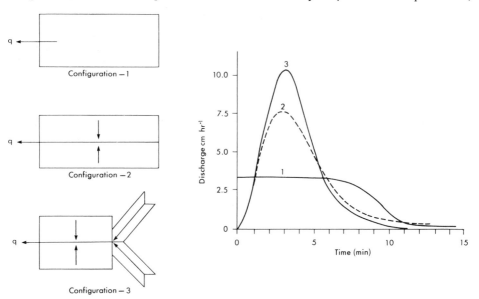

Figure 7.11 Simple hypothetical basin geometries and the corresponding hydrographs (Lane, Woolhiser and Yevjevich, 1976).

is not too easy to incorporate the complicated geometry of real catchments. On the other hand, the geometry may be varied to gain some idea of its effects (Lane, Woolhiser and Yevjevich, 1976). Another, sometimes serious, difficulty lies in the inability of the kinematic method to account for the back-water effects that are expected near channel junctions and at other changes in geometry.

Other workers have tried to evaluate the effect of channel network characteristics on the various flow properties. Much of the earlier work in this vein was based on the Strahler method of stream ordering which emphasizes topological characteristics of channel networks. Later work (Shreve, 1966; Jarvis, 1972) has attempted to provide a more rational basis for ordering. However, the fact remains that the network controls expressed topologically have little to offer in terms of hydrograph generation for process studies. As Kirkby (1976) has recently shown, the link frequency distributions and other basin characteristics based on topological measures differ appreciably from distance (i.e. geometrical) measures. He obtains an expression for the peak stream discharge from small catchments based on assumptions about the amount of area draining to a mean link length, which essentially reflects the growing contribution of area as one moves down channel. For individual channels this is a stepped function (Thornes, 1974) but empirical expressions have been obtained representing average increases in area for many basins. Two empirical equations incorporating the effects of basin area (A) and drainage density (D_d) are:

$$Q_{2\cdot33} \propto A^{0\cdot77} D_d^{0\cdot81} \quad \text{(Rodda, 1967)} \tag{7.9}$$

$$\text{and} \quad Q_{2\cdot33} \propto AD_d^2 \quad \text{(Carlston, 1963)} \tag{7.10}$$

Besides increase of water flow due to lateral inputs from the hill-slopes and from tributary channels, water enters from, and is lost to, the stream banks; water is also lost through the stream bed. These components have received much less attention from geomorphologists and yet they clearly have great importance in particular situations. Bank-storage is important for its effects on bank-collapse: as the water level rises, flow into the banks occurs and with falling levels water is released. In this way, a substantial weight of water may remain in the bank after the water level has fallen, leading to instability. This phenomenon has been studied in detail by Cooper and Rorabaugh (1963).

The loss of water through the bed is particularly important in semi-arid areas, but occurs to some extent in all channels where the bed is permeable. The volumes measured in semi-arid channels may be quite large. Keppel and Renard (1962), for example, showed that, over a 6·4 km reach, 50 per cent of the maximum discharge was lost and total flow was reduced by 35 per cent. Turner *et al.* (1943) stated that infiltration losses of 75 per cent in 25–40 km were common and even in perennial channels a study of 57 flood events on

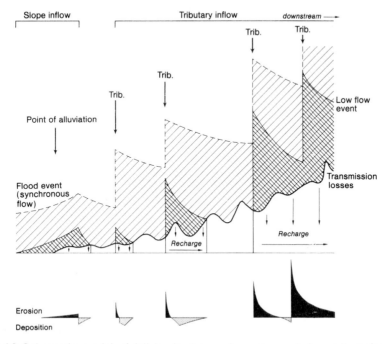

Figure 7.12 Schematic model of fall in discharge downstream below tributaries in a semi-arid channel resulting from transmission losses (Thornes, 1977b).

18 rivers in the Great Plains revealed an average loss of 40 per cent on an average channel length of 85 km, with as much as 75 per cent of the flood volume being lost. Transmission losses are generally related to the inflow to a reach, in which case, if there is no further lateral inflow, total volume will fall exponentially in a down-channel direction. This is shown diagrammatically in Figure 7.12.

Finally, one must bear in mind that there are few, if any, rivers in the world where flow conditions have not been affected by human activities. Long-term artificial regulation of flow by storage and regulation projects, and withdrawal of water for domestic and irrigation purposes, are the most common of these. In large rivers and estuaries, dredging and cut-offs are also important in determining the patterns of flow over time.

Dynamics of sediment transport

The basic reason for geomorphological interest in sediment transport lies in the fact that, if differences in input and output occur along a reach, the change in storage must have morphological implications. Conversely, if the form is unchanging the input and output over a reach should remain steady. The changes occur by scour and deposition; scour is removal of material, deposition is acquisition of material at any locality. With a completely movable bed, in non-cohesive sediments, scour and deposition may be modelled using laws for the transport of bed-load and suspended sediment based on such variables as velocity, slope or even simply discharge. Since many investigations of river morphology have dealt with this kind of situation, bed-load and sediment-load formulae of this type have prevailed. Three other important considerations must also be borne in mind: (i) the unsteady, localized behaviour of rivers; (ii) the fact that channels also exist in cohesive materials (especially solid rock: Plate XIV); and (iii) the fact that availability of sediment for transport varies spatially and through time. The last idea is expressed in terms of scour, a process that will be investigated in detail later in this chapter.

Sediment properties

Before considering the processes of transport, the properties of sediments need to be reviewed, since movement of particles depends on their physical characteristics. They can be divided into properties of individual particles and those of the sediments as a whole. The most important property of the particle is its size. In many studies, size alone is considered, but this can only give reasonable results if (as in controlled experiments) the variations in shape, density and size are fairly limited.

Sediments are usually divided into size classes or grades. The scale given in Table 7.3 has the advantage that sizes are arranged in a geometrical series and correspond closely to the mesh openings of sieves in common use. Because of variation in shape, any single length or diameter is arbitrary. Three measures are in common use: the 'sieve diameter', the length of the side of a square mesh opening; the 'sedimentation diameter', the diameter of a sphere of the same specific weight and the same terminal velocity as the given particle in the same sedimentation fluid; and the 'nominal diameter', which is the diameter of a sphere of the same volume as the given particle.

The settling velocity of a particle directly determines its reaction to flow and ranks next to size in importance. It depends on the size, shape and density of the particle and the density and viscosity of the fluid. Under steady-state conditions, the fall velocity is called the 'terminal velocity' and the drag on the particle is equal to its submerged weight. With no turbulence (laminar flow), the expression for fall velocity is known as Stokes's Law. Figure 7.13 shows

Plate XIV Potholes in former bed of Hvitá (river) in Iceland. Water standing in the pothole in the foreground is 2–3 m below the average level of the rock bed. (C.E.)

the fall velocity of a sphere for different temperatures, Reynolds numbers and diameters according to Raudkivi (1967). Attempts to define shapes that would give a unique relationship between shape and the other variables have not been successful. The fall velocities are also dependent on the concentration of sediments, the drag coefficient for a given fluid velocity increasing with concentration, though not as a simple function. If only a few grains, closely spaced, are in a fluid, they will fall in a group with a velocity that is higher than that of a particle falling alone. On the other hand, if particles are dispersed throughout the fluid, the interference is said to hinder the settling. The very finest part of the load consists of particles in which electro-chemical processes interfere with the simple fall velocity; these are called colloids. The electro-chemical processes tend to hold particles together once they come in contact. The agitation due to turbulence may also break up agglomerations of

Table 7.3　Commonly used particle-size classification

Class name	mm	phi units
Boulders	⩾ 256	⩾ −9
Cobbles	64 to 256	−6 to −9
Gravel	2 to 64	−1 to −6
Sand	0·064 to 2	+4 to −1
Silt	0·002 to 0·064	+4 to +9
Clay	⩽ 0·002	⩽ 9

particles. In this way the flocs tend to reach a size limit. When flocculation takes place by the particles coming together, the fall velocities increase—one expects particles to settle more slowly in a turbulent than a laminar fluid. This has been demonstrated by Field (1968).

Entrainment and transport

If one assumes an initially flat, uniform bed of non-cohesive sediments and if the velocity of the flowing water is increased, the power of the stream, the rate at which it can do work, is also increased. Eventually grain movement will be initiated, and particles will roll, slide or jump (saltate) across the bed. As the velocity continues to increase, the flow becomes turbulent and particles may be suspended in the flow. Up to a limit known as the transporting capacity of the stream, more and more material is put into motion.

Near the threshold of sediment movement, the bed is flat, but with a small increase in velocity ripples begin to appear with small slopes on their upstream faces, sharp crests, and steep downstream faces. At higher velocities and depths, the lengths of these forms may increase many-fold, still keeping sharp crests (Plate XV). They move downstream at a rate which is small compared with the velocity of flow. As the velocity increases, these ripples and dunes are replaced with a flat bed (Figure 7.14) and, with a further rise in velocity, a wave of sinusoidal shape may develop, called an antidune. This usually moves upstream, accompanied by waves on the water surface which are unsteady, moving upstream, increasing slowly in amplitude and then breaking by curling over in an upstream direction. Finally, at the highest velocities, chutes and pools occur; these are forms which, unlike ripples and dunes, do not occur in wind transport.

The loose particles that make up load are of greater density than the water, and hence differ in essential characteristics from materials which are chemically combined with the water and therefore move as the solute load. The essential features of granular flow have been described by Bagnold (1966). A shearing motion occurs in which successive layers are moved over one another. The motion has to be maintained by an impelling or tractive force. The solids are immersed in some pervading fluid (water, in this case) which itself is under shear and they are pulled down towards the lower boundary, or bed, because they are heavier than the fluid. In steady continuing motion the forces acting on every layer of solids must be in statistical equilibrium. Now if the carpet-like layer of immersed solids continues in motion, the layer must be supported by a stress equal to the immersed weight of the solid. This stress is maintained by the shearing motion and this, in turn, is derived from the applied tractive force (of the flowing water). There are two essential mechanisms for maintaining this stress. The first is the transfer of momentum from particle to particle by continuous or intermittent contact; the second is by the transfer of momentum from one mass of fluid to another and thence to the otherwise unsupported solid. The solid-transmitted stress arises

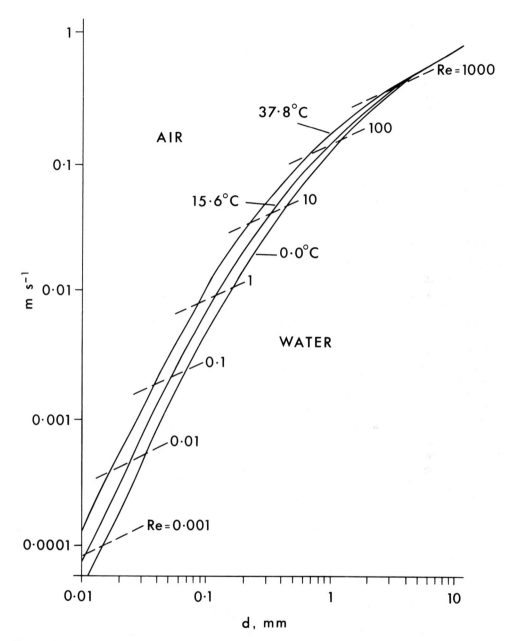

Figure 7.13 Fall velocity of a quartz sphere for different temperatures, Reynolds numbers and diameters (Raudkivi, 1967).

from the shearing of solids over one another. The fluid-transmitted stress arises from the shearing of the fluid, in the form of turbulence.

These mechanisms result in two basic types of solid-load transport. Suspension is the process whereby the excess weight of the solids is wholly supported by a random succession of upward pulses imparted by the eddy currents of fluid turbulence moving upwards relative to the bed. The solid may remain out of contact with the bed for an indefinite period depending on the essentially random nature of the turbulence. The other type of sediment transport is essentially unsuspended. This is where no *upward* impulses are imparted to the particles other than those attributable to successive contacts between the solid and the bed. The period

Plate XV Large dunes (amplitude about 2 m) developed on sand banks in the Xingu river, Matto Grosso, Brazil, exposed under very clear water. (J.B.T.)

between successive contacts depends on the force of gravity acting on the particle in motion. There are three types of movement: sliding, rolling and saltation. Rolling over a bed may be regarded as incipient saltation. Any body saltates if there is a constant body force having a component at right-angles to the boundary and a tangential motion exceeding a certain critical velocity. For this reason it can be observed in laminar flow (Francis, 1973). Bagnold (1973) likens saltation to pushing a wheelbarrow across a ploughed field; the normal force is the effect of gravity and the mass of the barrow, the tangential motion is that provided by the pusher.

In trying to obtain an expression relating the amount of transport to the characteristics of flow in the channel, the classical approach was to obtain the tractive force produced on the boundary by the flow. Du Boys (1879) and subsequently many others (e.g. Raudkivi,

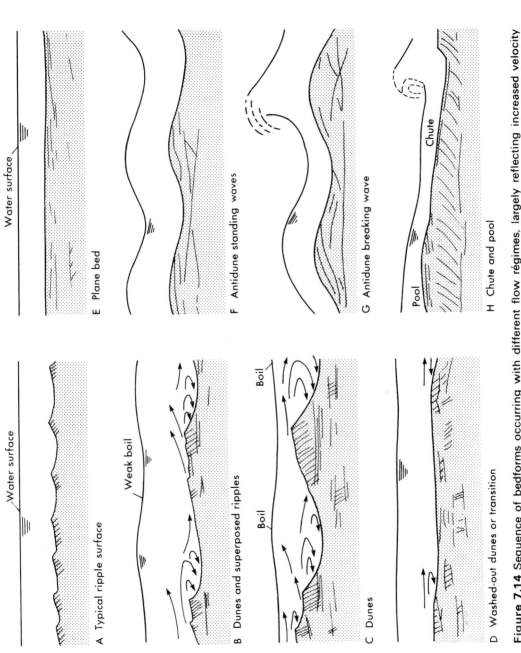

Figure 7.14 Sequence of bedforms occurring with different flow régimes, largely reflecting increased velocity and causing changes in roughness. Types A—D are representative of the lower flow régime where the Froude number is usually 0·4, E—H upper régime flow with the Froude number 0·7 (Simons and Richardson, 1966).

1967) have developed various equations to relate the rate of bed-load transport to the tractive force and the critical tractive force (for incipient motion). These developments have not led to a uniformly acceptable equation for sediment transport. Another approach is experimentally to compare sediment concentration against a set of other empirically determined characteristics, and obtain a multiple regression equation that 'explains' the former. Shen and Hung (1971) found that sediment concentration is a function of flow velocity, the energy slope and the fall velocity of the median sediment size of the bed sample. Because the migration process actually consists of a sequence of random steps and random rest periods, the third approach, initiated by Einstein (1937), relies on probabilistic descriptions of the mechanical processes.

Entrainment

The idea that, as velocity increases, a point will be reached when entrainment occurs is expressed in Brahms's (1753) equation:

$$v_{crit} = kW^{1/6} \tag{7.11}$$

In words, the critical velocity for entrainment (v_{crit}) depends on some proportion of the one-sixth power of the weight (W) of the particles; k is a proportionality constant. Unfortunately things turn out to be a little more complicated than this. Acting on the particle are two basic forces, the vertical or weight force moving it towards the bed, and the tangential force moving it along the bed. On a horizontal bed, the driving force comprises almost entirely the latter; on a steeply sloping bed, almost entirely the former. Their relative contributions depend mainly on the slope angle.

The usual model for the development of movement is outlined by Raudkivi (1967). Near the bed, velocity is a function of depth but there are also superimposed variations in velocity due to turbulence. As the fluid passes a particle, the streamlines in the flow are deflected, producing two components of force on the particle, the drag force, parallel to the direction of mean flow, and the lift force, normal to it. If the flow is laminar near the boundary (*cf.* Chapter 3) or if the thickness of the laminar sub-layer is much larger than the particle, the grain will not be affected by turbulent eddies and the drag is only that due to the viscous shear of the liquid. The drag is carried by the whole surface, and not by a few of the more exposed grains. As velocity increases, the boundary layer gets thinner and protruding particles develop a 'wake'. This produces a pressure difference across the particle which combines with the viscous drag. The pressure difference vertically across the particle is lift, and as flow increases this tends to lift the particle. Opposing both of these is the submerged weight of the particle. The condition of balance signifies incipient motion.

Instead of expressing this critical condition in terms of velocity alone, it is more meaningful to express it in terms of the shear stress or shear velocity at the boundary. The shear stress is a function of the depth of flow, the viscosity of the fluid and its velocity, and exists where the fluid is in motion relative to a fixed object or to a different fluid element. The shear velocity (μ_*) is the square root of the shear stress after division by the density of the fluid. Sometimes, the force is expressed in terms of dimensionless stress, θ_{crit}.

A plot of dimensionless stress (θ) against spherical-particle diameter (D) is shown in Figure 7.15 which Allen (1970a) developed from the work of Shields and Bagnold. The result shows that the relationship between the threshold stress and particle size of cohesionless material is not as straightforward as the simple resolution of forces described above would suggest. Moreover, the model does not take adequate account of the turbulent fluctuations along the stream bed, although for over 40 years a considerable effort has been made to obtain measurements of the character and spatial distribution of this turbulence. One clear result

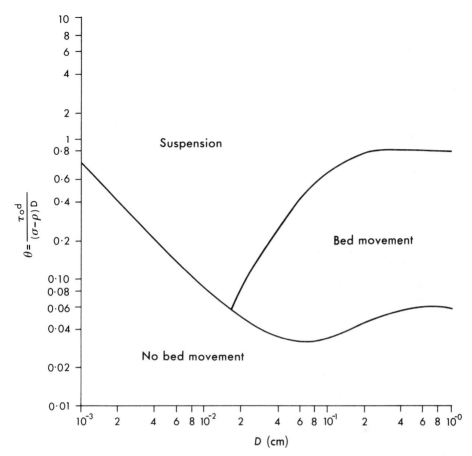

Figure 7.15 A plot of the entrainment curve in terms of particle diameter of quartz-density solids (D) and dimensionless threshold stress (θ) based on the work of Shields and Bagnold (Allen, 1970a).

of recent work, for example, shows a shift from the macro-scale to the micro-scale in turbulent boundary shear-stress and an increase in mean turbulent energy with increased Reynolds number (Blinco and Simons, 1974). As a result of this turbulence, the instantaneous shear-stress acting on the bed is up to three times the average shear according to Kalinske (1947). Raudkivi (1963) has, moreover, argued that the entrainment of particles and transport of bed-load can take place at values of temporal mean shear stress which are well below those expected by the Shields analysis on which Figure 7.15 is largely based, because of high instantaneous turbulence. He points out that turbulence can entrain particles in two ways: the particles may be moved by the drag exerted by a passing eddy or, if the eddy lowers the local pressure, the particles may be actually ejected from the bed by hydrostatic pressure. This kind of phenomenon was observed by Apperly (1968) who indicated that for lift there was a predominance of negative forces but that there were infrequent bursts of large positive lift forces. These large lift forces are apparently the ones that entrain particles.

Little attention has been paid to the critical conditions for the entrainment of cohesive sediments. The experimental plot of critical water velocities for quartz sediment as a function

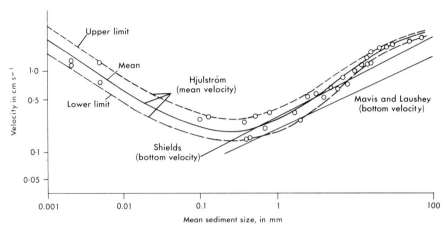

Figure 7.16 Critical water velocities for quartz sediment as a function of mean grain size (Vanoni, 1975).

of mean grain size shown in Figure 7.16 indicates that the finer the material, the greater the required entrainment velocity. Sundborg (1956) suggests that the cohesive force resisting entrainment of a grain is proportional to the shearing strength of the sediment. Actual shear strength was found by Smerdon and Beasley (1961) to be directly related to the plasticity index and to percentage clay, and Flaxman (1963) found that the compressive strength of unconfined saturated and undisturbed samples of the sediment was a good indication of the shear stress it would withstand without excessive erosion. He also noted that some channels in sediment with a small or negligible plasticity index were stable. The critical velocities are also related to chemical characteristics. Grissinger and Asmussen (1963) found that the resistance of clays increased with time of wetting and suggested that this is due to strengthening bonds with hydration. Obviously there is no critical shear stress for clays in the sense that one exists for non-cohesive sediments.

Transport equations

During the last century, many formulae have been presented based on what is known as the tractive-force approach. The oldest of these, still occasionally used, is the Du Boys formula. Although based on a simple model, it has considerably influenced many subsequent formulations. He imagined the bed-load material to be divided into n layers each of thickness d' and that the mean velocity increased upwards through the sediment at a rate ΔV per layer. In this way (Figure 7.17), with a linear increase in velocity with height above the layer of no movement, the surface layer has a velocity $(n-1)\Delta V$. The average velocity will be $[(n-1)\Delta V/2]$, the total thickness nd' and therefore the discharge

$$q_s = nd'(n-1)\Delta V/2 \qquad (7.12)$$

This is the volume of sediment moved per unit width of bed in a given length of time.

The force of the flowing water in a downstream direction is $\gamma y_0 s$ where y_0 is the depth of water, s the slope and γ the specific weight. This is balanced by the friction of the sediment–water mixture in the bed. The friction coefficient between the layers of the bed material (μ_s) is assumed constant so that the balance of forces is

$$\gamma y_0 s = \mu_s (\gamma_s - \gamma) \; nd' = \tau = \text{tractive force}; \qquad (7.13)$$

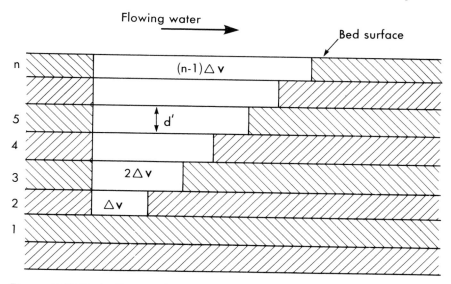

Figure 7.17 Basic diagram showing the development of the Du Boys equation.

γ_s is the specific weight of the sediment, so the term in brackets represents the specific weight of the water–sediment mix.

Now threshold conditions occur when the top layer just resists motion and the tractive force is critical. This is when

$$\tau_c = \mu_s(\gamma_s - \gamma)d' \tag{7.14}$$

and by substitution into the earlier formula, n can be represented by the ratio τ/τ_c, giving the classical Du Boys formula:

$$q_s = C_s\tau(\tau - \tau_c) \tag{7.15}$$

where $C_s = \mu d'/2\tau_c^2$ and is regarded as a function of sediment characteristics alone. If the Manning equation is used to replace the tractive force, the discharge of sediment per unit width can be obtained from the water discharge per unit width (q) by the equation:

$$q_s = \frac{C_s s^{1.4}\gamma^2 q^{0.6}(q^{0.6} - 1_c^{0.6})}{n^{1/2}} \tag{7.16}$$

The term q_c is the critical discharge (of water) on a critical slope s_c at which sediment transport will begin. This basic Du Boys-type formula is known as the bed-load type and similar equations are the Shoklitsch formula (Shulits, 1935), the Shields (1936) formula, the Meyer-Peter and Muller (1948) formula and the Einstein–Brown formula (Brown, 1950). They all indicate sediment discharge increasing as bed shear stress increases. Therefore these formulae show a decrease in sediment discharge for a given water discharge as the friction factor of the stream decreases and velocity increases (Vanoni, 1975). This is at variance with results for streams with sandy beds (Colby, 1964b).

Other approaches, for example the Inglis–Lacey formula (Inglis, 1968) and Colby's own work (1964a, 1964b) rely on mean velocity as a predictor of sediment discharge. In the Inglis–Lacey equation, the sediment discharge varies as velocity to the fifth power whereas in

Colby's work the exponent of velocity varies. For a given sediment size and water depth the exponent decreases continuously as velocity and sediment discharge increase. A comparison of the various methods is presented in Vanoni (1975) and an interesting evaluation and critique is to be found in Yang (1971).

An alternative to the more-or-less empirical approach to bed transport was initiated in the probabilistic work of Einstein (1950) which explicitly attempts to incorporate the effects of turbulence. In this approach the movement of any particle depends on the probability that, at a particular time and place, the dynamic forces exceed the resisting forces. Thus the probability of movement is a function of the size and immersed weight of the particle and a characteristic time (rest time) which is taken as a function of the particle size divided by its fall velocity. The jumping particles are received on an area of bed according to their size distribution, and for equilibrium, the number of particles eroded (N_e) must equal the number deposited (N_d). This situation is represented by

$$N_e = \frac{\mu*}{A_1 D^2} \cdot \frac{p}{t_1} = N_d = \frac{q_B}{\gamma_s A_2 D^3} \qquad (7.17)$$

where q_B is the transport rate in dry weight per unit time per unit width, γ_s is the specific weight of the particle, D the grain diameter, $A_1 D^2$ the projected area of the grain, $A_2 D^3$ the volume of the grain, p the probability of the particle being eroded and t_1 the exchange time. This yields a transport rate function

$$\phi = \frac{q_B}{\gamma_s} \left[\frac{\gamma}{\gamma_s - \gamma} \right]^{1/2} \cdot \left[\frac{1}{gD^3} \right]^{1/2} \qquad (7.18)$$

for the intensity of bed-load transport. The probability p is the probability that the lifting force is larger than the submerged weight. From ergodic considerations, p may also be interpreted as the fraction of the bed on which, at any given time, the lift on a particle is sufficient to cause motion. This equation of the lift force and submerged weight is then used to obtain an expression for ϕ. The interest in this work lies in its explicit recognition of the stochastic behaviour of transport, an idea that has been developed by Hubbell and Sayre (1964), Shen and Cheong (1971) and Hung (1971).

Another important restriction of the conventional bed-load equations, particularly for geomorphological investigations, is their failure to incorporate time-based effects of the sediment transport process. The case of channel armouring is used here to exemplify the problem. If an alluvial bed consisting of non-uniform coarse material is exposed to a constant, relatively small bed shear-stress, erosion of the finer fractions of the material is likely to occur before erosion of the coarser fractions. Eventually the underlying material will be covered by a stable armour coat, about the thickness of one grain, thus preventing further degradation. In this situation the concept of incipient motion, as well as flow-related transport, takes on a special character. The problem is an important one for geomorphologists, since the character of residual fluvial deposits (and current hillslope deposits) is a function of this differential removal and armouring process. The probability that a grain will survive could be expressed as a function of the ratio of $\tau_{crit}/\bar{\tau}$, the ratio of the critical to average bed shear stress. Gessler (1970) has shown experimentally that such a relationship exists. For a range of flows (or a flow distribution), the effects of this process on a pre-existing sediment mix should be computable. Figure 7.18 shows the change in grain-size distribution for different mean bed shear stresses calculated from experimental data by Gessler. Evidently the mean particle size should become so large that for, say, the annual distribution of flows there will be a limit to how

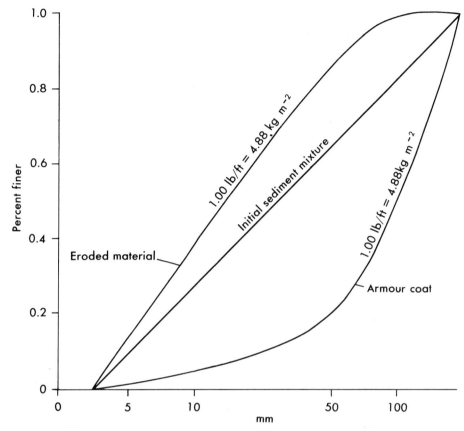

Figure 7.18 The change in grain-size distribution curves for different mean bed shear stresses calculated from Gessler (1970).

much scour can take place. If flow is also spatially variable, as for example in ephemeral channels, then the distribution of residual particle sizes should reflect this.

Suspended-load transport
The basic problem that exists in terms of suspended-load transport is its reliance on availability of material rather than the availability of power to transport it. Carson (1971a) argues that '... much of the suspended material in stream channels is so fine that almost any condition of flow will transport it provided that some mechanism exists for getting this material into suspension. The rate of suspended-sediment discharge is, therefore, probably best analysed in terms of sediment supply rather than stream capacity.' Sediment supply is largely related to the processes of bank collapse and hillslope erosion. However, that an upper limit to suspended-sediment load exists is indisputable. For otherwise, as Bagnold (1966) points out, rivers as we know them could not exist. No deposition of suspended material could occur and no suspendable bed material could remain unsuspended. In other words, there must be a limit to the efficiency of the work done by the internal energy of turbulence in supporting the excess weight of the suspension.

The usual approach is to assume that the downstream and cross-channel variations in velocity are unimportant and to obtain the vertical variation of suspended solids with depth. A flux through a small plane occurs with some currents moving up and some down. If the material in suspension were weightless, the concentration would be uniform. A gradient is produced by the continuous settling of the particles under their own weight. One method of analysis is to set the upward flux of sediment, analogous to the fluid shear-stress concentration, measured as momentum, equal to the settling of particles under their own weight. The upward flux (q_μ) is set equal to the concentration gradient (dc/dz) multiplied by a diffusion coefficient ε_s, giving

$$\varepsilon_s \cdot \frac{dc}{dz} = -cw \tag{7.19}$$

in which c is concentration, w the fall velocity of the particles and z the depth of flow. This is the Schmidt equation after Schmidt (1925) and its solution varies according to the assumptions made concerning the mixing or diffusion coefficient. (Note the similarity of the left-hand side to Darcy's Law, equation 7.4, p. 221). If ε_s is assumed constant (i.e. the turbulence is constant with depth), the expression can be integrated to give

$$\frac{C}{C_a} = \exp\left[\frac{-W}{\varepsilon_s} \cdot (z-a)\right] \tag{7.20}$$

where C_a is the concentration at a height a above the bed. If on the other hand ε_s is actually made proportional to the mixing coefficient for momentum exchange, say $\beta\varepsilon_m$ where β is a constant of proportionality, the solution is more realistic and ε_m is known from the momentum flux equation to be proportional to the rate of change of velocity with depth, i.e.

$$\varepsilon_m = K^2 z^2 \left(\frac{du}{dz}\right) \tag{7.21}$$

where K is the von Karman constant. Unfortunately von Karman's constant is not constant in turbid water. Moreover, when $z \to 0$ and $z \to D$, the predicted values are infinity and zero, both of which are unrealistic. Further manipulation leads to Rouse's equation, which takes the form

$$\frac{C}{C_a} = \left[\frac{a(D-z)}{z(D-a)}\right]^p \tag{7.22}$$

in which $p = w/K_{\mu*}$ where μ_* is the shear velocity (see p. 236). This equation relies on Stokes's Law, which in turn is viscosity-dependent (and hence temperature-dependent), and such a relationship is reported. Lane, Carlson and Hanson (1949) showed the graphical relationship (Figure 7.19) between suspended-sediment discharge and concentration, water temperature and water discharge against time for the Colorado River at Taylor's Ferry. The fluctuations due to temperature are seasonally important. Another effect of temperature through viscosity appears to be related to bed conditions producing a reduction in bed-form height and a diminished resistance to flow. This may be due to an increase in the depth of the laminar sub-layer as higher viscosities are offset by increase in friction, leading to more effective current attack on the bed (Vanoni, 1975).

Bagnold (1966) has expressed doubt about the procedure involved in deriving the Rouse equation on the grounds of the analogy with momentum transfer, where this is justifiably kinematic (there are no differences in density). It is equivalent to 'treating the solids as little fishes of zero excess weight swimming perpetually downwards relative to the fluid at a given

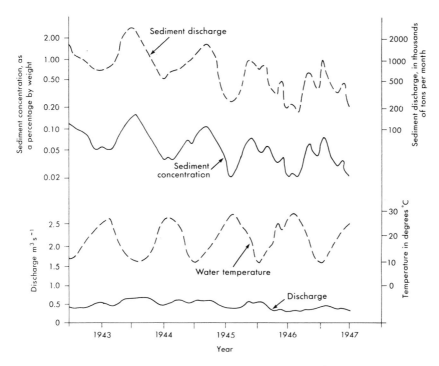

Figure 7.19 Graphical relationship between suspended-sediment discharge and concentration, water temperature and water discharge against time at Taylor's Ferry (after Lane, Carlson and Hanson, 1949).

flow velocity', and hence predicting the concentration of the fishes (=solids) with increase of distance from the boundary. The difficulty is that the Rouse procedure fails to predict the total weight of fishes which a given turbulent flow can carry in suspension! This is what we need to know. In order to overcome this difficulty, the flux of turbulent fluid momentum away from the boundary must exceed that towards it, and so be asymmetrical. Unfortunately the full implications of this criticism have not yet been fully evaluated.

Continuity approach

A more general approach, incorporating the typical bed-load equations (and some not-so-typical ones) is based on the continuity principle. This might also be regarded as the capacity approach. The concept of capacity of sediment transport (and the associated scour) has had a long history in geomorphology and formed a key part in arguments about channel morphology in the first half of the century. Laursen (1952) expresses the problem in terms of four general principles:

1 the rate of erosion will equal the difference between the capacity for transport out of the scoured area and the rate of supply;
2 the rate of erosion will decrease as the flow section is enlarged by erosion;
3 there will be a limiting extent of erosion for given conditions and
4 the limit will be approached asymptotically with respect to time.

The first principle may be expressed by the continuity condition (Vanoni, 1975):

$$\frac{df(B)}{dt}=g(B)-g(S) \tag{7.23}$$

in which $f(B)$ is a mathematical description of the boundary, t is time, $g(B)$ the sediment-output rate as a function of the boundary shape and position and $g(S)$ is the rate of supply to the zone of erosion. The remaining three conditions are boundary conditions on the continuity equation. If the local rate of transport is greater than the rate of supply, $df(B)/dt$ is positive and erosion is occurring; otherwise it is negative and deposition is occurring. When $df(B)/dt$ is zero, the bed is stable and transport steady. The application of this equation relies on the establishment of functions of flow conditions and time.

Foster and Meyer (1972) use the interesting model

$$D_F/D_C+G_F/T_C=1 \tag{7.24}$$

D_F is the net flow detachment rate and D_C the ultimate detachment capacity; G_F is the net transport rate and T_C the ultimate transport capacity of overland flow. This is the same as assuming that a particular flow which is carrying less than its capacity to transport will fill this capacity according to a first-order reaction, i.e.

$$D_F=G(T_C-G_F) \tag{7.25}$$

where G is the reaction rate coefficient. Bennett (1974) discusses some of the relationships between this approach and the solution to the equations of conservation of mass and momentum for flow and the equation of conservation of mass for sediment. Bennett concluded that solution procedures for these equations in two and three dimensions are so complicated that, in the light of present uncertainties concerning the physics of two-phase flows, the effort required for their solution in two and three dimensions is not warranted. Following the work of Simons (1965), Chen *et al.* (1975) have outlined the procedures available for solving the equations for the transport of material by unsteady flow. The basic equations are (i) the sediment-continuity equation, (ii) the flow-continuity equation and (iii) the flow-momentum equations. A major weakness of their analysis is in the sediment-discharge relations which have to be supplied empirically; the results obtained are close to reality, despite the approximations required in the procedure.

The Bagnold approach

Besides the continuity approach outlined above, the idea of the capacity of the stream to do work is also embodied in Bagnold's (1966) view of the fluid stream as a transporting machine expending power to do work. This has particular appeal geomorphologically because it enables the stream to be conceptualized at a scale closer to the geomorphologist's *modus operandi* than that of the hydraulic engineer. Bagnold uses the expression

rate of doing work=available power−unused power

or its equivalent form

rate of doing work=available power×efficiency.

The available power is the rate of energy supply to a stream and is equal to the rate of liberation of the potential energy as it descends the gravity slope s. This is called *stream power* and is found to be the product of the mean boundary shear stress (τ_o) and the mean velocity of the fluid stream (v), i.e.

$$W=\tau_o v \tag{7.26}$$

The work that is performed by the 'machine' is that of transporting sediments, both suspended and unsuspended. If this transport is statistically steady, the downward body force has to be counteracted by an upward acting force, perpendicular to the bed. On each unit area of bed there are, therefore, four forces: the two mentioned above, the boundary shear stress τ_o acting in the direction of motion, and the equal and opposite force τ exerted by the bed on the stream. For equilibrium, the upward force (p) equals the downward force and $\tau_o = \tau$. Resolving these forces, the rate of sediment-load transport (Θ) is given, per unit area of bed, as

$$\Theta = \frac{(\gamma - \rho)}{\gamma} \cdot mg \overline{V}_s \tag{7.27}$$

in which $\overline{V}_s$ is the mean velocity of the sediments, γ and ρ the density of the solid and the water respectively and m the dry mass of the solids. To convert this to sediment-load *work* rate it has to be multiplied by a conversion factor, A. The latter is defined as the ratio

$$\frac{\text{tractive stress needed to maintain transport of load}}{\text{normal stress due to immersed weight of the load}}$$

This conversion is necessary so that the load can be related to the boundary shear stress responsible for the transport. The source of the upward stress for this resolution of forces has already been discussed, and comprises the repeated frequent collisions between grains in fluid motion, which requires that the grains be closely spaced and that the normal-force component is directly transmitted from the bed. With the rate of suspended-sediment transport expressed in terms of the fall velocity of the sediment V_o, the expression for the rate of total load transport is

$$\Theta = \Theta_b + \Theta_s = W \left[\frac{e_b}{\tan \alpha} + \frac{e_s \overline{V}_s}{V_o}(1 - e_b) \right] \tag{7.28}$$

where e_b and e_s are the suspended- and bed-load efficiency and $\overline{V}_s$ the mean velocity of bed-load transport. This expression has been used by several workers (e.g. Smith, 1976; Armstrong, 1977) for modelling fluvial transport processes. Allen (1970a) describes the equation as 'remarkably accurate as such functions go'. The value of the bed-load efficiency factor is about 0·13 and is slightly influenced by the mean flow velocity and the calibre of the load. The efficiency factor for the suspended load is about 0·015 and α is approximated by the coefficient of solid friction. The equation is essentially valid only for uniform steady flow, but by expressing W as $\tau_o \cdot \overline{V}$, the variations in transport rate and stream power can be modelled according to unsteady fluctuations in velocity. This led Allen (1970a) to the expression for continuity of sediment, which takes the form:

$$\text{rate of erosion or deposition} = \frac{\delta\theta}{\delta x} + \frac{1}{\overline{V}} \cdot \frac{\delta\theta}{\delta t} \tag{7.29}$$

The first term on the right is the contribution from the non-uniformity of flow (change of power with length along the channel), and the second term is due to the unsteadiness of the flow (variations of stream power due to fluctuations through time).

Erosion and deposition

Static equilibrium

If a stream is flowing over a gentle slope with low velocity, there will be no transport of material because the stream power is insufficient to entrain the particles. As the power is increased progressively and if the particle size is homogeneous, a point will be just reached, at least theoretically, at which the grains on the bed and banks are on the threshold of movement. Because movement is produced by a combination of fluid drag and the downslope component of gravity, and because fluid drag is greatest in the centre, the channel slopes can increase up the sides to compensate for the decrease in drag. The resulting profile is roughly a parabola, similar to some natural channel forms, and is called a *threshold channel*.

In some respects this concept is idealized. It is optimally what some designers of canals would like to see, and it represents one end of the spectrum of channels in different stages of mobility. In practice, in alluvial streams with non-cohesive gravel or boulder banks and beds, this theory ought to be applicable. The largest flows in the long term should have cleaned out the finer materials and left channels which are in a threshold condition. In smaller flows, stream power is insufficient to have an effect and the boundaries of the channel system will be in equilibrium until subjected to discharge greater than that producing the threshold channel. The idea is that the coarse-bed stream morphology does not change between periods of extreme precipitation.

Many investigators have formulated the shape of threshold channels in homogeneous coarse alluvium. Lane (1953) developed the resolution of forces which is now most widely accepted. He assumed that:

1 At and above the water surface, the side slope is at the angle of repose of the alluvial material.
2 At all points on the periphery of the channel the particles are at a condition of incipient motion. The lift and drag forces of the fluid on the particle are exactly balanced by the friction force developed between the particles. The lift and drag force are directly proportional to the tractive force.
3 Where the side slope is zero, the tractive force alone is sufficient to cause incipient motion.
4 The particles are held against the bed by the submerged weight of the particle acting normal to the bed.
5 The tractive force acts in the direction of the flow and is equal to the component of the weight of the water above the area on which the force acts.

Figure 7.20 shows the basic formulation of the problem. The resultant of the drag force, F_d, and the downslope component of the submerged weight of the particle, $W \sin \theta$, are

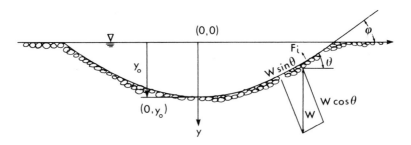

Figure 7.20 Basic terminology of the Lane model of the threshold channel.

balanced by the friction developed between particles. The angle θ is the local side-slope angle and W is the submerged weight of the particle. The friction is the product of the normal force and the tangent of the friction angle ϕ. The normal force is $W\cos\theta - F_i$, in which $W\cos\theta$ is the normal (V_0 the side-slope) component of the submerged weight and F_i the lift force on the particle. The balance of forces is then expressed as:

$$W^2\sin^2\theta + F_d{}^2 = [(W\cos\theta - F_i)\tan\phi]^2 \tag{7.30}$$

The drag force is given by $F_d = \delta\tau$ and the lift force by $F_i = F_d$, where τ is local bed shear stress and δ a proportionality constant. The term β is the ratio of lift and drag forces.

The maximum tractive force τ_0 occurs at the centre of the channel, where the slope is zero (i.e. $\theta = 0$) and, as before (p. 238), is given by

$$\tau_0 = \gamma\, y_0\, s \tag{7.31}$$

in which $\gamma =$ the unit weight of water, $y_0 =$ the maximum depth of flow, and $s =$ the slope of the channel.

One of the assumptions is that the local tractive force varies directly as the normal component of the weight of the fluid above the bed. Consider the cross-section at a distance x from the centre line, where the depth of flow is y. The weight of the fluid on the bed is γy; its normal component is $\gamma y\cos\theta$ and therefore the tractive force is

$$\tau = \gamma\, y\, s\cos\theta \tag{7.32}$$

With manipulation, this set of assumptions produces the shapes shown in Figure 7.21 if β is set equal to 0·85 and $\theta = 35°$ by the equation

$$\frac{y}{y_0} = \frac{1}{1-r}\left[\cos\left(\tan\phi\sqrt{\frac{(1-r)}{(1+r)}\cdot\frac{x}{y_0}}\right) - r\right] \tag{7.33}$$

The typical threshold channel, then, has a surface width of 4·49 times the maximum depth, an area 2·86 times the maximum depth squared and a wetted perimeter 4·99 times the maximum depth. This is regarded as the critical section or type B section. Two other sections (A and C in Figure 7.21) are also acceptable interpretations of the theory.

Henderson (1963) went on to introduce the size (D) of the bed material through the Shields entertainment function. He assumed it to be uniform in the section and found that there was a limiting value of slope and discharge at which the type B channel would remain stable. This value is defined by the expression:

$$S = 0.44D^{1·45}\, Q^{-0·46} \tag{7.34}$$

If the slope is *greater* than this value, the wide type (A) of channel, with less scouring capacity, would be needed for stability. With a lower slope the type C channel would be most stable.

This result is remarkably close to the empirical result relating to coarse channels found by Leopold and Wolman (1957). They presented a graph of channel slope and bankfull discharge on which is plotted the line discriminating between braiding and meandering. The line has the equation:

$$S = 0.06Q^{-0·44} \tag{7.35}$$

The implication is that the type A channel is the braided type while the type C channel is the meandering type, and that movement across this line will result in instability if there is an increase in slope or discharge over a period of years. Li, Simons and Stevens (1976) have recently extended the theory to derive the exponents of the equations relating discharge

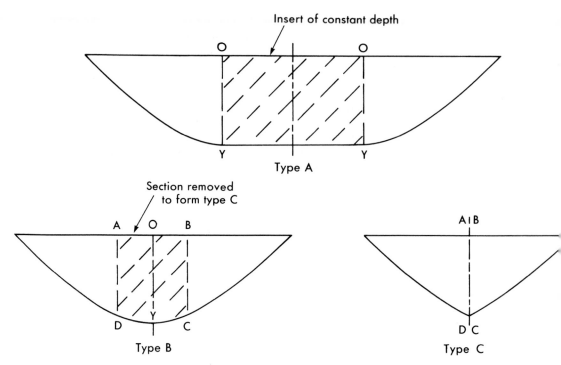

Figure 7.21 Characteristic profiles for stream channels.

and width, depth and velocity (i.e. hydraulic geometry) and have obtained results that are compatible with those of Leopold and Maddock (1953).

Dynamic equilibrium—bed forms

Unfortunately, the conditions outlined by Lane are not often met with in nature because the channels are transporting sediments and the bed comprises a non-uniform mixture. The channels migrate laterally by erosion of one bank, maintaining an average cross-section by deposition at the opposite bank. The form of the cross-section is in dynamic equilibrium. This was the cross-section sought after empirically by Lacey (1929) in his 'régime' theory. However, the general proposition that channel form is adjusted to the processes in operation, is revealed by other characteristics, though the extent and type of adjustment depends on the time-scales involved and is subject to complex lags (Allen, 1974). A well documented series of forms occurs in channels with sandy beds. As stream power increases beyond the critical threshold value of incipient motion the sequence shown in Figure 7.14 occurs. These bed-forms change from ripples to dunes (Plate XV) and ultimately to anti-dunes or chutes and pools. It is observed, with the simplest forms, that once one form exists it will be transformed gradually into an asymmetrical wave with a gentle upstream slope. The theory, due to Exner (1925), does not show, however, how the sinusoidal form is obtained from an initially plane bed.

One approach to the question is to treat it as an instability problem at the water–sand interface. If fluids of different densities are moving parallel to one another with different velocities, the common boundary is always unstable for perturbations of small wavelengths.

Observations in a glass-sided flume by Raudkivi (1967) show that the initial plane boundary begins to deform at some point and that this deformation gradually spreads downstream. The point of initiation of the ripple seems to be by chance. There is a tendency for the particles to 'pile up' and move intermittently when the velocity is only slightly beyond the threshold of particle movement. This could be due to a kind of 'traffic jam' effect whereby fast-moving particles move in on slower-moving particles, thus further slowing down the latter by a sheltering effect. Alternatively, it could be due to intermittent, strong turbulent eddies in the boundary region or to non-uniformity of the materials.

Once piling-up has occurred, the streamlines become curved causing dynamic pressure changes and accelerations. Once an inclined face has formed, the ripples appear with remarkable regularity. The particles slide up the upstream faces and roll over the crests of the ripples. There they come to rest on the downward slope at the angle of repose. This leads to the continuous downstream translation of material. An occasional grain is caught in the interface

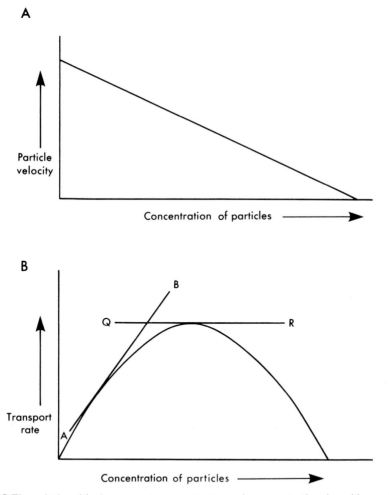

Figure 7.22 The relationship between transport rate and concentration in a kinematic wave.

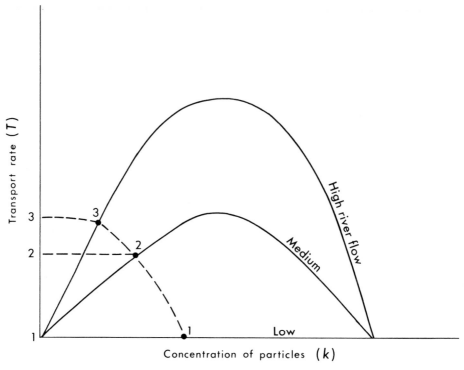

Figure 7.23 Flux-concentration curves for changing conditions in a river bar (Langbein and Leopold, 1968).

between the main stream and the wake downstream of the crest, and lands downstream at a distance of approximately six times the ripple height. There is here a critical point between backward motion (into the lee of the ripple or dune) and movement forward up the next form. This is the reattachment region.

Langbein and Leopold (1968) developed the analogy with the 'traffic jam' to explain the development of a pool-and-riffle sequence, in which there are alternate shallow and deep zones distributed regularly along the channel as steady-state forms. Like ripples, gravel bars form the riffles in the pool-and-riffle sequence as 'a group of non-coherent particles piled up in some characteristic manner'. The particles move through the pile at one speed whilst the bedform may be stationary or move up and down the channel at another speed. Occasionally, in very large floods, the whole assemblage including some of its largest particles may move downstream (Thornes, 1976). These ripples, dunes and bars are geometrical concentrations in a system of moving objects, like concentrations of cars on a motorway, and are called kinematic waves. As with kinematic wave theory (p. 228), dynamic properties of the material are not taken into account.

If the concentration of sediment on the bed increases, the velocity of individual particles will fall (Fig. 7.22), due to increasing interference between grains, until eventually no movement will take place at all. If total sediment transport is plotted against concentration, there will be a point at which maximum sediment transport can occur. Thereafter, increased concentration reduces the transport rate. It may be shown that the spread of individual grain velocities will lead to a grouping of grains, analogous to queues of moving vehicles, and that the velocity of the group (bar, ripple) is given by the tangent to the curve, AB, in Figure 7.22. The group speed is called 'celerity' to distinguish it from the individual particle velocity. The peak of the concentration-flux curve has a tangent of zero (QR) and Langbein and Leopold (1968) argue that the observed constancy of riffles during sediment movement is because sediment transport is occurring at a maximal rate. The flux-concentration for a river bar is suggested by Langbein and Leopold (1968) to take the form shown in Figure 7.23. As the river flow increases, the sediment transport increases, and since this is from the bars, the concentration decreases (moving from 1 to 2 to 3). The concentration between the riffles is zero: therefore, the flux-concentration change is zero and the riffles remain fixed.

In this hypothesis, once the perturbations are initiated, they themselves generate the flow conditions leading to regular distribution of the bedforms. Yang (1971) appealed to a slightly different mechanism which he considered capable of generating the classical pool-and-riffle sequence. He observed that in the Bagnold (1954a) model of dispersion of solid grains in a Newtonian shear flow, a repulsive pressure, P, should exist between grains of two layers:

$$P = c(\lambda D)^2 (du/dy)^2 \cos \alpha \qquad (7.36)$$

where P is a function of the velocity gradient (du/dy), the grain diameter D, the grain concentration λ, and a frictional property of the grains α. With a perturbation, he suggested that higher velocity gradients at the riffle result in greater dispersive stresses. This would tend to raise the bed at the riffle and, relatively speaking, lower it in the pools. Positive feedback would then accentuate the riffle until the water over the riffle became too shallow to sustain the shear stress being generated for a given flow. The dispersive stress pushes up the larger grains and the finer grains are washed out, leading to a sorting so that mean diameter on the riffles increases further.

Dynamic equilibrium—meanders

As with the threshold concept of static equilibrium, several models seek to obtain 'stable' cross-sections involving the transport of sediments in channels which are not straight in plan form. The fact that regular patterns of deep and shallow water exist, that meanders show a remarkable consistency in the relations of width and wavelength to discharge, and that there is a regular pattern in the cross-sectional geometry, leads to the conclusion that these are forms in dynamic equilibrium. Therefore some consistent and regular processes must be in operation. As before, it is easier to obtain a dynamic argument *sustaining* meanders than a reason for their initiation.

It has been argued that because meanders occur in supraglacial streams, in the waters of the Gulf Stream, and on beads of water flowing down clean glass plates, there is no need to invoke the presence of sediments to account for the meandering habit. In this view, meandering is a property of the flow, sediment transport serving only to make this instability visible on the bed. It is suggested that the role of sediment in meandering is similar to that of dye placed in turbulent water; the dye does not cause the turbulence, but simply makes it visible. Parker (1976), on the other hand, in a complex stability analysis, claims that, in river channels, sediment transport is essential for the formation of meandering although helical flow is not regarded as essential. He reconciles his view with those expressed above by arguing that, for meandering, there are three requirements, the first two being potential (inertial and gravitational) and friction effects. The third is the presence of sediment for alluvial streams, the Coriolis acceleration for oceanic currents, heat differences for glacial meltwater streams and surface tensions for water threads on glass plates. Most contemporary investigators of the phenomenon assume the need to incorporate sediments in their explanatory models.

Friedkin's (1945) description of meandering in a flume remains one of the most accurate of its type. In a whole series of experiments in a large sand-bed flume the basic processes were observed. Starting with a straight channel, with a velocity sufficient to move sand along the bed and erode the banks, it was observed that a local disturbance to the flow was created by a sand bar which resulted from bank erosion. Sand from caving banks overloaded the stream and deposition took place. These studies showed that, in the flume, the source of sand was the caving of the banks and that the sand travelled only a short distance to the first convex bar downstream, where velocities were low enough to permit deposition. Only minor amounts of sand crossed the channel to the convex bar opposite.

Besides the initiation of the meandering as a result of bar growth, the other most important process was thought to be the diversion of sand from the deep-water channel to the bar as the river enters a bend. This resulted in the building of the convex bars in the bends. Observations of this phenomenon showed that the sand was diverted to the bar because the slow bottom currents which carried the sand tended to divert from the thalweg to the bar.

Two other important observations were made. First, during high flows, the water surface slopes were very low and deposition took place in the crossings (riffles). During low flows, the slopes were gentle in the pools and steep over the riffles, which resulted in scour of the riffle. Secondly, during the low stage, the movement of sand was largely confined to the riffles where sand was scoured out and deposited in the pools below. In many tests there was low-water bank erosion in the upstream part of the bends and the eroded sand was deposited in the pools immediately downstream from the point of erosion. At bankfull stage, the path of the main current in the bends was generally across the convex bars, and the attack and erosion of concave banks was along the downstream part of the bends. In other words, the path of the main current changes with the stage, which results in changes of the loci of erosion and deposition.

Finally, it is worth quoting at length the concluding sentences for they seem to have been rediscovered in many later papers:

> The variables (discharge, load and rate of bank erosion) form a circle of dependency. These tests showed that a change in either discharge, slope, amount of sand entering a bend, or rate of bank erosion tends to bring about definite changes in the channel cross-section and patterns of a meandering river. However, the extent of these changes is often impossible to determine, because changes in one variable are opposed and limited by changes in another variable. This is the complex side of meandering which prevents set rules or formulae (Friedkin, 1945).

Although the initiation of meanders was observed to result from a perturbation, once generated this perturbation led in Friedkin's case to a very regular pattern. Anderson (1967) suggested a mechanism whereby this might occur. The initial irregularity causes a minor transverse component in the flow-velocity field which in turn causes the water surface at the banks to rise and thus initiate a transverse surge in the opposite direction. Because the main flow is still parallel to the channel banks, transverse oscillations will be eventually initiated in the whole mass. The period of this oscillation depends on the mass of the fluid involved and may be quite different from the period of the surface wave. The wave length λ is reflected in the construction of bars and shoals formed alternately at regular intervals on opposite sides of the initially straight channels. These bars are a means by which a more formal pattern is initiated. Mathematical models that determine the conditions under which such perturbations may propagate and under which meanders are initiated have been investigated by Callendar (1969), Engelund (1974), Parker (1976) and Ponce and Mahmood (1976).

Several variations on this basic process model have been observed in natural channels. Friedkin (1945) observed experimentally that 'pure' sinusoidal forms were distorted by variations in bed materials. Lewin (1972) observed that in real gravel-bed channels symmetrical forms are rather uncommon and that, particularly in the later stages of development, more complex geometries are involved. One important phenomenon was the development of several pools between cross-overs as the path length increased by meander growth. Hickin (1969) and Lewin (1976) have noted other variations on the theme of the initiation of the point bar in a straight channel.

Although there is still some debate about the precise type of circulation in bends once they are formed (see Chapter 3), the general depositional mechanism has been described by Allen (1970a). His description is followed here.

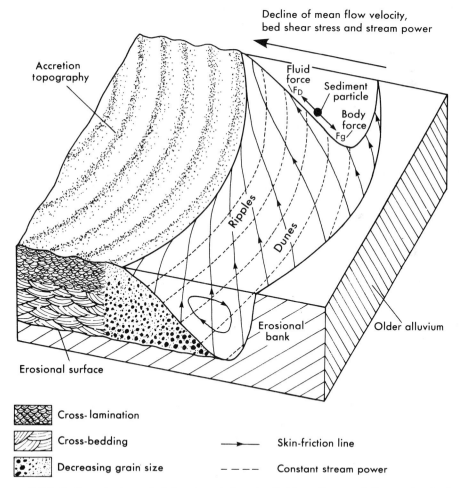

Figure 7.24 Sediment characteristics in relation to fluid motion and the major hydraulic variables in a river bend (after Allen, 1970a).

The flow of water in the bend gives rise to an excess of fluid pressure on the outer bank and a deficit on the inner bank (Figure 7.24). In the water a radial-pressure gradient is thus set up in balance with the centrifugal-pressure gradient. At the bed this radial-pressure gradient is imbalanced by frictional losses in the flow, and so water moves inwards over the bed towards the inner bank, down the radial-pressure gradient. This in turn requires compensatory flow from the inner to the outer bank, giving rise to the helical path. In the bend the local velocity and mean velocity all decrease from the outer to the inner bank. The particles of different sizes should then assume a radial position on the sloping bank so that the transverse up-slope component of the fluid force acting on the particles exactly balances the transverse downslope component of gravitational force. In this case the particles tend to travel parallel to the channel banks even though the flow is helical.

This model is similar to the analysis for the threshold concept, except of course that the bed is 'live'. The model has recently been 'broadly supported' by field observations made by Bridge and Jarvis (1976), and Bridge (1976, 1977) has outlined and modelled the interaction between flow and sediment in channel bends, extending the earlier work of Engelund (1974a, 1974b).

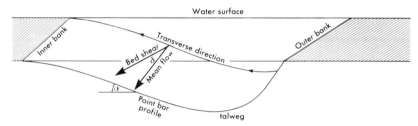

Figure 7.25 Terminology of the Engelund model of equilibrium form in a channel bend (Bridge, 1976).

In Engelund's model (Figure 7.25) helical flow causes the bed shear-stress vector to deviate by an angle δ from the mean flow direction. For the equilibrium in the channel bend, outlined earlier, in the direction of flow

$$F_D \cos \delta = (W - F_L) \tan \phi \cos \alpha \qquad (7.37)$$

where F_D is the drag force on a single particle in the direction of shear, W is the submerged weight of the particle, F_L the lift force, $\tan \phi$ the dynamic friction coefficient due to mutual collision between the grains and α the transverse slope of the point-bar surface. The mean bedslope in the downstream direction is assumed to be negligible. In the transverse direction the balance of forces is given as

$$F_D \sin \delta = (W - F_L) \sin \alpha \qquad (7.38)$$

This yields, for equilibrium (particle moving parallel to flow):

$$\tan \delta = \tan \alpha / \tan \phi \qquad (7.39)$$

Rozovskii (1961) showed theoretically that $\tan \delta = 11 y/r$ where y and r are the local flow depth and radius of curvature respectively. In combination with equation 7.39 and integrating, this yields

$$y = y_0 \left(\frac{r}{r_0} \right)^{11 \tan \phi} \qquad (7.40)$$

where y_0 and r_0 are reference depths and radius of curvature respectively. If the reference depth is the maximum water depth, then equation 7.40 becomes

$$y = h \left(\frac{r}{r_t} \right)^{11 \tan \phi} \qquad (7.41)$$

where r_t is the radius of curvature at the bed. Bridge (1976) also, following Allen (1970b), outlined the theoretical distribution of grain sizes and carried out comparisons of the theoretical profiles with field data. Both were predicted fairly well using these techniques.

Plate XVI Valley train (sandur) of the Hvitá, Iceland, down-valley from the Langjökull ice-cap, showing braiding over gravel surface. (C.E.)

Disequilibrium—braiding and scour

The stability of cross-sectional form and the repeated riffle-and-pool and meander sequences contrast strongly with unstable régimes. The former represent modal states that can be shown to represent minimum energy requirements. Kirkby (1977a) likens these modal states to depressions in an energy surface. The system spends most of its time at or near the bottom of some depression. If, therefore, a series of systems is observed at one time, most will lie in one or other of these depressions, which will be described as the modal states of the system.

With high slopes, high sediment concentrations and/or high discharges, the channel conditions shift from the stable forms outlined earlier to braided forms in which cross-sections include many channels, separated by islands, whose position is subject to change over quite short periods of time (Plate XVI). Leopold and Wolman in 1957 empirically described a discriminant function distinguishing between meandering and braided habits. Another, more recent version of the same idea is given in Figure 7.26 after Schumm and Beathard (1976).

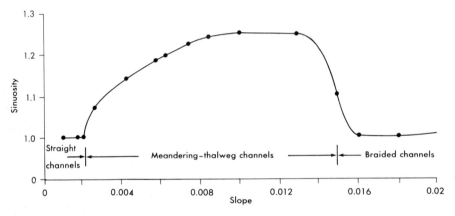

Figure 7.26 Relation of sinuosity to valley slope in an experimental flume (Schumm and Beathard, 1976).

Leopold, Wolman and Miller (1964) give a detailed description of the formation of a bar in a flume. A channel was moulded in moist uncemented sand, and water and poorly sorted debris was allowed to flow into it. After 3 hours a small deposit of grains, somewhat coarser than average, accumulated on the bed in the centre of the channel. This, they suggest, represented a lag deposit of the coarser fraction that could not be carried by the flow. The principal bed-load transport occurred over this bar, whereas grain movements in the deeper parts of the channel adjacent to the central bar were negligible at this stage. The central bar continued to grow until grains moving across it were actually breaking the surface of the flow. The flanking channels were deepened and then scoured causing the central bar to emerge. In rivers, vegetation on such a central island leads to trapping of fine material and stabilization by silting.

Several authors have described lateral bars whose growth is rather different from the sequence described above. These are essentially units of sediment attached to the bank and comprise a steep face of gravel ('riffle face' in the terminology of Bluck, 1974) which crosses

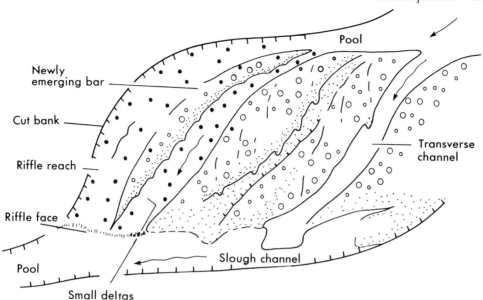

Figure 7.27 Braided channel forms in a riffle reach (Bluck, 1976).

Plate XVII Extensive boulder sedimentation and accompanying bar features in the Lower Guadelfeo river, southern Spain, resulting from the extreme flood of October 18th–19th, 1973. (J.B.T.)

the adjacent channel to cut the opposite bank at an acute angle. A shallow 'riffle reach' (Figure 7.27) is shallow and comparatively steep, and usually has water in it in a state of super-critical flow with standing waves and coarse deposition. Below the steep 'riffle face' at the end of the 'riffle reach', deltas are constructed. Gradual migration of the 'riffle reach' leads to construction of a bar platform above this and generally well above river level at its normal low stage. This is overrun and dissected at high stages of flow by channels that run obliquely to the bar. Repeated shifting of the channels often leads to dissection of older bars (Plate XVII) and rapidly fluctuating changes in stage, such as are common in semi-arid and glacial meltwater streams, contribute to the instability of the transport régime and the erosion of the banks.

The role of vegetation is crucial in these otherwise highly mobile channels. Where vegetative growth has already occurred, underground roots from dense growth of grass and shrubs reinforce the bank sediments and thus protect the channel banks. In aggrading river conditions, as Smith (1976) points out, vegetation roots accumulate rapidly and decay slowly, thus affording protection to the banks from erosion in the deeper parts of the channel. Smith found that bank sediment with 16–18 per cent by volume of roots and a 5 cm root mat for bank protection had 20 000 times more resistance to erosion than comparable bank sediment without vegetation. Nevins (1969) reports on how the Turandui River, New Zealand, with braided channels in shingle was changed over a period of years into a single meandering channel by planting willow shrubs at appropriate channel bends.

There remains debate on the role of scour and fill during rare events producing disequilibrium. Lane and Borland (1954) expressed the view that the role of scour had, relatively speaking, been much exaggerated, by the tendency to report mainly from narrow-gauged reaches where erosion was in any case likely to be intense. They computed the amount of sediment represented by the observed average depth of scour multiplied by the width and length of the alluvial reaches of the middle Rio Grande valley. This amount was much larger than that being deposited downstream (in the Elephant Butte reservoir) and they concluded that fill was taking place in many of the wider, unmeasured reaches.

Leopold, Wolman and Miller (1964), however, reported simultaneous scour over pools and bars and net losses during periods of high flow. They suggest that the actual process of scour involves a dilation of the bed down to the depth of scour. The grains involved in this dilation are in motion downstream but at a rate which, on average, is much smaller than the rate of movement of the water.

That there are specific loci of scour is difficult to dispute. Mosley (1976) describes the mechanics of converging flows where tributaries join the main channel, and outlines the expected distribution of scour at the junction. The field evidence for this in the rivers he examined appears to fit the model quite well, though there are many cases that do not. One problem is the concealment of scour holes by infilling with smaller material on the recession limbs. Neill (1976), investigating scour holes along the Athabasca River, found that the scour holes were located on bends and that they had been maintained over a long period of time in more or less the same location whereas the rest of the river bed (over a distance of 60 km) had undergone little change.

Unfortunately space does not allow further discussion of the many processes in river channels. Among these the biggest omission is the development of river flood plains. This question is especially important in terms of the analysis of Pleistocene and Holocene river-channel histories. Over-bank deposition may also be relatively important in lowland rivers in the humid tropics (Plate XVIII) (e.g. Blake and Ollier, 1971). The other major omission is a discussion of transport of dissolved loads. The basic principles involved here are outlined in Chapter 4.

Plate XVIII Horseshoe-shaped bars in the lower Rio Negro, near Manaus, Brazil. The forms are believed to represent the progressive change in channel conditions resulting from changes in climate during the late Quaternary. However, contemporary erosion and accretion in this environment is also intimately linked to the rate and character of vegetation succession on the islands. The channel section shown is about 20 km wide. (J.B.T.)

Hillslope responses to running water

Hillslopes respond to running (and falling) water by means of a complex group of processes generally classified under the heading of soil erosion. The latter includes, of course, the effects of wind, but these are considered separately in Chapter 10. Essentially, soil erosion involves detaching soil particles and transporting them downslope through the action of raindrop impact and runoff. The rate of erosion depends on the rain's erosiveness, the soil's erodibility, the steepness and length of slope, the cultural practices used, the stage of crop growth and the supporting conservation practices applied to the land. These various terms are expressed in the Universal Soil Loss Equation.

Ellison (1947) defined four sub-processes covering entrainment and transport: (i) detachment by rainfall; (ii) detachment by runoff; (iii) transport by rainfall, and (iv) transport by runoff.

Rain splash

The mechanics of detachment by rain splash (Plate XIX) have been investigated in detail by Smith and Wischmeir (1962), Mosley (1973), and Mutchler and Young (1975). The place of raindrop impact in water erosion began to emerge in the 1930s, when it was observed that the concentration of soil in runoff water was found to increase rapidly with raindrop energy. This in turn is related to rainfall intensities and several expressions for rainfall energy in terms of intensities have been obtained. Some of the relationships are shown in Figure 7.4. As a splash hits the ground it forms a small crater; the water bouncing back forms a parabolic curve which may be up to four times its height of rise. On level land the splashing particles tend to bounce back and forth, so there is no net loss of soil from any point on the field. But on a slope, the splashes move the soil downhill. Part of this movement is caused by the drops striking glancing blows that kick most of the particles to the bottom of the slope. Another part is caused by the fact that soil splashed in the downslope direction travels

Plate XIX Small pinnacles protected by large flat particles from the effects of rain splash. (J.B.T.)

farther in the air than that splashed uphill. On a 10 per cent slope the downhill movement is about three times that occurring in an uphill direction. The height and distance of splash depend upon the soil surface condition and the fall velocity of the drop. Milhara (1959) reported a maximum splash distance of 94 cm and a maximum height of 30 cm, and Ellison (1947) reported 4 mm stone fragments to have been splashed 20 cm.

Mosley (1973) showed that the amount (by weight) transported by splash to different distances decays exponentially and that 'the total weight of splashed material increased with

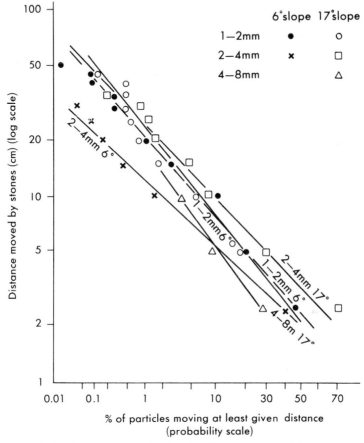

Figure 7.28 Cumulative frequency distribution for distances moved by small stones on 6° and 17° slopes in a 16 mm rainfall (Kirkby and Kirkby, 1974).

slope inclination', more sand being splashed downslope as the angle increased, giving an increase in the net downslope movement with increasing slope. The significance of rain splash under conditions of overland flow varies with the depth of flow; Mutchler and Young (1975) suggested that a film of water of about three drop diameters in depth effectively protects the soil from raindrop impact. In addition, the impact of raindrops enhanced the transporting capacity of flowing water and their erosive potential was greatest where a very thin layer of water is present. They concluded (i) that raindrop energy rather than surface-flow energy was the major force initiating soil detachment on rill areas, (ii) that detachment and transport from non-rilled areas is by raindrop impact but that (iii) splash *alone* accounts for a very small proportion of the movement into the rills and that, finally (iv) raindrop impact was the driving force transporting soil in thin surface flows to rills.

Kirkby and Kirkby (1974) observe that rain-splash transport distances are probably best described probabilistically and that, for particles in the range 1–8 mm, the distribution is approximately log normal (Fig. 7.28). Within the range there is an 'effective grain size' that is carried most effectively; this they found to vary from about 2 mm on a 3° slope to about 12 mm on a 12° slope. In attempting to model the process, Meyer and Wischmeir (1969) differentiate the two processes of splash detachment and transport. They argue that splash detachment is proportional to the square of rainfall intensity. For moderate to intense storms, kinetic energy per unit of rain per unit of area varies approximately as $I^{0.14}$, where I is intensity. For steady-state conditions, rain amount per unit area is proportional to I, so total energy E is proportional to $I^{1.14}$. For a uniform rainfall intensity, the EI (erosion index) parameter is then approximately proportional to $I^{2.14}$. The transporting capacity due to splash is a function of slope steepness, amount of rain, soil properties, microrelief and wind velocity, but they chose the simplified expression $T_R = S_{TR}s.I$ in which T_R is transport due to splash, S_{TR} a coefficient, and s the angle of slope.

Detachment by running water

Detachment of soil by running water is essentially the same problem as the initiation of sediment transport in alluvial channels except that (i) the width is usually great compared with depth; (ii) the roughness elements are, relatively speaking, much larger, particularly vegetation, and (iii) the effects of ground infiltration are more significant, approaching the situation found in ephemeral channels.

The Horton model of overland flow provided the theoretical basis of sheet erosion for many years and is still thought to be applicable in some situations (Pearce, 1976). Recall that the Manning equation can be expressed as

$$v = \frac{r^{2/3}s^{1/2}}{n} \qquad \text{(3.9, p. 48)}$$

where r is hydraulic radius, s is slope and n the Manning roughness coefficient. Since $v = Q/dw$, and q is the discharge per unit width, then $v = q/r$ and so

$$q = \frac{r^{5/3}s^{1/2}}{n} \qquad \text{(7.42)}$$

and therefore

$$r = \left[\frac{q.n}{s^{1/2}} \right]^{3/5} \qquad \text{(7.43)}$$

With a very wide channel (a slope plane), the hydraulic radius is practically the same as the depth, so equation 7.4 expresses the depth of flow as a function of the discharge, the roughness and the slope. In this simple analysis, however, many other factors have to be borne in mind. For example Parsons (1949) found that raindrop impact increased depths from 8 to 28 per cent over theoretical depth, the average increase being 17 per cent. Horton

(1945), formulating discharge as a function of rainfall excess (intensity minus infiltration capacity) and the length over which flow occurred, attempted to delimit a critical belt of no erosion. This was the zone over which the tractive force of the flow was insufficient to overcome the resistance of the soil and the vegetative cover. In the model, the resistance is assumed to be constant (τ_0). In other words, with γ_w as the specific weight of water and d as flow depth:

$$\gamma_w \, d s = \tau_0 \tag{7.44}$$

This leads to the expression of a critical length of overland flow (x_c) for a soil of known resistance given as

$$x_c = \frac{1}{nI}\left(\frac{\tau_0}{\gamma_w s}\right)^{5/3} \tag{7.45}$$

where I is the rainfall intensity. Carson and Kirkby (1972) express slope as $\sin^{0.7} \beta \cos^{0.4} \beta$, where β is slope angle, which is close to the Horton expression $s = (\sin \beta / \tan^{0.3} \beta)$.

Using a slightly different approach, Pearce (1976) expresses the erosive potential as stream power following the ideas of Bagnold (1966). Stream power (W) is then

$$W = \tau_0 V = \gamma_w \, d s v = \gamma_w q s \tag{7.46}$$

In his measurements of stream power (energy expenditure), Pearce (1976) found values ranging from 30 mW m^{-2} at a runoff intensity of 80 mm hr^{-1} to 0·2 mW m^{-2} at a runoff intensity of 0·5 mm hr^{-1}. These values are very low. With an assumed n of 0·35, the boundary shear stresses were found to be in the range of 0·1—2·0 N m^{-2} which are well within the range of stresses found by Parthenaides (1971) to cause erosion of a 50 per cent clay, 50 per cent silt mixture. Increasing the value of n to 0·5 increases the boundary shear stress by about 25 per cent; decreasing it to 0·2 decreases it by about 30 per cent. Total sediment loss and mean surface lowering were found to be highly correlated with both total rainfall kinetic energy and total runoff kinetic energy, but efficiency of energy use was found to be low. In the magnitude and frequency relationships, the greatest amount of erosion was carried out by the moderate intensity (5–15 mm hr^{-1}) runoff events of 1–6 hr duration, with return periods of about one year.

Several other workers have generally expressed overland-flow discharge as a function of roughness, slope and depth of flow in the general form

$$q = c d^b \tag{7.47}$$

where $c =$ a function of slope and roughness with fixed parameters. Chow (1959) and Carson and Kirkby (1972) express the exponent b as a function of the depth of flow relative to the grain diameters which control surface roughness. They find the range of values for b to lie between 1·62 for deep flows and 2·5 for shallow flows. Experimental results range from 1·0 to 3·7, the lowest values being for flow through vegetation. Whereas Pearce (1976) uses the Du Boys-type equation to determine stream power, Carson and Kirkby (1972) use simplified versions of a suspended-sediment-load equation to obtain an expression for sediment discharge, and this takes the general form

$$\Theta \propto q^L K^M s^P \tag{7.48}$$

where Θ is sediment discharge for a particular value of the overland flow (q), grain roughness (K) and slope (s), with L, M and P as constants. This result may be compared with others of a similar type, for example the equation that Meyer (1971) gives for the transport due to runoff (T_Q)

$$T_Q = C\,q^{5/3}\,s^{5/3} \tag{7.49}$$

where C is a constant.

In recent years, doubts have been expressed concerning some of Horton's ideas. The development of runoff on slopes above the channel as a result of throughflow and saturation has already been discussed. The question has been raised, however, as to whether discharge will continue to increase with distance from the divide, as Horton and others have suggested. Yair (1973) found experimentally that volume of runoff and weight of sediment load collected (in desert areas) were highest on the divides that belong to the Horton belt of no erosion, and Yair and Klein (1973) in a similar environment found an inverse relationship between slope angle and slope erosion due to spatial variations in the texture of the surface materials. Likewise, Nortcliff (1976) found virtually no correlation between depth of regolith and distance from the divide in a semi-arid area of Spain. Finally, de Ploey and Moeyersons (1975) argued, from a small experimental soil-erosion flume, that discharge would quickly reach a constant value which depends on the inflow to the slope section and the input of rainfall because infiltration into the bed is a function of total discharge. There is much speculation in this paper, but the argument is very similar to that applied to semi-arid channels. For example, Murphy, Diskin and Lane (1974) demonstrate a high correlation between transmission losses and the volume of the inflow hydrograph. The function is linear, of the form

$$f = kV + C \tag{7.50}$$

where V is the inflow volume, f the infiltration and C a constant. The implication of de Ploey and Moeyersons' work is that a limiting discharge occurs because below it, infiltration decreases and above it, infiltration increases.

Another source of problems with the general thesis is that it tends to deal with undivided (i.e. unrilled) flow. The relative role of rill-work and sheet-flow depends on the extent of the inter-rill areas. Foster and Meyer (1975) argue that the relative importance of these processes depends on whether one is dealing with inter-rill areas, where flow is unconcentrated, or rill areas where flow is essentially concentrated into channels.

A third major problem in explaining transport on slopes in terms of discharge alone is that it fails to take adequate account of the longer-term effects of the concentration of coarse debris to produce 'armouring' of the slope. Coarse debris protects the soil immediately beneath it from splash but tends to increase the volume, the turbulence and the erosive power of the flow on the uncovered patches. Wash tends to be slightly increased on the downslope sides of most of the pebbles as long as the Reynolds number does not exceed a certain limiting value, depending on such factors as runoff volume and the nature of the subsoil. According to de Ploey, Savat and Moeyersons (1976), when the critical value of the Reynolds number is exceeded, wash is accelerated by vortex erosion on the upslope sides of the pebbles. This can cause the particles to move upslope. For laminar flow, the critical R value is about 50, for turbulent flow about 3000, on sands. On clays and loams, slope wash is concentrated along scar lines where the eroding force of the runoff is enhanced by heavy turbulence and a sort of plunge-pool effect. This process they call *runoff creep*. Thornes (1975) has described similar processes in coarse materials on semi-arid debris slopes in Spain. These processes lead to the production of fairly coarse debris sheets.

Very little has so far been said about the effects of vegetation. In the Universal Soil Loss Equation, this was recognized by a 'cropping factor'. The role of vegetation is essentially that of the interception of rainfall, the decrease in velocity of the running water, the effects of roots on the granulation and porosity of the soil and the use of moisture by transpiration. These effects have been widely observed experimentally (Smith and Wischmeir, 1962) but, to a large extent, have been modelled principally through the roughness coefficient. This is clearly an area where much more thought must be given to modelling the processes. Another area that has been largely neglected is the influence of soil biota in preparing material for transport by slope wash, and the actual erosion by splash of mounds built by animals (Imeson and Kwaad, 1976).

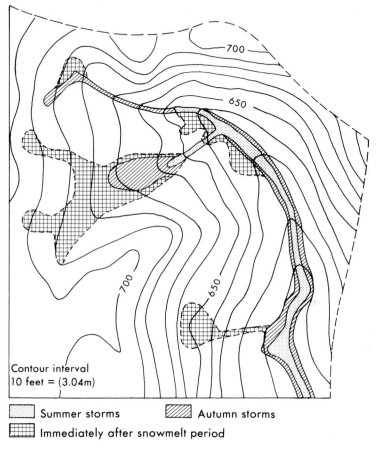

Figure 7.29 Seasonal variations in pre-storm saturated area for a small catchment with steep, well-drained hillsides and a confined valley bottom (Dunne, 1978).

The procedures discussed so far regard steady flows as a function of depth and slope in a downslope direction. Recently, as mentioned earlier, attention has been paid to the plan form as well as to the downslope physiographic characteristics. Plan-form characteristics are important for they control spatial variations in the intensity of operation of the hillslope processes outlined above, and even determine the likelihood or otherwise of the actual operation of the processes. Thus, for example, the likelihood that overland flow will occur is in part a function of the soil moisture content. A map of the change of soil moisture through

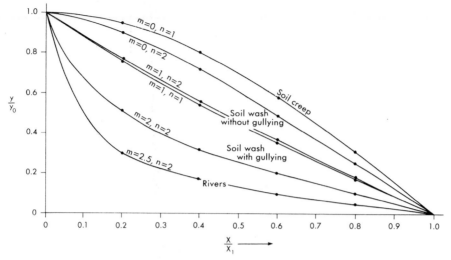

Figure 7.30 Dimensionless graph showing approximate characteristic-form profiles for a range of processes according to Kirkby's analysis (Kirkby, 1970).

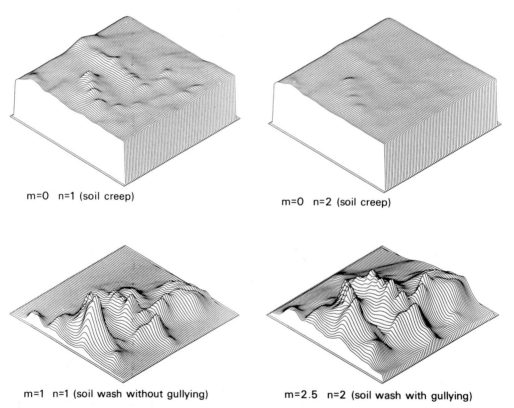

m=0   n=1 (soil creep)   m=0   n=2 (soil creep)

m=1   n=1 (soil wash without gullying)   m=2.5   n=2 (soil wash with gullying)

Plate XX Three-dimensional computer simulation of soil erosion under different process laws, with parameters as in Table 7.4.

an individual storm (Figure 7.29) reveals the area in which overland flow is likely to occur. Empirical evaluations of the effects of spatial topographic variations are also reported in the literature (Young and Mutchler, 1969).

Three models of hillslope erosion by running water

In the final part of this section, we outline three attempts to model the effects of the operation of the processes described above. They represent attempts to model slope change (in two dimensions) over three different time-spans. The model of Kirkby (1971) is long-term

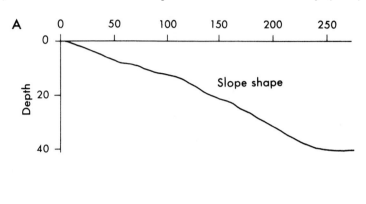

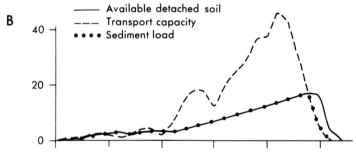

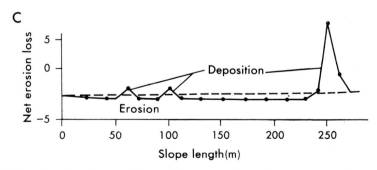

Figure 7.31 Simulated erosion and deposition on an irregular slope according to the model of Meyer and Wischmeir (1969).

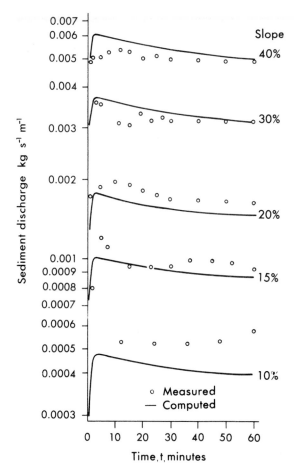

Figure 7.32 Effects of slope angle on time-dependent sediment discharges for an experimental situation, according to a computer model (Li, Simons and Carder, 1976).

(several thousand years?); that of Meyer and Wischmeir (1969) for the intermediate term; and that of Li, Simons and Stevens (1976) short-term. They represent the development of corresponding equilibrium, steady and unsteady forms.

Kirkby's model starts with a continuity equation expressing the height loss from a small section of slope as a result of input and output of material and weathering. A further continuity equation defines the increase in soil depth as the increase of elevation of the land surface and the reduction of the elevation of bedrock. These are combined with a transport law and solved in an approximate analytical fashion to yield a characteristic form that is determined by the transport law and not by the initial conditions. Each profile approaches a form $y = y_0 \cdot f(x)$ in which the lowering through time is applied equally over the profile so that only y_0 is time dependent. The essential process law for wash processes is defined as

$$\Theta \propto x^n s^m \tag{7.51}$$

in which Θ is the rate of sediment removal and n and m are the exponents of distance and

Table 7.4 Variation of exponents m and n for Kirkby's transport limited case when the process law is of the form $\Theta \propto a^m \cdot (\text{slope})^n$ when Θ is sediment load per unit width of flow and a is area drained per unit width of flow

Process	m	n
Soil creep	0	1·0
Rain splash	0	1–2
Soil wash	1·3–1·7	1·3–2
Rivers	2–3	3

slope. A set of values for m and n is given in Table 7.4 and the corresponding profiles are shown in Figure 7.30 (see also Plate XX).

A digital simulation approach was used by Meyer and Wischmeir (1969) to model soil erosion. They define the four terms

$$D_R = C_1 I^2 \qquad \text{where } I = \text{intensity}, \ D_R = \text{rainfall detachment}$$
$$D_F = C_2 s^{2/3} Q^{2/3} \ \text{where } Q \text{ is the runoff, } s \text{ the slope}$$
$$T_R = C_3 s I$$
$$T_F = C_4 s^{5/3} Q^{5/3}$$

D_R and D_F are detachment due to rainfall and flow; T_R and T_F are transport capacities of rainfall and flow; and C_1, C_2, C_3, C_4 are constants. The model is evaluated by iterative techniques using a digital computer. Starting at the top of the slope, the four sub-processes are calculated for each slope-length increment. If the total quantity of soil detached is less than the transporting capacity, all the material is carried downslope. If the transporting capacity is *less* than the volume of soil detached, sediment load is equal to transport capacity.

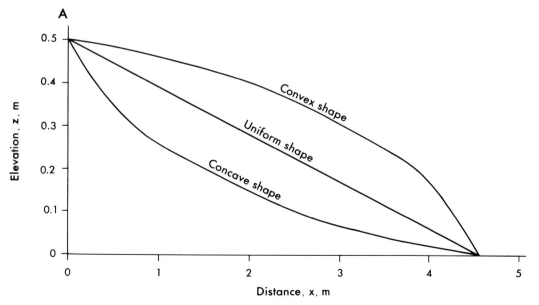

Figure 7.33 Effects of slope shape on time-dependent sediment discharges for an experimental simulation. A: Slope shape types; B: *(see p. 270)* (Li, Simons and Carder, 1976).

Unfortunately, values for the four constants were not available and had to be intelligently guessed. Some results of the applications of these models are shown in Figure 7.31. A more sophisticated attempt to obtain a closed-form analytical solution, again based on the relationship between capacity and detached load, was published by Foster and Meyer (1972), and Armstrong (1977) has attempted a three-dimensional computer simulation along similar lines. The Meyer–Wischmeir model is based on assumed rainfall intensities for individual events, but because of the need to iterate a large number of times to obtain meaningful results it should be thought of as an intermediate-term model. It relies on the assumptions of the Horton type of runoff model whereas, in fact, the essential character of soil erosion at the level of individual storm events is that it is controlled essentially by unsteady flow.

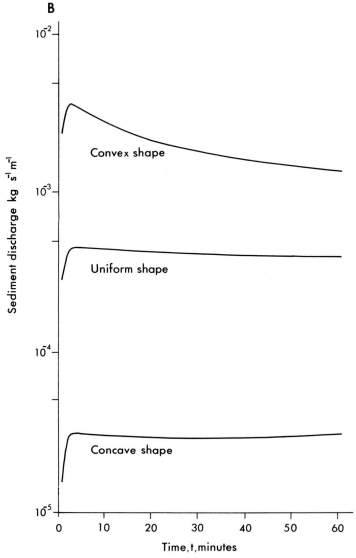

Figure 7.33B: experimental simulation (Li, Simons and Carder, 1976).

Li, Simons and Carder (1976) have attempted to use a finite-differencing approach to the problem and simulate the sediment outflow hydrographs over a sandy soil. In particular, their model draws attention to the effects of slope shape, especially since overland flow is routed kinematically (see p. 228). They assume that slopes are less than 25 per cent and that the effects of raindrop impact are negligible. The water is routed down the slope using a non-linear kinematic wave scheme and to these discharges a sediment-transport law is coupled. The following equation is involved:

$$q_b = B_1 (\tau - \tau_c)^{B_2} \tag{7.52}$$

a typical Du Boys-type equation, with τ the boundary shear-stress determined by $\tau = \gamma y s_0$ with γ the specific weight of water, s_0 the ground slope and y the flow depth. τ_c is the critical shear stress; it is given by $\tau_c = 0.047 \, (s_g - 1) \, D_{50}$ in which s_g is the specific gravity of the sediment. D_{50} is a function of the size of sediment on the bed, representing the value at which 50 per cent of the sediment is finer by weight. Suspended load is defined by a complex differential equation related essentially to the settling velocity of the particles and the shear velocity (*cf.* p. 242). With the passage of a water wave, the occurrence of erosion or deposition, defined by the sediment-continuity equation and these transport laws, is determined, subject to fixed upper and lower boundaries. Iterations in this model are based on one-minute intervals. Li, Simons and Carder found that the model operated successfully except at low rainfall intensities. The modelled and observed results are shown in Figure 7.32 and the effects of slope and time in Figure 7.33. Erosion rates on convex slopes were found to be nearly five times greater than those on a uniform slope, whereas the erosion rates on concave slopes were much *less* than those on rectilinear slopes (Figure 7.33). They observed that on concave slopes, rate of slope erosion actually increased slightly with time. This is similar to the results observed by Meyer, Foster and Romkens (1975).

8

Glacial processes

Clifford Embleton

Within the last one or two million years, 32 per cent of the Earth's land surface has been glacierized. Although two-thirds of the bulk of the Pleistocene ice has now melted away, evidence of its erosive and depositional activity dominates the present landscapes of northern Europe and more than half of North America as well as other smaller areas.

The temperature of an ice mass plays a fundamental part in its movement and morphological activity. Two basic types of glacier may be distinguished: temperate (sometimes termed 'warm') and cold.

Temperate glaciers are those in which the ice temperature is approximately at pressure melting point throughout their thickness, except for the uppermost few metres which may be temporarily chilled to lower temperatures each winter. Below the depth to which any winter cold wave penetrates, temperatures will decrease because of the increasing pressure of ice overburden, at about $6 \times 10^{-4}\,°Cm^{-1}$ (Shreve and Sharp, 1970). Owing to this reversed temperature gradient, geothermal heat reaching the base of the glacier, and any heat generated by glacier movement, cannot be dissipated upwards but must be used in melting the lower surface of the glacier. Thus meltwater exists in varying quantities at the bases of temperate glaciers, and, as shown in Chapter 3, is vital in the basal sliding mechanism.

Cold glaciers are characterized by temperatures below pressure melting point throughout their thickness. A normal temperature gradient (that is, temperature increases from the surface downwards) will permit the upward escape of geothermal and frictional heat, so that cold glaciers are normally frozen to bedrock. This in turn means that basal sliding at the ice–rock interface cannot occur, though there will be rapid shearing in the lowermost layer of ice. Ahlmann (1935) distinguished between two types of cold glacier, *high polar* and *sub-polar*. In the former, ice and air temperatures remain below freezing point even in summer and melting is virtually absent except where rock outcrops absorb solar heat. In the sub-polar type, surface melting occurs in a short summer ablation season, and marginal and supra-glacial streams temporarily appear. Summer meltwater will, however, be confined to the surface or shallow depths, below which impermeable cold ice will prevent it penetrating to the glacier base. Exemplifying high polar characteristics, Figure 8.1A shows the temperature profile of the Camp Century borehole in north-west Greenland. From the surface where the temperature was $-24°C$, the temperature rose to $-13°C$ at the bedrock contact at a depth of 1387 m (Budd, Jenssen and Radok, 1971). The temperature distribution in a glacier depends on several factors:

1 surface temperatures and their fluctuations both seasonally and over the longer term,
2 the thickness of the ice,
3 its conductivity,
4 the rate of geothermal heat flow at the base of the ice, and
5 heat generated by sliding, shearing or deformation of the ice.

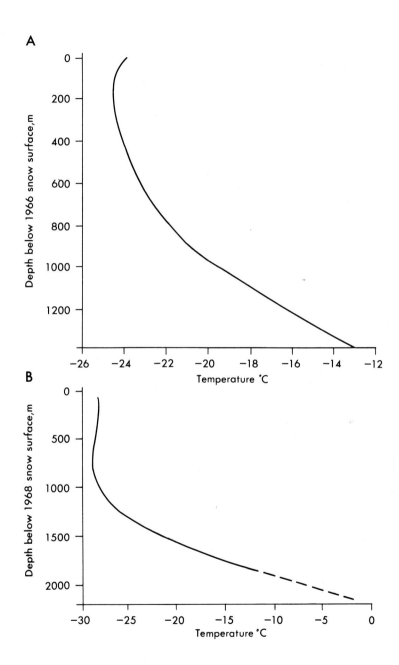

Figure 8.1A: Temperature profile for the Camp Century borehole, Greenland.

B: Temperature profile for the Byrd Station borehole, Antarctica.

Because of the great variability of each of these factors (except 3), the thermal régimes of actual glaciers vary widely, and many glaciers do not fall neatly into one or other of the two main categories, temperate and cold. If, for instance, we could imagine the ice at Camp Century being roughly twice as thick as it is now, and holding the other factors constant, the temperature at the base of this cold-ice profile would be close to pressure melting point. This is, in fact, the general situation in the case of the Byrd Station drill hole, Antarctica (Figure 8.1B) where basal melting beneath 2164 m of cold ice was revealed in the meltwater that rose 60 m up the newly-completed drill hole. There are also glaciers where thicknesses of temperate ice underlie cold ice. Recent data discussed by Robin *et al.* (1969) suggest that, in parts of Greenland, there may be a basal layer of temperate ice 100–200 m thick beneath cold ice. The accumulation zones of some sub-polar glaciers consist of cold ice in their upper layers, overlying temperate ice below, but in the ablation zones of the same glaciers where the ice is thinner, the whole thickness of the ice may be cold and frozen to bedrock. It is also possible for a glacier to be completely cold in a high-altitude accumulation area and

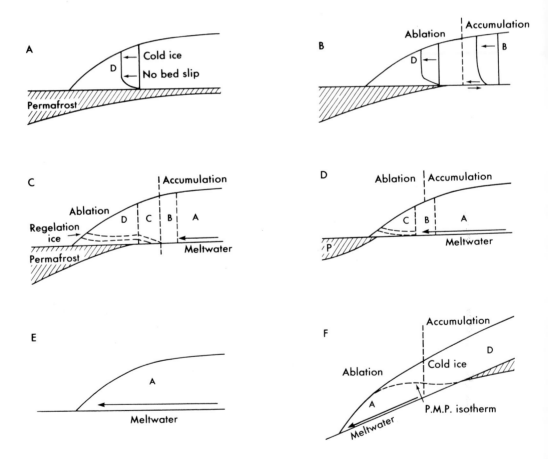

Figure 8.2 Idealized examples of thermal zonation in glaciers and icesheets (for explanation and discussion, see text) (after Boulton, 1972a).

to flow to much lower levels with a more temperate climate where the glacier may warm up to pressure melting point throughout. Figure 8.2, after Boulton (1972a), summarizes some of these and other possible thermal régimes. The implications for processes of erosion, transport and sedimentation are fundamental and will be examined in subsequent sections of this chapter.

Erosion processes

There is now general agreement that, under certain circumstances, moving ice can be an effective instrument of erosion of the underlying bedrock. The evidence is partly morphological, based on studies of glaciated bedrock surfaces and the assemblages of landforms unique to glaciated regions; partly sedimentological, in that the vast bulk of glacial deposits, particularly till, can only be the product of glacial and glacifluvial erosion of new rock; it is partly based on studies of rates of erosion by present-day glaciers; and there is evidence from laboratory and sub-glacial studies of rock wear by moving ice. Processes of abrasion which scratch, scour or polish bedrock surfaces can be usefully differentiated from quarrying processes in which fragments are broken off or removed from the bedrock mass; a third category is that of meltwater erosion which, although no different in principle from fluvial erosion processes considered in Chapter 7, is so closely associated with the sub-glacial erosional activity of temperate glaciers that its relative contribution must be discussed here.

Abrasion

The micro-relief forms of glacially abraded surfaces provide some of the most persuasive evidence in favour of rock wear by moving ice. These forms include scratch marks parallel with the direction of ice motion (Plate XXI), ranging from the fine almost microscopic markings on polished surfaces, through striations of a millimetre or few millimetres in depth. Individual striations and shallow grooves are more persistent. The giant grooves described by Smith (1948) in northwest Canada attain lengths of as much as 12 km, with depths up to 30 m and widths approaching 100 m. Friction cracks represent another family of micro-relief features on glaciated rock surfaces. Harris (1943) proposed the term 'friction crack' to cover a variety of rock-fracture features, including gouges from which crescentic chips of rock have been removed, and semi-vertical fractures with crescentic outcrops. Other forms include chattermarks and conchoidal fractures. Some of these features are not strictly the result of abrasion and will be considered separately under the heading of crushing and rock fracturing on page 281. Another group of micro-features, even less easily classified and described, falls under the heading of 'p-forms' (Dahl, 1965). This category covers complexly moulded (Dahl 'plastically-moulded') hard rock surfaces, often smoothly rounded and partly polished. Round-edged grooves and bowls, *sichelwannen* (small trough forms of sickle-shape) and convexly rounded bosses intermingle with more sharply cut channels, the whole assemblage often being additionally marked by striations and friction cracks. Although some elements are almost certainly the result of fluvioglacial erosion (Dahl, 1965), others are closely related to ice movement and glacial abrasion processes.

Abrasion processes have also been invoked in part to explain the modelling of larger-scale forms. The stoss (up-glacier) sides of roches moutonnées, for instance, usually carry all the hallmarks of glacial abrasion and fracture in the form of striations, friction cracks and other micro-relief, and the features as a whole, considering the regularity of their distribution and orientation, and their interdependence of lithological variation, strongly suggest glacial sculpturing. At a still larger scale, the macro-forms of glacial erosion such as cirque basins and glacial troughs are modelled by a combination of processes of which abrasion is one, though not necessarily the most important quantitatively.

Field observations of abrasion processes are necessarily limited because of the difficulty of access to sub-glacial sites. In practice, this means that field data are derived from a few tunnels into glaciers reaching bedrock, from natural sub-glacial cavities, and from studies of processes at the sides of glaciers near to the surface. As long ago as 1843, Forbes observed how the basal ice of Alpine glaciers was armed with rock fragments causing scratching and grinding of bedrock surfaces and producing rock flour by the pulverizing action (p. 47). Other pioneer workers attempted to estimate the wear of rock surfaces after glacial advances

Plate XXI Glacially abraded and moulded rock outcrop near the margin of Breiðamerkurjökull, Iceland. Directon of ice movement from behind the photograph, parallel to the shaft of the hammer. (C.E.)

by studying the reduction in depth of previously engraved marks (e.g. Simony, 1871). Carol (1947) discussed the varying pressure exerted by the ice on basal rock fragments, following his observations in a cavity beneath the Grindelwald glacier, and other workers have shown how debris-laden 'soles' beneath sliding temperate glaciers are the essential agent of bedrock abrasion (e.g. McCall, 1960; Kamb and LaChapelle, 1964). Some more detailed quantitative experiments have been reported by Boulton (1974) (Table 8.1), which show that glacial abrasion can be a significant erosional process. Such a view is reinforced by measurements of the silt loads carried out from under temperate glaciers by meltwater streams. The silt or rock flour seems likely to be almost entirely the product of glacial abrasion, and the quantities carried in suspension can be immense. Østrem (1975) shows how the corresponding erosion rates for a series of small Norwegian glaciers range up to $1 \cdot 18$ mm year^{-1} in the period

Table 8.1   Some experimental measurements of glacial abrasion (Boulton, 1974)

Locality	Average abrasion rates mm yr^{-1}* Marble plate	Basalt plate	Ice thickness (m)	Ice velocity (m yr^{-1})	
Breiðamerkurjökull	3	1	40†	9·6	Plates flush with bedrock surface
Breiðamerkurjökull	3·4	0·9	15	19·5	
Breiðamerkurjökull	3·75	—	32	15·4	
Glacier d'Argentière	up to 36	—	100‡	250	Plates project above bedrock

* Estimate of measurement error = 0·3 mm.
† Probably underestimates actual pressure at bed.
‡ Probably overestimates actual pressure at bed.

1967–71, suggesting that abrasion alone is currently lowering the floors of the glacier basins by about a metre every 1000 years. Other glaciers, larger and more active, may be lowering their beds by two or three times this amount. It should be noted that all these figures should probably be doubled if the bedload of the meltwater streams is taken into account, but such coarse material is derived from non-abrasional processes of fracture and plucking.

What factors control the rates of glacial abrasion?

1   A prerequisite is that there must be rock debris in the basal ice. Observations show that clean temperate ice, rather than abrading, is itself grooved and shaped by the rock with which it is in contact (McCall, 1960). There is debate and uncertainty over the relative quantities of rock debris being carried by different types of glacier at the present day (see, for instance, the discussion between Andrews and Boulton, *Journal of Glaciology*, 1971, 1972). In theory, cold glaciers frozen to bedrock, unless they are fed with debris derived from other areas where the ice is not frozen to bedrock, are likely to be relatively clean in their basal layers, which would indicate lower potential rates of bedrock abrasion. Parts of the Antarctic ice sheet probably fit this description; Warnke (1970) comments on the lack of debris seen in many Antarctic glaciers and icebergs today and suggests that the ice is now playing a relatively protective role, contrary to Linton's (1963) claim for bulk removal by glacial erosion. In the Byrd Station bore-hole, only the lowermost 5 m out of 2164 m contained significant rock debris, mostly small-size fragments and of limited potential usefulness for abrasion. (In this borehole, it should be noted, pressure melting point was reached near the base.) As well as the presence or absence of debris, the nature of the debris must also be considered. Its hardness relative to that of the bedrock is clearly critical, and the shape of individual fragments will affect the type of abrasion. Sharply angular fragments will cut striae, but as the pulverized material accumulates, it may build up to form a 'carpet' over which the fragments move, so that abrasive polishing begins to replace the cutting of striae (Boulton, 1974). The size of the fragments is even more important since this partly controls the pressure that can be exerted on the bedrock. If ice is taken to be a perfectly plastic substance with a 'yield stress' in compression of about 2 kg cm^{-2} at pressure melting point, the maximum downward force that could be applied to a rock fragment with a bearing surface of 100 cm^2 would be 200 kg. A column of ice 22 m high would be sufficient to generate a loading of 2 kg cm^{-2}. In reality, ice is not perfectly plastic and has no constant yield stress: its rate of deformation varies with the level of applied stress, as discussed in Chapter 3. However, there is no doubt that the larger the rock fragments in the glacier sole, the greater the force that they can transmit to individual points on the bedrock beneath. Thus, fine striations are not carved in bedrock simply by grains of matching size embedded in the under-surface of the glacier but by those grains being pressed onto the bedrock by much larger rock fragments in the glacier sole, or by the sharp corners of the larger fragments themselves. The most effective tools of abrasion are undoubtedly large, resistant and angular boulders, provided that the tractive force is large enough to overcome their friction with the bed. The tools need to be renewed as they become worn or fragmented and there are several ways in which new debris can reach the glacier bed. Some may fall down the glacier sides, or reach the bed by way of deep crevasses if the glacier thickness is less than about 30 m. In the case of temperate glaciers, basal melting will be an important mechanism for bringing englacial debris to a sub-glacial position, while for both cold and temperate glaciers, englacial debris if present may encounter bedrock obstacles rising in the path of ice movement.

2   A second major group of controls on abrasion is glaciological: temperature régime, thickness and the speed of movement of the ice. It is now well established that abrasion is minimal beneath cold ice: and in many cold glaciers lacking in basal debris, abrasion may be wholly absent. Since the adhesion of cold ice frozen to a solid surface that is wettable can resist shear stresses of the order of 10 bar, whereas the normal shear stresses at the bases of actively moving glaciers rarely exceed 1·5 bar, basal sliding of cold ice will not occur. Creep of the ice will, however, be possible. On the basis of Glen's flow law, it will be greatest close to the bedrock where the applied shear stress is highest; more rapid deformation of the lower-most layers will also be assisted by the characteristic temperature gradients of cold-ice masses which show increasing temperatures with depth, so that the basal ice is least cold (values of B, the temperature-sensitive constant in Glen's equation, 3.15, will be higher here). Thus, even though the actual base of a cold glacier will be immobile, rates of motion will increase rapidly upwards through the lowermost few centimetres of the ice where shearing is concentrated. Particles of debris contacting the bedrock and yet of sufficient size to be substantially enclosed by this shearing ice may be rolled forward or dragged across the bedrock, so that some abrasion is possible in theory, though quantitatively it will be far less than below actively sliding temperate ice. The moss-covered boulders described by Goldthwait (1956, 1961) from beneath cold ice at Thule, Greenland, show that basal sliding and abrasion are at present negligible here. Andrews (1972) compares rates of erosion for Pleistocene glaciers (? cold ice) in east Baffin Island with those for Pleistocene temperate glaciers in Colorado, based on rates of cirque excavation. Although processes other than abrasion will have operated in addition, the contrast is considerable: 50–200 mm per 1000 years as against 650–1300 mm per 1000 years respectively.

Ice thickness (h) will determine the normal pressure (ρgh) on both bedrock and basal fragments. For temperate glaciers, it is important to consider $(\rho gh-p)$, where p is the basal water pressure. Basal meltwater is also important in facilitating the sliding process and therefore the movement of abrading debris. Frictional drag (F) between the abrading debris and the bedrock will be given by

$$F = A\mu(\rho gh - p) \qquad (8.1)$$

where A is the area of contact and μ is a coefficient of dynamic friction. As effective basal pressure rises, for any given speed of sliding, the abrasion rate will also rise initially (Figure 8.3), but as Boulton (1974) shows, there will come a point where friction (F) between particle and bedrock will retard particle movement – the ice will flow over the particle instead of dragging the particle along. Thus in the sector of the graph shown as zone B (Figure 8.3), abrasion rates will decrease with further increase of effective basal pressure, until the condi-

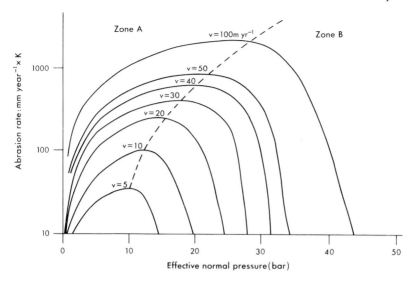

Figure 8.3 Theoretical abrasion rates plotted against effective basal pressure for different ice velocities (Boulton, 1974). On the vertical axis, values of *K* will depend on the relative hardness of abrading fragments and bedrock and on the amount of debris in traction.

tion is reached where there is no abrasion and deposition or lodgement ensues. This concept is important, suggesting a continuum between erosion and deposition, and the depositional implications will be discussed in greater detail later in the chapter. At present, it should be noted that the ideas still need testing, and that field data to enable this to be done are still wholly inadequate.

The speed of glacier sliding enters into the abrasion process mainly because it controls the amount of abrasive debris passing over a given area of bedrock. For cold ice, as described already, the amount of abrasion will be small and may be zero, even if the basal ice contains debris. Temperate glaciers, on the other hand, may exhibit basal sliding rates that represent the greater part of total ice motion: for example, 87 per cent on one cross-section of Athabasca glacier (Raymond, 1971) and 65 per cent below the margin of Østerdalsisen (Theakstone, 1967).

3 There are, thirdly, the characteristics of the bedrock itself to be taken into account. Lithology of the bed must be compared with that of the abrading fragments in the glacier sole, in terms of hardness and liability to fracture. Permeability of the bedrock is another consideration: this will affect the quantity (and pressure) of meltwater beneath temperate ice. Highly permeable lithologies will reduce the water flow at the glacier base and limit the removal of rock flour, encouraging the accumulation of clay-rich debris and hindering abrasion.

Laboratory experiments to study abrasion processes have been carried out in two main fields: first, to investigate ice-sliding behaviour over different rock types, and secondly, to attempt to reproduce glacial bedrock markings on other substances such as glass. Barnes *et al.* (1971), for instance, studied the friction values of ice sliding over granite specimens with surface asperity roughness of about 1·5 μm. Varying ice temperatures and sliding speeds were used (Figure 8.4). For low or moderate sliding speeds and temperatures below −10°C,

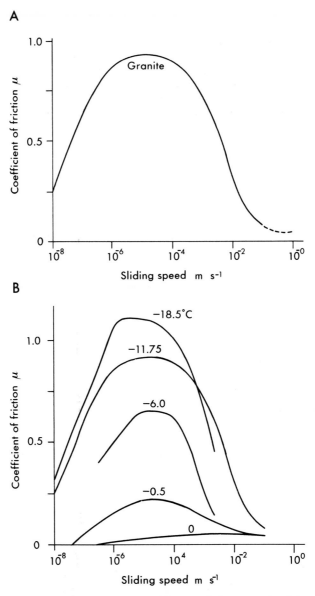

Figure 8.4 The friction of ice over granite at different sliding speeds. **A**: at a temperature of −11·75°C; **B**: at temperatures between 0° and −18·5°C (Barnes *et al.*, 1971).

high values of friction show effective adhesion between ice and rock. Such experiments confirm that glacial abrasion, while not necessarily wholly absent beneath cold glaciers, is nevertheless dominantly related to actively sliding temperate glaciers armed with basal debris. The other critical condition is the effective basal pressure, though this should not exceed a certain threshold in order that the debris should move across, rather than stick onto, the bedrock surface.

Crushing and fracturing

As well as scratching and grooving bedrock surfaces, basal debris may also be competent to shear off or crush bedrock protrusions, or to gouge out chips of rock producing friction cracks. In theory, there is little difficulty in providing adequate forces (Figure 8.5). A cubic fragment of basal debris A with dimensions x cm is held up in its sliding over the bedrock surface by a smaller-sized cubic protrusion B (dimensions y cm). Estimates of basal shear stress (τ_B) have been derived for many glaciers (Table 8.2), suggesting that, in the case of

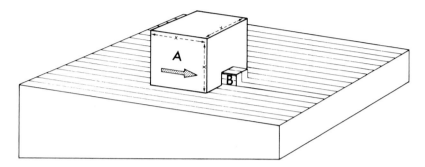

Figure 8.5 Diagram showing a cubic block of basal debris (A) arrested in sliding over bedrock by a smaller protrusion (B).

Table 8.2 Values of calculated basal shear stress (τ_B) and related values of ice thickness (h) and surface slope (α)

	h(m)	α (degrees)	τ_B (bar)	Reference
Blue Glacier	26	28	0·7	Kamb & LaChapelle (1964)
Salmon Glacier	490	2	1·4	Mathews (1959)
Athabasca Glacier	209	6·3	2·0	Savage & Paterson (1963)
Athabasca Glacier	310	3·5	0·85	Reynaud (1973)
Barnes Ice-cap (near centre)	463	1	0·71	Orvig (1953)
Barnes Ice-cap (near edge)	32	6·5	0·33	Orvig (1953)

actively moving ice, values are likely to fall in the range 0·5–1·5 bar. Taking a mean value of 1 bar $\approx$ 1 kg cm^{-2}, the tractive force exerted on the up-glacier face of block A would amount to x^2 kg. Additionally there would be varying amounts of drag exerted on other faces lying parallel to the direction of ice flow, depending on the levels of adhesion between the ice and the block. For cold ice, adhesion would be greater than the shear strength of the ice, but for temperate ice, adhesion might be negligible, depending on the roughness of the rock surface and the amount of basal meltwater. Whether the total tractive force exerted by the ice on block A would be sufficient to shear off the obstruction B would depend, then, on

1 the relative dimensions of A and B; the greater the size of A, the greater the tractive force that can be exerted on it by the ice.
2 Thermal régime, governing the degree of adhesion between the ice and block A.
3 The shear strength of B; if this is of the order of 200 times that of ice, then the critical dimension (y) of B is likely to be about $x/\sqrt{50}$ in the case of cold ice adhering perfectly

to the sides of A, but perhaps only $x/\sqrt{200}$ for temperate ice with no adhesion to the sides of A.

As McCall (1960) points out, the situation would be complicated by the tendency of block A to roll over. In the basal layers of a cold glacier, flow increases rapidly upwards from the bedrock and this would encourage rotation of basal rock fragments. If the obstruction at B is large enough to resist the forward motion of A, there can be several possible outcomes. The first is that more debris carried by the flowing ice will lodge against A, effectively increasing the area of A until fracture of B occurs. A second possibility is that accumulating debris around A might become anchored to other neighbouring bedrock obstructions, forming an immobile mass around which the ice would flow, possibly streamlining it into a drumlinoid form.

The process of gouging by moving blocks is described by Glen and Lewis (1961) at the side of Austerdalsbreen, Norway, where boulders are jammed in the gap between the ice and bedrock and prevented from moving by protruberances of the latter. Although the boulders here were not wholly encased in ice and although the latter was at pressure melting point, there were many signs of gouging and grooving of the sidewalls. Boulders jammed temporarily against bedrock obstructions and then breaking free are a possible explanation of the jerky ice motion described not only by Glen and Lewis but also by many other workers such as Vivian and Bocquet (1973, base of the Glacier d'Argentière) and Kamb and LaChapelle (1964, base of Blue Glacier).

Attempts to reproduce experimentally the friction cracks and gouges induced by sliding ice are described by Harris (1943). In none of his experiments did he attempt to reproduce conditions to scale in respect of time, pressure, hardness or character of material or cutting tool. Most successful in forming crescentic gouges or fractures was a pointed knife or file on glass. Chattermarks were relatively easier to simulate when speed and applied pressure were properly adjusted. MacClintock (1953) outlined experiments using a steel ball on glass: when allowed to rotate under pressure, crescentic fractures formed with the horns pointing in the reverse direction to that of the ball motion, but when the ball was held rigid, the horns of the fractures developed pointing forward. Gilbert as long ago as 1906 likened the fractures associated with crescentic gouges to conoids of percussion, i.e. conchoidal fractures obtained by extreme pressure on substances such as flint or obsidian. Pressure must be sufficient to exceed the elastic limit of the bedrock.

There is also the possibility that ice may itself exert sufficient stress directly on bedrock to cause elastic deformation and fracture of the latter. The matter of dilatation jointing and its possible association in some cases with static loading by ice will be taken up in the next section, but there has recently been some discussion about the stresses imposed by moving ice. Trainer (1973) found some correlation between directions of ice motion and the patterns of near-vertical bedrock jointing at a wide variety of locations in the USA, but the nature of any causal relationship is not clear and many more investigations are needed. Boulton (1974) argues that pressures at the glacier sole will be locally intensified where ice passes over obstructions, and that shear stresses induced in the surface of the bedrock will be greatest where the normal pressure is least, that is, on the down-glacier sides of obstructions. If a sub-glacial cavity exists here, the maximum shear stress will be developed just before the point of closure of the cavity. Whether the rock will yield or not will depend on its cohesive strength, on its pre-existing joint structure, and on the possibility of fatigue due to fluctuations in the levels of stress imposed. Reinforcement of existing rock weaknesses seems an important possibility and very relevant in preparation for joint-block removal (plucking).

Joint-block removal

Many workers have considered this process to be more important quantitatively than abrasion, though it is agreed that much depends on bedrock lithology and structure. Matthes (1930) contended that, for tough unjointed rocks, slow abrasion only was possible, but that for jointed rocks with favourable spacing of fractures, massive removal of blocks by 'quarrying' was possible. He suggested that joint spacing of 5 to 25 feet (1·5–7·5 m) provided optimum conditions, wider spacing producing too large blocks for the ice to move and smaller spacing not producing as large a bulk of material. He and many other workers succeeded in demonstrating a general relationship between rock type and structure on the one hand and depth of glacial excavation on the other (e.g. Crosby, 1945; Zumberge, 1955).

There are two aspects of the process to be considered: one, the provision of fractures in the bedrock, and second, the manner of extraction of loosened blocks by moving ice. The second aspect will be considered later in this chapter under the heading of entrainment and transport of debris. The origin of bedrock jointing can be wholly non-glacial, resulting from previous tectonic or igneous events. There is also dilatation jointing of non-glacial type, caused by unloading of bedrock by denudation. In addition, it has been argued in the so-called hypothesis of 'defonçage périglaciaire' (Boyé, 1950, 1968; Cailleux, 1952; Lliboutry, 1953) that pre-glacial cold-climate conditions could be important in the preparation of the bedrock for glacial quarrying. Ground freezing prior to ice advance could stress and deepen any existing fractures in the rock; subsequent advance of temperate ice would cause ground thawing, and the loosened blocks would be available for glacial transport.

Apart from jointing of pre-glacial origin, there is also the possibility of jointing developing contemporaneously with glaciation. Trainer's (1973) claims have already been mentioned. Many more workers have looked into the question of sub-glacial dilatation processes. Lewis (1954) was the first to suggest a link between pressure release and glacial erosion, arguing that the normal pressure of ice in a thick glacier could load the bedrock (approximately 9 bar for every 100 m of ice) sufficiently to cause shallow fractures to develop parallel to the surface of the bedrock after glacial retreat. Linton (1963) has supported such a view, and claimed that the process might occur without the need for deglaciation, for as glaciers erode into bedrock, the load on the latter is effectively reduced, glacial ice (density 0·9) replacing rock of density, say 2·5. Evidence for dilatation joints developing as a direct result of glacial erosion and enabling joint-block removal to continue, is given by Battey (1960). The cliffed faces of the cirques containing Vesl-Skautbreen and Veslgjuv-breen in Jotunheimen, Norway, show extensive joint systems whose surfaces run parallel to the cirque outlines and dip at angles conforming approximately to the slopes of the cirque walls. There is no doubt that rocks can fracture after release of extreme confining pressure and even 'burst' spontaneously (see, for example, the description of Bain, 1931 and Jahns, 1943), but whether the pressures exerted by even the thickest glaciers are sufficient to create new joints, or whether the process is merely one of reinforcement of existing joints, is still open to debate. It has also been pointed out that sheet joining alone is not enough for joint-block removal: there must also be near-perpendicular intersecting joint sets reflecting horizontal rather than vertical stresses in the rock (Birot, 1968b) which are more likely to be non-glacial in origin. Addison (1975) working on the detailed joint patterns of Snowdonia, North Wales, finds that most glacial landforms here are strongly controlled by jointing and other structures, but is not persuaded that any dilatation mechanism of glacial origin has operated.

Once joints or other fractures are created, the way is open for another set of processes to operate in extending and widening them. These comprise various freeze–thaw mechanisms over which controversy has ranged since the beginning of the century. Johnson (1904) proposed freeze–thaw action as the major process involved in the shattering of cirque headwalls; the process, he claimed, was one of diurnal temperature changes across the freezing point in bergschrunds, so that meltwater soaked into existing rock fissures, froze at night and widened them until the joint-blocks became loosened. Lewis (1938, 1940) elaborated on this idea, suggesting mainly external sources of meltwater to wet the rock face exposed in bergschrunds or to seep to even lower levels behind the glacier until it refroze. Measurements of temperatures in bergschrunds by Battle (1951, 1960) and McCall (1960), and beneath the Glacier d'Argentière (Vivian and Bocquet, 1973) have provided little support for these ideas. The deeper one penetrates behind or below a temperate glacier, the more closely do temperatures approximate to that of the isothermal ice; external meltwater reaching such a position will not be refrozen because of the lack of any means for removing latent heat. There is just some possibility that refreezing can occur in the upper parts of temperate-glacier bergschrunds where cold air can penetrate in winter, or in sub-glacial cavity systems where passages connect directly to the glacier surface. Small changes of air temperature have in fact been measured in cavities and stream tunnels 100 m below the surface of the Glacier d'Argentière. Here, sub-glacial streams rushing through the passages appear to aid air circulation, and while in summer warm air and meltwater are dominant, freezing conditions occur at times in winter. Temporary refreezing of meltwater, mainly in winter, in the upper parts of open bergschrunds is also known to occur (e.g. McCall's observations of conditions in the headwall gap behind Vesl-Skautbreen), so that the hypothesis of freeze–thaw action in such locations in the case of temperate glaciers should not be entirely dismissed, though the extent and importance of this process is probably much less than Johnson and Lewis thought.

Freeze–thaw action in closed sub-glacial cavities is another possibility to be considered. Carol (1947) postulated that in sub-glacial cavities on the downglacier faces of roches moutonnées, the release of pressure could encourage refreezing, possibly forming regelation ice on the bedrock or in fissures in it. Pressure-induced freeze–thaw, discussed theoretically a few years previously by Holmes (1944), has been confirmed by more recent observations in sub-glacial cavities by Kamb and LaChapelle (1964) and by Vivian and Bocquet (1973). The main problem is whether the small oscillations of temperature around the freezing point are of sufficient amplitude to have any disintegrating effect on bedrock—experiments on freeze–thaw action in the laboratory (see Chapter 4) suggest otherwise. In the case of cold glaciers frozen to bedrock, sub-glacial freeze–thaw is absent. Some sub-polar glaciers where englacial temperatures are negative to a considerable depth may, however, experience a short summer melt season when surface melting may be sufficiently extensive to cause runoff and marginal streams to develop, as Maag (1969) describes in Axel Heiberg Island, Canada. Marginal streams tend to work their way under the ice edge as the melt season progresses, before freezing up with the onset of winter. The conditions are thus temporarily favourable for freeze–thaw action, but this will be confined to the glacier margin zone.

There remains the interesting case of those glaciers where cold ice is present in the upper layers but temperate ice exists at depth (see p. 274), or where temperate ice passes downglacier into a cold-ice zone. Fisher commented in 1955 and 1963 on the significance of such a situation for potential freeze–thaw action. In the cold-ice zone, the ice will be frozen to bedrock; in the zone of temperate ice, the bedrock will be unfrozen and wetted by basal meltwater. The 0°C isotherm will approximately mark the junction of the two zones, and its position will be controlled basically by external climatic influences and changes in glacier mass balance. As it migrates in response to these changes, temperate ice may be brought into contact with frozen bedrock, or cold ice may come into contact with unfrozen bedrock. In other words, alternate thawing and freezing of bedrock on a major scale will be induced. The time-scale in this hypothesis may have to be reckoned in terms of hundreds of years, rather than the seasonal or even daily fluctuations sought for by Battle, Lewis and others, but the efficiency of the process may be far greater because of the opportunity for greater oscillations in rock temperature and for greater depths of penetration of these oscillations. Considering the varied climatic conditions of the Pleistocene, and the climatic oscillations since, the hypothesis probably represents one of the most important applications of freeze–thaw action in glacial erosion. So far, though, much of the discussion is conjectural because of the lack of field data. With few exceptions, such as the Breithorn tunnels described by Fisher in 1963, and the Byrd Station borehole in Antarctica, no tunnels or borings have penetrated to bedrock near the critical junction between cold and temperate ice nor has it been possible to measure any temporal changes of thermal régime that the hypothesis demands.

There are, then, certain circumstances in which freeze–thaw action can be an important agent of joint-block loosening, and others, in contrast, where its importance is negligible. Thermal conditions at the ice–bedrock junction provide the critical controls.

Sub-glacial meltwater erosion

The powerful erosive ability of channelled meltwater streams is well documented. Vivian (1970), for instance, describes how the bursting of the Gornersee, Switzerland, in 1967 released a torrent escaping from the ice that moved more debris in 3 days than was normally shifted in a year, and that was responsible in those 3 days for 10 cm of erosion in granite bedrock. When meltwater flow occurs in tunnels beneath the ice, even greater velocities of flow may be encountered. The rates of discharge of sub-glacial streams are well known to be erratic, possessing high peaks when sub-glacial or englacial water bodies are suddenly tapped. Together with the frequently heavy loads of angular debris picked up from the basal ice, bedrock corrasion can be rapid. The gorges and channels cut by such sub-glacial streams have been extensively described. With high velocities of flow, cavitation erosion becomes a significant possibility, borne out by the distinctive *sichelwannen*, bowls and other micro-forms on rock surfaces subjected to such action. Another very important aspect of sub-glacial water discharge is the removal of debris produced by glacial processes. There is some uncertainty about the possible corrosive role of sub-glacial meltwater, mainly because of the lack of data, but there is no reason to think that, fundamentally, meltwater streams behave any differently from non-glacial streams in this respect. At times of high discharge, mechanical erosion will be overwhelmingly dominant, but it is possible that, at other times of low flow, when the rock surface is being merely washed by meltwater, maybe moving more as sheetflow than channelled flow beneath the ice, chemical attack of certain rocks and minerals, such as clay-rich sediments and feldspars, may be significant. Vivian and Zumstein (1973) show an increase in the pH values of meltwater samples taken at low flow from a stream below the Glacier d'Argentière, compared with high flow conditions.

Erosion processes: summary and evaluation

1 Erosion processes can be differentiated as abrasion, crushing, fracturing and block removal (plucking). Their relative quantitative importance is uncertain but varies with conditions at the ice–rock contact.

2 Thermal conditions at the ice–rock contact provide a fundamental set of controls on erosion processes. Cold glaciers are capable of limited abrasion because of lack of basal slip and because, on the whole, they seem to be poorly provided with basal debris; on the other hand, superior adhesion to bedrock may enable plucking to operate more effectively, especially

Plate XXII Contact of glacier with bedrock, Austerdalsbreen, Norway. Jointing of the bedrock, including probable dilatation jointing, together with melt-water flowing over the rock, provides favourable conditions for freeze–thaw weathering. Rock fragments jammed in the gap between ice and bedrock assist in gouging and abrasion of the rock. Measurements of differential motion between ice and bedrock at this locality amounted to $0.26\,\mathrm{m\,day^{-1}}$. (C.E.)

where the bedrock is incompetent, fractured, or frozen only to shallow depths below the glacier base. Temperate glaciers exhibit varying degrees of basal sliding (and therefore abrasive potential); moreover, basal melting can bring englacial debris into contact with the bed, while regelation can remove debris into the ice. It is most important to note that glaciers can change their thermal characteristics both spatially and temporally: for instance, a transition from temperate to cold conditions due to climatic deterioration can result in the freezing and break-up of wetted bedrock, and debris can be removed by freezing onto the glacier sole.

3 The debris content of the basal ice is also critical. Without debris, erosion (apart from any effects of direct ice pressure on bedrock) is impossible; and abrasion needs tools to operate. Cold glaciers often seem to possess little or no basal debris, and it is difficult to see how their basal layers can produce or acquire much debris. There is, however, also the problem that the bases of some temperate glaciers seem to be relatively clean. These problems will be considered further in the next section.

4 Pressure conditions at the ice–rock contact are also critical for erosion. Where the ice is sliding, there is friction between the ice and any rock debris resting on bedrock, and between the bedrock and the debris that is being dragged across it. Above certain pressure thresholds (allowing for basal water pressure, if any), the debris will tend to stick to the bedrock rather than slide across it; thus abrasion will decrease, eventually to reach zero when deposition or lodgement will take over.

5 The condition of the bedrock also needs to be considered. A glacier may be dealing with consolidated or unconsolidated rock, with fractured or massive rock, with frozen or unfrozen rock. Fracturing of the bedrock prior to or during glacierization will facilitate block removal by the ice: various freeze–thaw or dilatation mechanisms have been proposed for such fracturing (Plate XXII).

6 Although the major controls on erosion are now becoming clearer, there is still a dearth of data on such matters as rates of abrasion, effective basal pressures, the relative abundance of debris in the soles of different glaciers, and conditions near the boundary between cold and temperate ice. There is an especially great need for more borings to test temperature profiles and the dispersion of debris in a wide variety of glacier conditions.

Entrainment and transport of debris

Removal of the debris produced by erosion is clearly a prerequisite for continuing effective erosion of the bed. Some removal will be effected by sub-glacial meltwater beneath temperate ice, but the processes involved are fluvial and are dealt with in Chapter 7. This section considers purely glacial mechanisms of transport.

Debris is acquired by glaciers in two ways—from material that falls onto the ice surface (*supraglacial debris*) from nearby nunataks, valley sides, cirque headwalls, talus slopes, etc; and from material available at the glacier bed (*sub-glacial debris*), whether produced by glacial erosion or left over from preglacial weathering processes. *Supraglacial debris* falling on to ice in the ablation zone will remain on the surface and be carried forward with any ice movement, except for minor quantities that may fall into crevasses to travel englacially. In the accumulation zone, debris will become gradually buried by subsequent snowfall. In the case of temperate glaciers, this is often a regular seasonal process leading to the formation of a series of dirt bands or sedimentary layers (Plate XXIII). The debris thus moves to successively deeper englacial positions (Figure 8.6).

Englacial debris may be emplaced by this method, or it may be derived from sub-glacial sources as will be shown later. As it travels with the ice, it may return to a surface position by ablation of the overlying ice; it may sink to a basal position with basal melting or by encountering bedrock obstructions; or it may maintain an englacial position up to the glacier snout. It is analogous to the suspended load of streams, except that, because of the vastly greater viscosity of ice (10^{12} to 10^{14} poises), the particles are maintained in suspension not by turbulence but by the ice itself. Englacial debris may be carried in the ice in two ways. First, it may be disseminated throughout the whole or part of the ice thickness. In this case it is unsorted, it may include large boulders, and the proportion of debris to ice rarely exceeds

Plate XXIII Ice-cliff margin of Vesl-gjuvbreen, Norway, showing annual sedimentary layers (dirt bands), representing englacial transport of originally supraglacial debris. The thickest layer, about 2·5 m above the fallen debris from melt-out, probably represents a period of several years when there was little winter accumulation and the glacier surface became blanketed in debris. (C.E.)

15 per cent by volume. Secondly, it may be concentrated in bands sometimes as much as 5 m thick, separated by relatively clean ice. The debris in such bands is mainly fine-grained and may attain concentrations up to 80 per cent. Boulton (1968) estimates that the snout of Dunerbreen, Spitsbergen, contains enough banded englacial debris to give a layer of till 5 to 8 m thick on melting. The bands dip at various angles depending on glacier flow, and their possible origins have been widely discussed. Some may be of supraglacial origin in the accumulation zone as already noted (Figure 8.6); some may mark the infilling of crevasses, either at the surface or by the squeezing-up of basal debris; some may mark shear

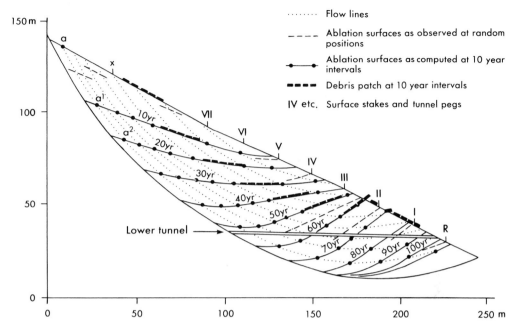

Figure 8.6 Longitudinal section of Vesl-Skautbreen, a cirque glacier in Norway, showing sedi-mentary layers (ablation surfaces), computed from velocity data, at 10-year intervals (McCall, 1960).

planes in the ice; but probably most represent sub-glacial debris frozen onto the glacier sole and carried upwards along flow lines. These ideas will be further considered later in this chapter.

Sub-glacial debris may travel throughout in a basal position, or it may become englacial, while in the ablation zone it may reach the surface of the ice. Whereas englacial particles are subjected only to hydrostatic pressures in the ice, particles on the bed are subjected to pressures ($\rho gh-p$) that depend on ice thickness and basal water pressure (if any).

A series of processes affects the vertical and lateral dispersion of debris.

1 In the accumulation zone, already noted, supraglacial debris will sink relatively to the surface by the addition of new snow.

2 Flow lines in the ice will control the movement of englacial debris; as it passes through the glacier system it can be raised or lowered, spread laterally or confined. Zones of longitudinal compression will be of the greatest significance in producing upward movement of debris in the ice.

3 Diffusion of debris, sub-glacial or englacial, can occur. The larger particles may collide and be pushed over one another, though the extent of vertical mobility possible is probably limited to a metre or so.

4 In the case of temperate glaciers, general basal melting can help to concentrate debris in the lowermost layer where the bed is smooth.

5 If, however, temperate ice is flowing over obstructions, the regelation mechanism identified by Kamb and LaChapelle (1964) comes into play. Obstructions cause pressure melting, but on their downglacier flanks, refreezing of meltwater occurs, enclosing and attaching

small-sized debris to the basal ice. The resulting debris-rich 'sole' (McCall, 1960) is limited to the thickness of the regelation layer, of the order of a few centimetres, the thickness being controlled by the heat flux through the obstacle and therefore related to the size of the obstacle. In theory, a series of obstacles could produce alternate debris-rich and clear ice layers, depending on the sizes and arrangement of the obstacles; however, the debris-laden regelation ice produced by one obstacle will tend to be destroyed and then merely reformed beyond another obstacle, downglacier, of the same size, so that it is difficult to see how thick basal debris sequences can be built up by this process.

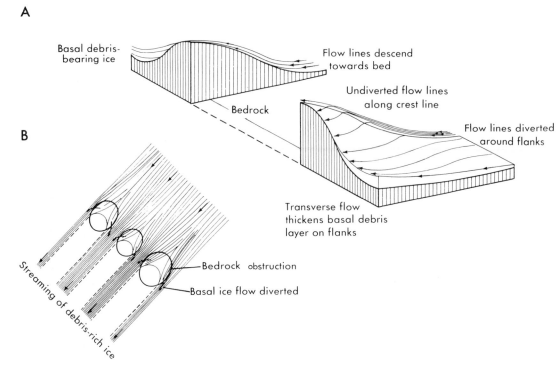

A

Basal debris-bearing ice

Flow lines descend towards bed

Undiverted flow lines along crest line

Bedrock

Flow lines diverted around flanks

B

Transverse flow thickens basal debris layer on flanks

Streaming of debris-rich ice

Bedrock obstruction

Basal ice flow diverted

Figure 8.7A: Flow of basal ice around a bedrock obstruction, showing thickening of basal debris-rich ice in troughs and its thinning over the summits of obstructions.

B: Streaming of basal debris-rich ice around bedrock obstructions (Boulton, 1975).

6 Much thicker debris layers can form at the ice base if temperate ice moves into a cold ice zone (Figure 8.2). Basal meltwater moving into such a zone can be refrozen in much greater quantities than the simple regelation mechanism in (5) permits. For some distance into the cold-ice zone, the actual base of the ice could be maintained at pressure melting point if the supply of meltwater were adequate: this would depend partly on the permeability of the underlying bedrock (Boulton, 1975). Boulton (1970a) has suggested that such a situation could account for the thickness and sequence of debris-rich layers in the terminal parts of Makarovbreen, Spitsbergen, for example.

7 For obstacles larger than about 1 m in length, the process of regelation slip becomes inefficient, and a creep mechanism governed by the flow law of ice comes into operation (Weertman, 1964). This mechanism, unlike that outlined under (5), is applicable to both cold and

temperate glaciers. If the basal ice layers already contain debris, the latter will be forced laterally as well as vertically around the obstacles, so that streaming of the basal debris and moulding of the bedrock obstacles by erosion can result (Figure 8.7). As the debris-laden ice is forced into the hollows between obstacles, thickening of the basal debris-rich layer will occur. The creep mechanism also allows new debris to be incorporated in the basal ice if loose fragments are already available (bedrock prepared by previous freeze–thaw or by pressure-induced jointing). Beneath the Glacier d'Argentière, Boulton (1974) notes how fragments are entrained by both regelation and creep: some fragments become frozen in on the lee sides of obstacles, others become pressed in to the basal ice.

Plate XXIV Margin of Breiðamerkurjökull, Iceland. Water-soaked till appearing beneath glacier margin, probably being squeezed out to the right. Just to the left of the two figures, till can be seen to penetrate a fissure in the ice narrowing upwards. (C.E.)

8 The entrainment of very large erratics (e.g. the Schollen of north Germany, consisting of huge masses of relatively unconsolidated sediments up to 4 km long) can probably only occur beneath cold ice where the 0°C isotherm lies at shallow depth (up to about 100 m) below the glacier base. The rafts of sediment frozen to the glacier base are torn away from the parent bedrock along any lines of weakness at or just below the depth of ground freezing (see p. 302).
9 Some basal debris may be entrained beneath temperate glaciers by upward squeezing into basal cavities or crevasses (Plate XXIV). Okko (1955) noted evidence for this in the terminal parts of some glaciers in Iceland; Hoppe (1952) invoked such a process to account for hummocky moraine. The process is unlikely to be significant quantitatively in terms of entrainment, but will be considered further in connection with processes of deposition.

10 A theory of entrainment to which much importance has been attached in the past is that of over-thrusting, carrying basal or englacial debris to higher levels in the ice. Gripp in 1929 identified debris bands in Spitsbergen glaciers, cutting across the ice foliation, and concluded that these represented thrust-planes, up which the over-riding ice carried sub-glacial debris. Goldthwait (1951) similarly attributed the formation of marginal moraines around the cold ice of the Barnes ice cap in Baffin Island to debris being brought up from the base along thrust-planes. There is no doubt that englacial thrust-planes do exist at certain positions in many glaciers. Commonly they develop where ice motion is obstructed at the margin by accumulation of moraine, the actively moving ice shearing up and over a wedge of stagnant ice. Thrust-planes may also develop in response to bedrock obstructions, and not only at the ice margins but also in zones of strongly compressing ice flow (see p. 59). A section (Figure 8.8) provided by a natural ice cliff at the southern terminus of Makarov-breen, Spitsbergen (Boulton, 1970a) enables a major thrust-plane to be traced from the ice

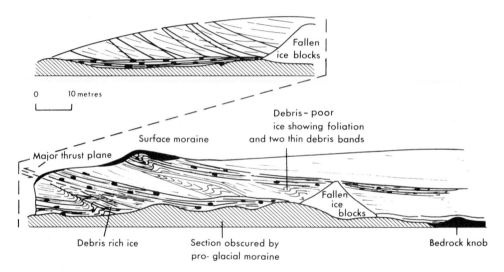

Figure 8.8 Section along the southern ice-front cliff of Makarovbreen, Spitsbergen (Boulton, 1970a).

surface down to the glacier sole. The role of thrust-planes in raising debris from sub-glacial to englacial or supraglacial positions is an important one, but their role in entraining new material at the glacier bed has probably been overemphasized. More probably, the debris is already entrained in the basal ice (e.g. by a regelation mechanism) and over-thrusting is simply a means of transport of debris-laden ice to higher levels. Indeed, the thrust-planes may develop preferentially *along* existing debris-rich bands since these may have a lower shear strength than the surrounding ice.

11 It has recently been suggested (Clapperton, 1975) that surging glaciers (Chapter 3) may acquire relatively great amounts of basal debris, owing to the high velocities of forward move-ment, the increase in the size and number of sub-glacial cavities, and the possible increased supply of basal meltwater involved in the surging mechanism. Clapperton's hypothesis is that these conditions may favour accelerated regelation and incorporation of debris in the lee of obstacles. If the surge at the same time carries the lower part of a glacier over outwash or other sediment-rich areas, the load of basal debris thus acquired may be considerable.

The hypothesis needs further testing: it would imply that differences in sub-glacial (and englacial) debris content between glaciers may be not only the result of differing thermal régimes but also related to surging or non-surging behaviour.

These, then, are the main processes of entrainment and redistribution of debris within a glacier. It is evident that thermal régime and, possibly, surge behaviour play a fundamental role in determining which process or group of processes functions at any given locality. The conditions under which the greatest bulk of sub-glacial debris is entrained are probably associated with those parts of cold sub-polar glaciers where meltwater from adjacent zones of basal melting is freezing in quantity to the glacier sole. In these glaciers, debris moves up into the glacier as more ice and debris is added at the sole, and together with the effects of compressive flow or internal thrusting, transport of englacial debris tends to occur at a relatively high level, especially in the terminal zones. Temperate non-surging glaciers will acquire relatively thin debris-laden soles through the regelation mechanism; on the other hand, debris of supraglacial origin may be relatively more important. Cold glaciers wholly frozen to bedrock will be unable to acquire sub-glacial debris loads except when extraction of large blocks or rafts of frozen sediment takes place; and it should be noted that such glaciers do not exhibit the phenomenon of surging.

Where the glacier base is sliding over bedrock, the speed of movement of particles at the ice–rock junction will depend on their shape, their size, the frictional drag (F: see p. 278), and the nature of the bedrock. Frictional drag will be strongly affected by ice thickness. For a thick glacier with small or zero basal water pressure, friction between the particle and the bed may be too great to allow any differential movement. For smaller thicknesses of ice, the speed of movement of small particles will be relatively slow because the regelation-slip mechanism is relatively efficient for small obstacles. Thus there will tend to be greater differential movement between the ice and the particle than between the particle and the bed. Similar conclusions apply to large particles when the creep mechanism becomes relatively more efficient. Boulton (1975) suggests that traction at the bed is most effective for particles of intermediate grain-size. For effective basal pressures of less than about 20 bar, particles in the range -3 to -7ϕ show relatively little retardation—they tend to move at 80–100 per cent of basal ice velocity. Particle shape must also be considered, however, as this will affect the tendency of the particle to slide or to roll along. Basal particles will suffer attrition through breakage; crushing will occur where large particles are pressed on smaller ones, and this will increase the content of fines. Wherever the primary source of a particular rock-type can be identified, it is found that the roundness of the derived particles increases rapidly in the first kilometre or so of transport, but thereafter there is less change from a dominantly sub-angular shape. The nature and roughness of the bedrock are other factors. Deformable bedrock will allow harder particles to plough through it, though eventually their movement may be arrested (Figure 8.9). Thus, bed particles will be moved at different velocities depending on a range of parameters; collisions between particles will occur (Figure 8.9E), leading to accumulation of aggregates or to further diffusion of debris.

The processes of entrainment and transport impart distinctive fabrics to some glacial sediments. Miller as long ago as 1884 noted how elongated stones in till tend to become orientated in the line of ice flow, and the significance and analysis of till fabrics has become a major field of investigation, beginning with the work of Holmes in 1941. Here we are concerned only with an outline of the processes causing distinctive fabrics to develop. Glen, Donner and West (1957) point out that elongated pebbles placed at random in a flowing liquid will quickly adopt an orientation in line with the flow, but that, after a time, the particles turn to adopt a transverse position, theoretically the position of minimum energy requirement, provided the ratio of the a/b axes is less than about 15:1. In glacier ice, which

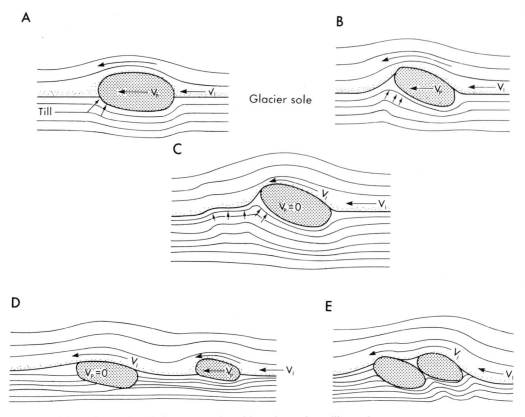

Figure 8.9A: A basal rock fragment ploughing through a till matrix.

B–C: Stages in the arrest of movement of the rock by progressively deeper ploughing and consolidation of the till.

D–E: Lodgement of the rock fragment against another one already embedded in the till (Boulton, 1975). V_p—particle velocity; V_1—ice velocity.

is not a linear viscous liquid, the theory may not apply exactly, but a preferred orientation does develop following the line of flow. Such a parallel orientation will also develop, as Holmes pointed out, if stones in the ice interact with the bed by dragging over it or if stones are being dragged along shear-planes in the ice. A secondary transverse orientation develops over a longer period, because of the tendency of stones totally immersed in the ice to assume a position in which they rotate about their long axes. Collisions between stones, especially prolate stones, can also favour a transverse orientation. Much, then, depends on stone shape and degree of roundness, and many till fabrics show not only great variability within short distances but also considerable scatter around the preferred values. A fuller discussion is given in Embleton and King (1975a).

Processes of deposition

The sands, gravels and clays deposited either directly or indirectly by Pleistocene ice cover practically 8 per cent of the world's land surface, including one-third of Europe and one-quarter of North America. They include material of both glacial and glacifluvial origin. In this section we are concerned largely with the former, the sediments (termed 'glaciogenic' by Francis, 1975) laid down directly by ice; the fluvial processes governing the accumulation of glacifluvial material are dealt with in Chapter 7 though some reference to depositional processes operating in an ice-contact environment will be made in this section. Essentially, we are concerned with the origins of till (often termed ground moraine or boulder clay in the past).

Till is one of the most varied of all deposits. It may be disposed in sheets or layers, but except for occasional signs of lamination produced under hydroplastic or sub-aqueous conditions, it does not usually possess any internal stratification. It is essentially an unsorted deposit, containing a great variety of grain-sizes, minerals and rock types. For any specific mineral or mineral group, the particle size distribution is often bimodal, one mode in the clast size and one in the matrix material. Dreimanis and Vagners (1971) have shown that the clast mode is dominant in tills that have undergone only a short distance of transport, but that with increasing distance, attrition and fracture increase the importance of the matrix mode until this becomes dominant. The matrix consists mostly of mineral grains, and beyond certain limits (about 5 to 8 ϕ) the grains of each mineral do not become further comminuted by glacial transport.

Table 8.3   Classification of depositional processes

Basal processes:	Melting (undermelt)	Lodgement	Flowage
Supraglacial processes:	Melt-out (ablation)		Flowage
Marginal processes:	Melt-out (ablation)	{ Pushing { Dumping	
Structural changes:		Deformation	Collapse, consolidation

Till fabrics have already been touched on in the last section. They have proved of great value in the analysis and interpretation of tills, but their complexity has at the same time raised many problems and differences of opinion. Fundamentally, as Francis (1975) emphasizes, their complexity reflects the fact that many different processes are involved in glacial transport and deposition. Several genetic types of till are therefore distinguished. Boulton (1971), for instance, suggests a division into sub-glacially formed lodgement till, sub-glacial melt-out till, and supraglacial till produced by ablation (supraglacial melt-out till) and affected, in some cases, by supraglacial flowage (giving flow till).

Many of the processes contributing to the formation of till are imperfectly identified or understood, for the sub-glacial processes, especially, are rarely susceptible to direct examination. Goldthwait (1971) comments in respect of sub-glacial tills that 'several mechanisms have been proposed, but none has been demonstrated and none excluded by actual observation'. Table 8.3 presents an attempt at classification, but it should be emphasized that there are rarely any clear dividing lines between one process and another, and there is controversy about their relative importance. The separation of marginal and supraglacial processes is arbitrary, and structural changes may also act penecontemporaneously with sedimentation. The term 'lodgement' has been in use since the early days of glacial geology (e.g. Chamberlin, 1894) and consequently has frequently been employed in various senses, often to include more than one process. Dreimanis (1971) distinguishes between (i) lodgement of debris at the base of a moving glacier through friction with the bedrock and pressure melting, the glacier sometimes shearing over the basal debris-rich layers of ice; and (ii) basal melting beneath a moving glacier. Sugden and John (1976) use the term lodgement to include basal melting, basal drag and shearing, and the 'plastering' of till on bedrock. 'Lodgement' will be used in this chapter only in the latter sense (i.e. plastering).

Undermelt

This term, adopted by Carruthers (1953) to refer to bottom melting of a stagnant ice mass,* can be usefully applied to conditions beneath any glacier whose base is at pressure melting point, i.e. temperate glaciers. The process will not, of course, act below cold ice. In ideal conditions, debris in the ice will melt out slowly and will be gently deposited on the surface beneath, whether this consists of bedrock or existing drift. If the basal meltwater can escape without disturbing the melt-out deposits (for instance where the sub-glacial bedrock is permeable), and if the ice is not moving, then the fabric may preserve basal or englacial structures. Pressure of the overlying ice will help to expel water from the deposits, the till becoming more compact, but if there is any accumulation of basal meltwater, flowage or injection of the melt-out till will occur, modifying or destroying the original fabric. Basal melting can occur beneath both active and stagnant temperate ice. If the ice is moving over an irregular surface, pressure melting on the up-glacier faces of obstacles can be additional to overall basal melting. There are three possible sources of heat at the base of a temperate glacier:

1 Geothermal heat. This will be trapped beneath both active and stagnant ice. Mean world heat flow is about $47\,\mathrm{cal\,cm^{-2}\,yr^{-1}}$ ($198\,\mathrm{J\,cm^{-2}\,yr^{-1}}$), which would be capable of melting about 6 mm of ice at 0°C annually.
2 Heat of friction due to glacier sliding, which will vary with the speed of sliding. For velocities about 20 m per year, heat production might be of the order of $45\,\mathrm{cal\,cm^{-2}\,yr^{-1}}$.
3 Increased pressure due to any obstructions of a rough bedrock surface.
  1 and 2 together could result in the melting of about 12 mm of ice a year. If the debris content was 20 per cent, the rate of deposition would theoretically amount to about 2·4 m per 1000 years. Mickelson (1971) referring to Burroughs Glacier in Alaska, has calculated actual rates of 5 to 28 mm per year for the accumulation of basal melt-out till.

Basal lodgement

A process that has been widely invoked to explain the deposition of clay-rich basal till is that of lodgement. Dreimanis (1976) summarizes it as the forcible pressing of rock flour into voids between larger particles; the resulting clay till is smeared or plastered onto the bedrock surface or existing drift, and it will be subjected to shearing in the process. Thus actively moving ice is a prerequisite, unlike basal undermelt. The process has been described as one of accretion; once clay is smeared on the bedrock, further clay-rich till will tend to be added as clay sticks to clay better than it does to ice.

Boulton (1975) gives the following as the main controls on the lodgement of individual particles:

1 Effective basal pressure ($N=\rho gh-p$): lodgement will be encouraged by higher values which will increase friction.

---

* Its adoption here should not be taken to mean that Carruthers's views on the undermelt drift sequence are acceptable. Carruthers was mistaken in his belief that melting of stagnant ice takes place mainly from the bottom upwards.

2 Basal ice velocity (V): lower sliding speeds will assist the process up to a point, though the lack of sliding beneath cold glaciers will clearly be unfavourable to lodgement (see below).

3 Particle size: in the case of temperate glaciers, both large and small particles will be more liable to lodgement than particles of intermediate size, since the ice can flow relatively efficiently over the larger clasts by plastic flow and over the smaller particles by regelation slip. The lodgement of small particles (less than 5ϕ) will also be assisted by cohesion. Beneath cold glaciers, it is possible that minor amounts of lodgement may occur as creep carries any particles in the lower ice layers against obstructions projecting up into the ice.

4 Particle shape: plate-shaped particles will lodge more readily than rounded particles which will tend to roll.

5 Bed roughness: this will promote more frequent contacts between particles and the bed.

Actual tills consist, however, of aggregates of particles of various sizes. Using actual till samples, Boulton attempted to identify thresholds separating conditions of lodgement from those of transport. The boundaries will be strongly governed by factors (1) and (2) above, and lodgement will occur when

$$\frac{N}{V^m} > L_c \qquad (8.2)$$

where m is a constant (value about 0·3) depending on the mode of ice flow around obstructions, and L_c is the critical lodgement index depending on grain-size distribution and particle shape in the debris and on the nature of the bed. Values of L_c for several modern and Pleistocene tills are found to range from 6 to 25. The lower the value, the easier it is for lodgement to occur. As till is transported, comminution of the debris occurs (down to certain limits: see p. 295), adding fines to the matrix and reducing the modal clast size. This will reduce the value of L_c; and as the matrix mode becomes progressively more important with increasing distance of transport, more material falls within the small particle-size category that is less easily transported by the ice because of the efficiency of the regelation-slip mechanism (this, of course, applies only to temperate ice).

It has already been noted that significant lodgement of till is unlikely to occur beneath cold glaciers. As regelation slip is absent, small particles will move with the ice; moreover, the stationary boundary layer frozen to bedrock will separate moving englacial debris from the bedrock unless obstacles project upwards into the moving ice. The adhesion of cold ice will effectively entrain any loose debris.

Clay-rich till resulting from lodgement often has a high degree of compaction (see p. 305), relatively high shear strength and low porosity. Foliation, fine laminations or thin lens-shaped structures may in part reflect shearing by overriding ice. Friction between the debris-rich sole and the bed may be greater than between the sole and cleaner ice above, so that shearing below the latter will occur. Shearing of thin layers and structural foliation of basal till has been observed by Boulton (1970b) in Spitsbergen, and plastering on the underside of a shear-plane may be important. Folds, overthrusts and flame structures are also sometimes seen, but these may also be of post-depositional origin. The fabric of lodgement till, if undisturbed, usually shows strong preferred orientation of elongated stones parallel with the direction of ice flow: Glen, Donner and West (1957) point out that the plastering-on process is probably effective in eliminating any transverse peak.

Some workers claim that undermelt is the major (or even the only) process in the formation of so-called lodgement till. Price (1973) rejects lodgement by 'plastering' and claims that basal till is intimately associated with melting. It seems more likely, however, that both processes, undermelt and lodgement, operate simultaneously beneath temperate glaciers, and that lodgement by the mechanisms described above takes place only beneath moving ice. As, however, ice slides across the surface beneath, friction will generate heat, which will in turn cause undermelt. Because of lack of field data (very rarely has basal-till deposition been observed), only tentative conclusions are possible: that nearly all basal till is deposited by temperate ice: that beneath stagnant temperate ice, undermelt is the mechanism of deposition; and that beneath active temperate ice, undermelt and lodgement work in combination.

Basal flowage

Fluted or drumlinoid till surfaces are evidence of the ability of moving ice to streamline unconsolidated deposits. Although this may be partly an erosional process, the streaming of basal till by plastic flow into leeside cavities downglacier from bedrock obstacles, the deformation of dilatant materials under certain stress levels (Smalley and Unwin, 1968), and the squeezing of watersoaked debris into basal crevasses or out from under the margin of the ice, are processes that have also been postulated. Gripp (1929) and Hoppe (1952) referred to the squeezing of basal debris into fissures in the lower part of the ice; Stalker (1960) describes a variety of ridge and plateau features that can result from disintegration of stagnant ice, the underlying saturated till being pressed into basal cavities of various shapes. Okko (1955) and Price (1969, 1970) show how semi-liquid till can be squeezed out from under a glacier to form marginal moraines (Plate XXIV). Andrews and Smithson (1966) have suggested that the process may be an annually recurring one beneath some temperate glaciers: meltwater will penetrate in quantity to the base of the ice soon after the beginning of the melt season, the basal till will then become incompetent because of high pore-water pressures, and ice pressure will then squeeze it out towards the margin. In winter, the process may cease. This annual cycle may be in part the explanation of cross-valley (De Geer) moraines. Strong fabrics resulting from squeezing characterize the proximal slopes of the moraines. The process clearly demands temperate ice lying over unfrozen till, and will be favoured by conditions in which basal meltwater cannot easily escape, for example, impermeable bedrock or closed bedrock basins.

Supraglacial melt-out

The processes of ablation at an ice surface comprise melting, evaporation, wind erosion, and removal of ice or snow by avalanching, together with calving at an ice front ending in deep water. Melting is the chief form of ablation in terms of quantity of ice lost. It characterizes the ablation zones of temperate glaciers. Glaciers with cold ice at the surface show only local and short-lived melting, for instance near to rock outcrops heated by radiation, and direct evaporation assumes greater importance, though small quantitatively.

The largest amounts of melting occur on the snouts of temperate glaciers. Those in Iceland, Norway and Alaska can lose as much as 12 m of ice a year—Nigardsbreen in Norway has been known to lose over 20 m in one summer. At the other end of the scale, ablation in the Antarctic is extremely small away from the coast, probably as little as a few centimetres a year in places. Even this, however, is greater than the average rates of basal melting, and many temperate glaciers have rates of surface ablation of the order of 10^3 times the rate of basal melting. Debris in the upper layers of the ice can be liberated relatively quickly in such cases, resulting in the formation of supraglacial melt-out till. If it melts out in a stable situation, this till may preserve elements of the parent englacial structures. The englacial fabric will never be wholly undisturbed, however, as melting will bring stones into contact with each other, giving some change in orientation or dip, and if meltwater cannot escape from the till, pore-water pressures will rise and the till will become liable to deform under overburden pressure or simply by gravity flow. On some glaciers, melt-out till from englacial debris may be supplemented by debris that has travelled wholly on the ice surface; this supraglacial addition will usually lack any clear fabric. Once a cover of melt-out till or any other supraglacial debris has begun to accumulate on the ice, rates of ice melt will be considerably retarded. For temperate glaciers experiencing a warm summer melt season, melting will still be possible even beneath a few metres of till, but in sub-polar regions, accumulation of only a few decimetres may be enough to insulate the ice and prevent further melt-out.

Differences in the character of supraglacial melt-out till and basal undermelt till will depend very much on local circumstances. For a given debris content, the former will accumulate much more rapidly because of the greater heat flux. Boulton (1967, 1972b) describes how thick covers of supraglacial moraine characterize the snouts of many Spitsbergen glaciers, the englacial debris content sometimes exceeding 60 per cent. As supraglacial melt-out tills will be underlain by ice, their deposition on bedrock will post-date final melting of the ice, which may take hundreds or even thousands of years if a thick till cover forms. During down-melting, the supraglacial deposits will be liable to disturbance such as collapse and mass-movement, and also erosion by meltwater or wind. Undermelt tills develop beneath an overburden of ice (which may cause them to deform if meltwater is abundant); supraglacial melt-out tills similarly develop beneath an overburden which may be outwash or flow till (see below). The fabrics, if undisturbed by deformation, will be similar and related to the movement and structures in the ice; together, these melt-out tills mark 'the stratigraphic position of the vanished glacier' (Boulton, 1970b). Supraglacial melt-out tills can form at the surfaces of both temperate and cold glaciers, provided that in the latter case there is a summer melt season. Sub-glacial undermelt is restricted to temperate glaciers, or polar glaciers where a layer of temperate ice underlies the cold ice.

Supraglacial flowage

Mass movement of supraglacial debris is a widespread phenomenon on many glaciers, especially those carrying thick loads of till at the surface. Lamplugh described such tills in Spitsbergen as long ago as 1911, noting that flow of till downslope into hollows allowed accumulation of up to 30 m. Hartshorn (1958) introduced the term 'flowtill', and several papers by Boulton (1968, 1971, 1972b) have discussed the processes involved and shown how significant this form of till can be in Pleistocene glacial sequences as well as in present-day accumulations.

Flow tills represent the products of supraglacial melt-out and of any other supraglacial debris subjected to downslope movement in an unstable situation. Factors controlling movement include slope angle and direction, the roughness of the surface, the water content of the till and its grain-size characteristics, and the permeability of the underlying material (whether ice, till or bedrock). The slip-plane at the base of the flow may be the surface of the glacier or of bedrock, or it may be the junction between frozen and unfrozen till. For widespread development, flow tills depend on a high englacial debris-content melting out at the surface. In glaciers characterized by mainly basal transport and little material at the surface or melting-out, flow tills will be absent; they will also be absent from cold glaciers not subject to any melt season.

Rates of movement vary from rapidly moving mud streams to barely perceptible creep. Boulton (1971) suggests a rough division into (i) thin mobile liquid flows, capable of carrying

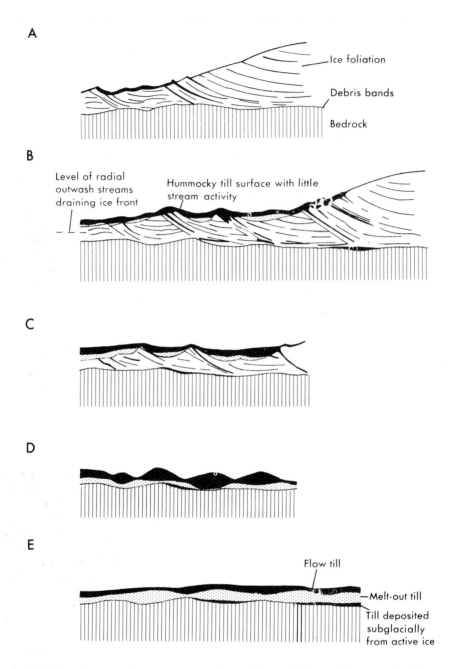

Figure 8.10 Development of a continuous sheet of flow till by mainly supraglacial processes (after Boulton, 1972b).

stones up to 0·3 m, (ii) semi-plastic flow which is just visible, and (iii) creep on well drained slopes where motion is only just measurable. The fabrics of flow tills will differ somewhat accordingly, and there may be folds or shear-planes developing in the more rapidly moving flows. Single flows rarely travel more than a few hundred metres, but coalescence of many flows can produce an apparently continuous surface till (Figure 8.10).

Marginal dumping

Several processes may be responsible for the formation of end-moraines: basal squeezing of water-soaked till has already been mentioned. Dumping is another process, in which supra-glacial, englacial and sub-glacial debris is carried forward by ice flow to the edge of the ice and there deposited by ice melt. Substantial linear ridges mark positions where the ice margin existed in a steady state for some time, forward ice movement being approximately balanced by ablation. Supraglacial flowage may contribute material that has melted out farther back up the glacier. Andrews (1975) cites the example of the Arapahoe cirque-glacier moraine in Colorado. Its volume is estimated at 1800 m³. Over the last century, estimates of ice supply and wastage indicate that 1200 m³ of ice per metre width of the ice front have melted. This in turn suggests a very high debris content for the ice, some 66 per cent, which seems unlikely considering the present appearance of the glacier. While many of the data involved are uncertain, Andrews concludes that a considerable part of the morainic debris is derived directly from rock falls from the cirque walls. Rock falls are also important in adding material to lateral moraines, but dumping is likely to be a major process accounting for transverse valley moraines and moraines around the edges of temperate ice caps. Fabrics of dump moraines are complex and often show no strong preferred orientation.

Pushing An advancing ice margin may be capable of overriding and bulldozing loose deposits in front of it. Both glacial and pre-glacial sediments may be involved in this process, recognised by Penck in 1879. Some geologists have held that the majority of terminal moraines originated in this way, including massive examples up to 100 m high at Kvíárjökull, Iceland, but such a view is being increasingly questioned. Boulton (1972b) states that ice-pushed moraines have long been considered typical of Spitsbergen glacier margins following the classic work of Gripp (1929), but contends that most of these features are in reality of supraglacial origin, the internal structures being the result of collapse and supraglacial flowage as ice-cores melt. Price (1973) accepts the validity of pushing as a possible mechanism, but points out that the ice front should not advance over too great a distance: he considers it unlikely that individual moraine ridges would survive a prolonged period of pushing. Kalin (1971) describes the Thompson glacier push moraine in Axel Heiberg Island, one of a considerable group in this part of the Canadian Arctic. It attains an average height of 45 m and shows a complex series of thrust planes and anticlinal folds. The glacier is frozen to its bed and shearing extends deep into the sediments beneath. Examples such as this show that pushing does occur, but true push-moraines may be rarer than was once thought, and the process should not be invoked without adequate evidence from internal structures. The fabric on the proximal side of a push-moraine should show preferred orientation parallel to the direction of ice movement, the stones dipping up-glacier more steeply than the proximal slope. On the steeper distal side, the fabric may not show any particular tendency, since on this side the debris will be falling or sliding down from the ice edge.

Glacio-tectonic deformation

The formation of push-moraines just outlined is part of a much broader topic concerned with deformation structures resulting from ice movement and pressure. Such structures have, like push-moraines, been recognized for over a century, but partly because the structures are mostly concealed, the widespread nature of glacio-tectonic deformation has only relatively recently been appreciated, and there are many aspects of the process, especially the mechanics, that are still not well understood. In part, the process represents incomplete erosion of bedrock, Pleistocene or recent sediments, caused by drag of moving ice, ice push, or freezing to the sole of a moving glacier, but it is more convenient to discuss the processes here rather than under glacial erosion. In so far as many basal tills are laid down beneath moving ice, glacio-dynamic structures are commonly present on a small scale in them. Dreimanis (1976) speaks of 'deformation till' and points out that it may thicken and merge into glacio-tectonic deformations on a much larger scale. The dividing line between lodgement till deformed during accretion and glacio-tectonic deformations is therefore rather arbitrary.

Glacio-tectonic structures are of two main types. First, there are the small folds, shears, lineations and faults developed during deposition of some tills. Secondly, there are larger-scale folds, thrusts, and other faults developed within a sequence of deposits or in bedrock, and the transport of massive rafts (Schollen). The first group occur in sub-glacial tills and in supraglacial flow tills. They may be caused by ice movement, the collapse of buried ice and gravity sliding. Fault throws rarely exceed 3 m and may be normal or reversed. Episodes of deposition interspersed with melting, erosion and collapse can produce complex sets of small-scale structures and intraformational unconformities (Figure 8.11).

The second group requires some rather different considerations. Rotnicki (1976) notes that, in our present state of knowledge, there are many difficulties in formulating exact theoretical models of such glacio-tectonic features. The complex mechanical properties of the materials, the thermal régime of the ice, the amount and duration of ice pressure, the occurrence and thickness of permafrost, and the rate of increase of stress in the substratum are all crucial variables, about which little is known. Banham (1975) considers some properties of ice, rock and permafrost relevant to glacio-tectonics. Ice is a relatively weak substance, but its adhesion to a wettable rock surface is greater than its own strength in shear if temperatures are below pressure melting point. As temperatures rise near to melting point, however, the strengths of ice and permafrost are considerably reduced. The weight of ice is only about half that of wet sediment, but even so the loading represented by thick ice is considerable: 300 m of ice will give a normal loading of about 270 t m^{-2}. Sediments subjected to ice loading and ice movement possess very varied properties, but one of particular significance here is the drastic reduction in the shear strength of clay with increasing water content. Rapid loading by ice of water-saturated sediments, particularly clays, can induce excessive pore-water pressures in them and lead to failure at quite shallow depths. Boulton (1972a) has discussed the process whereby rafts of frozen sediment may be entrained by freezing to a cold moving glacier above, shearing off at the base of the permafrost or along planes of weakness just

Figure 8.11 The development of typical structures in ice-contact sediments. In each row, the left-hand diagram shows the situation at the start of ice decay, the right-hand diagram the situation after the supporting ice has melted. **A–B**: post-depositional melting; **C–D**: melting during deposition; **E–F–G**: local melting causing flexuring and lowering of deposits on to the sub-glacial floor in **F**, whose support produces the flat-top of **G**; **H–I**: sub-glacial floor exposed by erosion before deposition, again giving flat-topped deposits; **J–K**: lowering on to an irregular floor (after Boulton, 1972b).

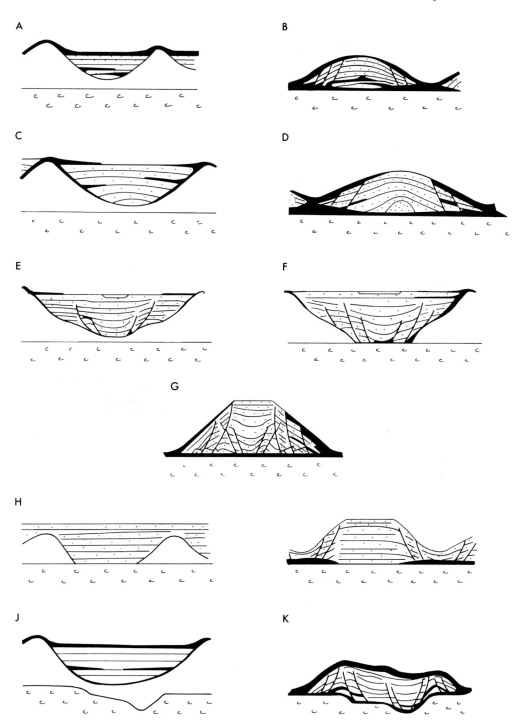

A

B

C

D

E

F

G

H

J

K

below this (see p. 291) if the shear strength of these materials is less than the shear strength of ice. Permafrost helps by sealing in pore-water under pressure and therefore reducing the shear strength in the unfrozen sediments below. Banham (1975) puts forward two general models of glacio-tectonic deformation (Figure 8.12). The first shows the expected results of the loading of a water-saturated clay or similar rock (e.g. chalk), and may be used to

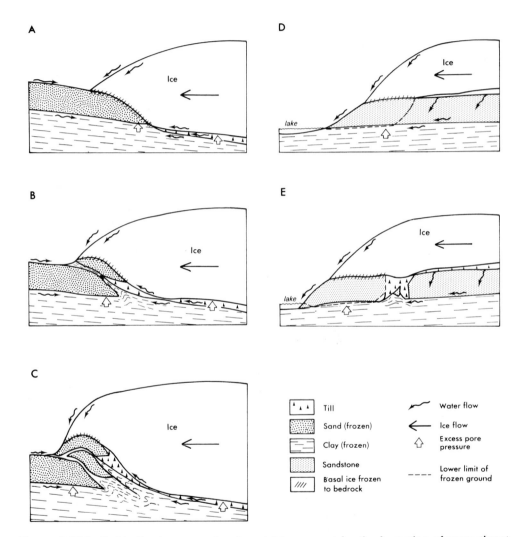

Figure 8.12A–C: Idealized compressional model to account for the formation of some thrust rafts. Vertical scale (up to hundreds of metres) exaggerated. **A:** before significant deformation;

B: loading of wet clay and lateral push of the ice cause movement of a frozen sand raft; **C:** the detachment of a second raft.

D–E: Idealized tensional model. **D:** before significant deformation; **E:** loading and ice push cause rifting of the sandstone (after Banham, 1975).

explain, for instance, the valley-side thrusting of Veluwe in the Netherlands. Rafts of frozen sand have become detached and carried forward, and may be incorporated in the basal till as large-scale 'block inclusions' (Moran, 1971). The second model shows how loading and ice push can cause rifting of a stratum under tension.

Consolidation and compaction

Most, but not all, basal tills show a high degree of consolidation. Ice loading is one factor in this if the ice is thick enough, and the vertical pressure may cause some flow and creep in the basal till, affecting the till fabric. Small-scale low-angle shear-planes and high-angle transverse joints develop, splitting the till into a series of shear-lenses. The pressure will also help to expel pore-water, and this is a very important process in consolidation. It seems that the moisture content of a till is the principal factor determining its compactness and penetration resistance. The softness of some basal tills may be caused by excess pore pressure during deposition (Adams, 1961).

Glacio-aqueous deposition

One of the problems in discussing glaciogenic sediments is that of definition. The variability of till is notorious, and a rigid distinction between material formed directly by or from glacier ice, and material entrained, transported and deposited by water, is frequently impossible in practice. It is possible for material to be transported by ice but to melt out into running or standing water; it is also possible for fluvial material to be transported and deposited by ice. Although fluvial processes are not the concern of this chapter, some reference must be made to the glaciofluvial environment, its distinctive characteristics and the processes active there.

Waterlain till

Francis (1975) includes waterlain tills under the heading of glaciogenic sediments, defining them as ice-transported debris that melts out or is otherwise released into water. According to the environment of deposition, waterlain tills may be distinguished as glaciomarine or glaciolacustrine. Many other terms have been used for such deposits (sub-aqueous till, sub-marine moraine, etc.) and there is considerable ignorance of the precise nature of the processes and conditions of deposition. Some may originate by melt-out under floating ice shelves, glacier tongues or ice bergs, the material falling from the underside of the ice and becoming mixed with non-glacial sediments. Some crude stratification and sorting may result, but the fabric is often random and unrelated to ice movement. Flowage may occur in the water-saturated deposits, and they may show other deformation structures where the ice locally grounded. Evenson *et al.* (1977) propose the term 'sub-aquatic flow till' to describe deposits emplaced by semi-plastic flow under water in an ice-margin environment, giving laminated tills and interbedded glaciofluvial sediments.

Glaciofluvial processes

As the processes are essentially fluvial (Chapter 7), this section will only indicate the main ways in which the glacial environment will affect their operation (for fuller accounts of melt-water erosion and deposition, see Embleton and King, 1975a; Sugden and John, 1976). Melt-water may flow at the surface, inside or beneath a temperate glacier; in the case of cold glaciers frozen to bedrock, meltwater is absent englacially and sub-glacially, but may appear temporarily and locally at the surface and margins if there is a summer melt season. Supraglacial streams are guided by channels melted in the ice; for a given gradient, rates of flow are more rapid than for a channel cut in rock or in sediment because of the lower coefficient of friction. On temperate glaciers, such streams soon disappear into crevasses or moulins to become englacial and, eventually, sub-glacial if ice pressures permit. Tunnel systems represent a delicate balance between ice pressure tending to close the tunnel by plastic deformation on the one hand, and water pressure and temperature on the other. Flowing water will melt the ice walls at a rate depending on its temperature and on the amount of frictional heat generated by water movement. Change in diameter of a tunnel by these processes can in turn affect rates of flow, velocities being reduced, for instance, if the tunnel widens by melting. Rates of flow are also affected by position with respect to any englacial water-table, such as may be present in a stagnating glacier, water movement becoming sluggish or static below the water table whereas streams above may be flowing rapidly and carrying large volumes of sediment. Ice collapse can cause tunnel blockage and local cessation of flow; the blocked tunnel may become choked with sediment and the water diverted elsewhere.

Sources of sediment for meltwater streams range from surface moraine to englacial and basal debris. Sediment deposited in supraglacial channels and englacial tunnels will be liable to disturbance (slumping and collapse) as the ice floor subsequently melts, while eventual melting of tunnel walls will cause lateral collapse of all englacial or sub-glacial sediments.

Depositional processes: summary

The range and variety of depositional processes explains the complexity of the resulting tills or moraine. Many processes are still inadequately identified or understood—and observations of sub-glacial deposition processes in particular are, to say the least, inadequate as a basis for theory. However, it is now becoming clear that the fundamental controls are two-fold—dynamic and thermal—and Table 8.4 attempts to summarize these relationships.

Table 8.4   Dynamic and thermal controls on glacial deposition

		Thermal conditions	*Dynamic conditions*
Sub-glacial processes	Undermelt	Temperate only	Active or stagnant
	Lodgement	Mainly temperate; limited amounts of lodgement by cold ice possible if debris already entrained	Active only
	Basal flowage	Temperate only	Active or stagnant
Supraglacial processes	Melt-out	Temperate; and to a limited extent depending on external temperature range, cold ice	Active or stagnant
	Flowage	Temperate and sub-polar ice	Active or stagnant
Marginal processes	Dumping	Mainly temperate ice	Active only
	Pushing	Temperate or cold ice	Active only

9
Nival processes

Clifford Embleton

The action of snow on the landscape is a field of study that has been relatively neglected by geomorphologists. Although its importance was recognized early this century by Matthes (1900), Ekblaw (1918), Ahlmann (1919) and others, and although many detailed studies have since been made on the properties of snow and on the nature and occurrence of avalanching, specifically geomorphological studies are few. The subject has also often been largely ignored in general geomorphological texts (e.g. Cotton, 1945; Thornbury, 1954; Sparks, 1960), though more recently greater attention has been paid to it in more specialized works (e.g. Tricart and Cailleux, 1962; Jahn, 1975; Embleton and King, 1975b; Washburn, 1973).

In considering nival processes, an important distinction must be drawn between those associated with snow remaining largely motionless on flat or moderately sloping surfaces, and those associated with snow moving rapidly, sometimes catastrophically, as avalanches on steep slopes. An intermediate category of movement in which the snow slides gradually over the ground surface can be distinguished (Table 9.1), while an entirely separate group of processes concerned with wind-blown snow (niveo-aeolian processes) needs to be briefly considered.

Processes associated with stationary or slow-moving snow

Weathering

Matthes (1900) introduced the term 'nivation' to describe the set of processes responsible for the breakdown and removal of rock around and beneath relatively immobile snow patches. Essentially, weathering mechanisms, especially freeze–thaw action but possibly also chemical weathering, have been thought to be involved, together with any available trans-porting agents to remove the weathering products. These processes will be considered in turn.

Freeze–thaw action
The principles of freeze–thaw action are examined in Chapter 4; here, we are concerned with the process acting on the ground beneath a snow cover or around the borders of snow patches. The first detailed studies were undertaken by Lewis (1936, 1939) in Iceland and by McCabe (1939) in Spitsbergen. Both were convinced that freeze–thaw action was the main component of nivation but their work showed that a simple hypothesis of intermittent refreezing of snow melt is untenable. Although snow provides an important source of melt-water to wet the ground, the snow itself acts as an efficient insulating medium to protect the ground from atmospheric temperature changes. Thus frost cycles (Figure 4.14, p. 120), of sufficient amplitude to promote rock disintegration where the ground is exposed, will

Table 9.1 Classification of nival processes

Snow movement characteristics	Erosion	Weathering	Transport	Deposition
Stationary	—	Mechanical; Chemical	Meltwater runoff; Gelifluction Debris sliding on snow surface	Melt-out;
Slow-moving	Abrasion by snow			Dumping
Fast-moving	Avalanching: Abrasion by snow, by debris, by mixed avalanches	—	On or within avalanche Slushflow	(formation of protalus etc.)
Niveo-aeolian	Corrasion	—	Aeolian transport	Melt-out (formation of niveo-aeolian deposits)

become progressively less effective the thicker the snow cover. The degree of ground protection will be a function of the thickness of the snow and its thermal conductivity, and of the duration and amplitude of the frost cycle. The thermal conductivity of snow is less than that of ice, and is dependent on temperature as well as on density (Table 9.2), so that these two variables will also affect the degree of penetration of atmospheric frost cycles.

Harris (1974) found that snow thickness was of major importance in controlling freezing of the ground beneath a snow patch when atmospheric temperatures fell below zero. In the Okstindan area of Norway, an area of sporadic permafrost and patchy snow cover, freezing rates beneath snow more than 1 m deep were less than half those beneath a snow cover 5 cm thick or less. Snow was observed to delay spring thawing of the underlying soil frozen in winter, and to protect the ground from any short-term frost cycles. Figure 9.1 shows a record of 4-hourly temperature changes in autumn and spring for two sites at Okstindan. Up to late October, the ground is generally exposed, and air-temperature fluctuations are mirrored by ground temperatures at the surface (0 cm) though, at a depth of only 5 cm, soil temperatures show only minor corresponding fluctuations. With the development of even a thin snow cover, ground temperatures remain relatively constant even when frost cycles

Table 9.2 Thermal properties of snow samples (Weller and Schwerdtfeger, 1971)

Snow density	Temperature, T, °C	Thermal conductivity (W m^{-1} deg^{-1})	
		at T	at 0°C
0·42	−60	0·71	0·50
0·57	−17	0·91	0·82

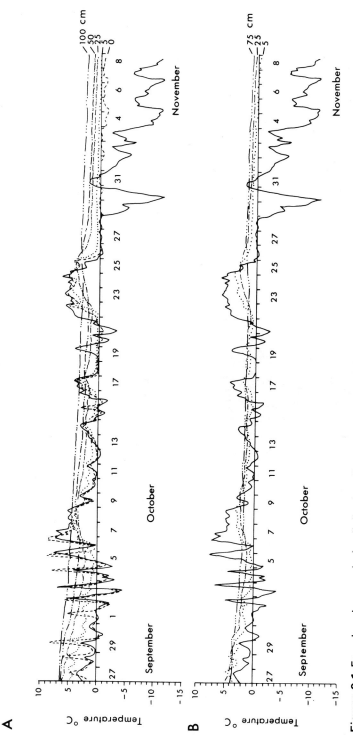

Figure 9.1 Four-hourly record of soil (broken lines) and air (continuous lines) temperatures for two sites at Okstindan, Norway (Harris, 1974).

carry overnight air temperatures down to $-11°C$. In spring, too, ground temperatures remain around zero, showing no freeze–thaw alterations, until thawing of the snow re-exposes the ground in late May. There is then a time lag before the soil at depth thaws out. Harris (1972) points out that geothermal heat as well as solar heat contributes to ground thawing; geothermal heat may also maintain sub-snow temperatures at melting point in non-permafrost areas.

The duration, amplitude and frequency of frost cycles will also have a bearing on the efficacy of sub-nival freeze–thaw action. Minor fluctuations of air temperature will be absorbed by only a few centimetres of snow, so that diurnal frost cycles less than 5°C in amplitude are probably unimportant for freeze–thaw action. Conditions in which air temperatures remain well below zero for at least several days followed by a period of thaw are likely to be more effective, for the greater duration of freezing will allow atmospheric cold to penetrate a greater thickness of snow. The total number of frost cycles per year varies greatly from one area to another, as shown in Chapter 4, and this will also clearly affect the extent of disintegration of the ground material by freeze–thaw. Gardner (1969) has measured temperatures in the clefts behind two snow patches banked against steep rock walls above Lake Louise, Alberta. Observations were made for one month in summer up to the time when one of the snow-patches disappeared. The presence of snow had the effect of maintaining minimum temperatures in the cleft close to melting point, or even fractionally lower, at a time when external air temperatures, as recorded on the exposed rock wall above the snow patches, did not show mean minima lower than 6°C. The sub-zero temperatures recorded at night behind one of the snow patches (mean minimum $-1·4°C$) probably represent residual cold from earlier spells of colder weather and provide another instance of the insulating capacity of a snow cover. Gardner notes signs of freeze–thaw disintegration of bedrock, and that external meltwater was seen to refreeze on contact with the cold rock behind one of the snow patches, but the fact that temperatures in the cleft fell only fractionally below zero suggests that the period of observation was not one of effective freeze–thaw action.

Lewis (1939) considered the possibility that sub-snow tunnels or caves opened up by streams (Plate XXV) might allow cold air to penetrate at times beneath the snow and cause refreezing of snow-melt. The very limited degree of air circulation afforded by caves open at only one end makes it unlikely that nocturnal or short-period freezing conditions in the air outside would have any significant effect beneath the snow. The main function of such caves is to act as a zone of thawing because of the concentration of snow-melt drainage.

The possible presence of permafrost beneath a thin snow cover is a relevant factor in potential freeze–thaw action. Here again, snow thickness is critical. Annersten (1966) in studies at Schefferville, Quebec, showed that a cover of snow about 40 cm thick can begin to inhibit permafrost formation or lead to degradation of existing permafrost. Further work (Nicholson and Granberg, 1973) suggests that 65–75 cm is likely to be a better estimate of the critical thickness in the Schefferville area. Thorud and Duncan (1972) show that a snow cover as thin as 13 cm can be very significant in affecting depths of soil freezing, and it is clear that surprisingly small depths of snow cover can act as an efficient insulating medium. If permafrost is present beneath snow that is no more than, say, half a metre thick, it is possible that the refreezing of snow melt may be encouraged to a limited extent.

To sum up, freeze–thaw weathering beneath a snow cover seems unlikely to be effective if the snow exceeds a metre or so in thickness, if the snow is new and therefore of low density, if there are only a few frost cycles per year and if their amplitude and duration are small. Thick snow plays essentially a protective role in terms of freeze–thaw action, keeping the ground surface beneath it at a temperature close to melting-point, unless atmospheric temperatures drop well below zero over a prolonged period. Geothermal heat trapped by thick

Plate XXV Cave under thin snow patch, Devon Island, Canada. Melting of the underside of the snow occurs in daytime (August) as warm air penetrates from outside; freezing may occur at night-time but is more likely during prolonged cold spells and in winter. Freeze–thaw conditions such as these only obtain during a small part of the year. (C.E.)

snow will help to maintain melting-point temperatures, and it is clear that any external meltwater seeping beneath such a snow cover cannot be refrozen in the absence of severe or prolonged winter cold. In general, the efficacy of freeze–thaw action beneath snow seems questionable. On the other hand, the margins of discrete snow patches represent more favourable locations, for here snowmelt is available to wet the soil around them, and atmospheric frost cycles can act directly on the exposed ground surface. Many workers (e.g. Tufnell, 1971) have also noted the overnight development of needle ice in the ground wetted by the previous day's snow melt. The growth of the needles raises particles from the soil surface, and if the latter is sloping, subsequent thawing and collapse of the needles lead to downslope displacement of the debris. The ground most liable to disturbance by needle ice and freeze–thaw action will be that lying around the edges of snow patches, and as these change in size according to seasonal or other periodic weather changes, so the zone of destructive action will migrate accordingly.

Hydration shattering
The question whether some, or even a large part, of the rock disintegration apparent in many arctic or alpine regions is caused by hydration shattering rather than freeze–thaw action has been reviewed by White (1976). Hydration is examined in Chapter 4; it involves the addition of complete water molecules to mineral surfaces, so as to build up layers of adsorbed water. White notes that the expansive force generated may be as much as

2000 kg cm^{-2}, very similar to the theoretical figure deduced by Bridgman (1912) for the maximum pressure generated by freezing water. Snow melt, wetting the ground surface beneath and around snow patches, can provide conditions for hydration; and it may well be that the difficulties of accepting a hypothesis of sub-nival freeze–thaw action can be overcome by a hypothesis of sub-nival hydration. More work is needed to investigate this possibility.

Chemical weathering

The extent to which chemical weathering processes contribute to nivation is still poorly understood. Williams (1949) asserted that snow banks might encourage CO_2 weathering of the underlying bedrock, and showed that samples of air taken from inside a snow bank contained twice as much CO_2 as normal. Williams explained this in terms of freeze–thaw alternations in the snowbank. The optimum temperature for absorption of CO_2 by water is near 0°C, so that its concentration in snow melt might be expected to be relatively great. When water freezes, it liberates CO_2 which, being a relatively heavy gas, will tend to remain in the snow bank ready to be absorbed again by any percolating meltwater. But the effectiveness, as a weathering agent, of water containing even high amounts of CO_2 is doubtful, except where the bedrock is calcareous (Boyé, 1952). Moreover, it should not be concluded that snow meltwater always contains high levels of dissolved CO_2, for much depends on the porosity of the snow (and therefore the quantity of CO_2 in the voids) and on the extent of freeze–thaw action, if any, in the snow bank. Clement and Vaudour (1968) discuss the variations in pH of snow melt in the southern French Alps. Snow pH is mainly a function of CO_2 content, and their data suggest varying degrees of acidity for melting snow, newly fallen snow being the most acid with pH about 5·4, whereas older snow exposed to freeze–thaw action apparently tends to be slightly less acid, contrary to Williams's view. Miotke (1968) working in the Cantabrian Mountains of Spain finds that snow melt there has actually less potential for limestone solution than rainwater. In Somerset Island, Arctic Canada, Smith (1972) also concludes that rates of limestone solution are small compared with those in lower latitudes of more temperate climate. Analyses of snow melt in this area do not suggest that the snow banks contain abnormal concentrations of CO_2, and it appears that rates of solutional lowering of the limestone floors of supposed nivation hollows can only account for a relatively insignificant amount, about 0·01 per cent, of material removed from the hollows since the area was deglaciated.

Salt weathering is a process recently advocated as of significance in arctic as well as other areas (Chapter 4). In the relatively dry McMurdo Sound area of Antarctica, Selby (1972) views it as an important aspect of chemical weathering, in which salt crystals, mainly NaCl, are carried by wind from the sea, or derived from ground efflorescences, and incorporated in accumulating snow. Snow melt will then form solutions which later re-crystallize, helping to wedge-out rock crystals. The relative contribution of salt weathering to nivation will vary widely according to climate and location; so far, few studies are available.

Other forms of chemical weathering that might be enhanced in the presence of snow have not yet been identified, though this is a field in which much work remains to be done. In our present state of knowledge, the role of chemical weathering in nivation is still an uncertain one. Although hydration may replace freeze–thaw action as the major process, salt weathering probably plays only a minor part, restricted to dry or coastal environments, and the possibility of enhanced CO_2 weathering seem to have been greatly exaggerated.

Transport

Several processes have been invoked to explain the removal of weathering products from around or beneath immobile snow patches. The evidence for removal is both direct, in the

Plate XXVI Sliding of fallen blocks over snow slope, Austerdalen, Norway, leading to formation of a protalus rampart at its foot. On the opposite valley wall, funnel-shaped avalanche chutes regularly discharge snow and debris during each winter. (C.E.)

form of observations of active transporting processes, and indirect, resting on observations of the landforms. It has often been claimed that snow patches deepen the hollows in which they lie, and the term nivation applied to this action must involve both weathering and transport of the weathered debris.

The main agencies available for transport are running water and gelifluction, and opinion has supported one or the other at times. Ekblaw (1918) favoured sub-nival gelifluction; Boch (1946) observed that gelifluction terraces were frequently present downslope from snow patches, and argued that the weight of the snow would itself encourage the process of gelifluction in the moistened layer of debris beneath the snow patch. Watson (1966) argues from evidence in central west Wales that the scarps of major gelifluction terraces may mark the downslope limits of snow patches at various stages of growth or dissipation. Lewis's observations in Iceland (1936) led him to support both gelifluction and meltwater runoff as transporting agents. The relative importance of these two processes will probably vary with local climatic conditions, meltwater runoff being dominant in more humid sub-arctic climates, while colder drier conditions might favour gelifluction, especially frost creep. Sub-surface piping (suffosion) may also contribute to the removal of fines and solutes from beneath a snow patch. Tufnell (1971) notes how the washing-out of fines by snow melt leads to the undermining and collapse of larger fragments in the steeper back-slopes of nivation hollows. Further experimental work in different areas is needed to establish the relative importance of the various transporting agencies.

On the surfaces of sloping snow patches, debris sliding is an important process of rock transport (Plate XXVI), though the debris involved is likely to be derived from sub-aerial weathering processes that may have no connection with the presence of snow. Frost action, for instance, may cause debris to fall onto snow beds lower down a slope, the debris sliding over the snow surface to accumulate at its foot as a 'nivation ridge' (Behre, 1933) or a 'protalus rampart' (Bryan, 1934). Such deposits are only incidentally related to the presence of snow and should not be described as nivation forms.

Snow sliding

Although the term nivation does not include the effects of rapid snow movement or avalanching, there is some evidence that slow mass movement of snow can cause surface erosion, as well as disturbance of any loose material produced by weathering. Dyson (1937) found clear signs of bedrock striation associated with snow sliding, and many observations refer to evidence of slow movement of snow beds such as minor crevasses (e.g. Lewis, 1939) and deformation of the snow layers. Both basal sliding and internal creep are involved, but there are rather few data about how effective the basal sliding component is likely to be in terms of erosion. Costin *et al.* (1964, 1973) have studied the area around a semi-permanent snow patch on Mt Twynam, New South Wales. Fresh abrasion marks on granodiorite bedrock here were clearly caused by recent movement of stones, some of which were still in position at the ends of their tracks, and both the stones and their tracks in places bore traces of fresh rock flour. Moving snow was considered to be the only possible agency for impelling the stones and producing rock flour. Avalanches are virtually unknown in this area, and would be unlikely to leave stones resting in precarious positions; therefore slow mass sliding of snow was postulated as the motive force. But although striation of bedrock surfaces was occurring, the resulting amount of erosion is probably quite small. Patches of peat still adhered to the scratched surfaces, though it is possible that, if these were frozen prior to winter snowfall, snow would slide easily over them. Estimates of the range of shear stresses likely to be exerted on bedrock protrusions by the moving snow were obtained by drilling mild steel rods into the bedrock. The rods, left projecting from the rock surface, were notched to give effective diameters ranging from 2·5 to 7·9 mm and were placed in positions where snow sliding had been previously detected. Their breaking-points and bending resistances were first determined in the laboratory. Observations over 8 years showed that snow sliding here was able to generate stresses ranging from 1·6 to over 15·9 bar, ample to account for the evidence of snow-patch erosion.

The ability of snow sliding to abrade rock surfaces is, then, not in doubt, though more field experiments are urgently needed. Quantitatively, however, erosion is probably small in comparison with other nivation processes, and insignificant compared with the erosive capacity of avalanches.

Avalanching

Avalanching can be defined as the rapid mass movement of snow on slopes caused by instability of the snow cover. The avalanche may be of dry or wet snow, it may collect varying amounts of debris in its movement, and it may slide in contact with the ground or underlying snow surfaces, or it may be airborne, losing contact with the ground. Instability of the snow pack initiating an avalanche may be related to many factors but the most common causes of instability are overloading (after heavy snowfall), structural weaknesses in the snow (for example, former surface layers that refroze before further snow was added), and percolating snow melt. Tremendous energy may be released in a major avalanche—Allix (1924) estimated 300 million HP (2.24×10^5 MW) in a major Italian avalanche of 1885—and for too long the geomorphological role of avalanches has been inadequately investigated and appreciated. On the other hand, there has been a great deal of research on the causes, prediction and prevention of avalanches and on protection from them, especially since 1931 when the Swiss Commission for Snow and Avalanche Research was founded. An important characteristic of avalanches is their variability both in time and space. In some localities they represent infrequent events; in others they occur regularly, re-using the same routes year after year. They may range in size from minor slips to events of catastrophic proportions that may be ranked with floods, hurricanes and earthquakes as major natural hazards. Their

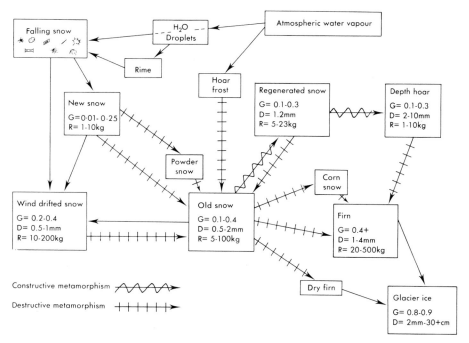

Figure 9.2 Origin and transformation of snow, and links with other phases of the hydrological cycle (US Department of Agriculture, Forest Service, Handbook No. 194 (1968), p. 9).

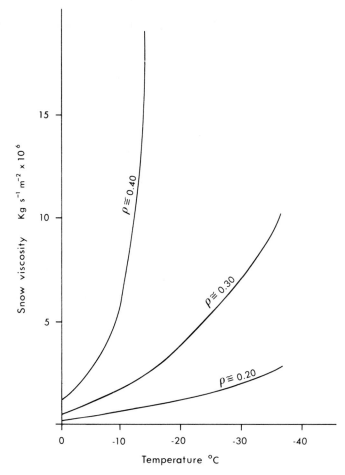

Figure 9.3 Dependence of snow viscosity on density (ρ) and temperature. As density increases, viscosity becomes very sensitive to changes in temperature (US Department of Agriculture, Forest Service, Handbook No. 194 (1968), p. 12).

timing can often now be successfully predicted in areas where there is regular surveillance, but their precise location usually can not.

Just as glacier motion is intimately related to the physical properties of ice, so avalanching depends on the characteristics of the snow cover. However, in dealing with snow, one is dealing with a substance that is much more variable than glacier ice (Figure 9.2). The density of snow ranges from less than 0·01 in the case of flakes deposited under cold calm conditions, to more than 0·25 under temperatures close to melting point. Strength is even more variable: that of old frozen snow is probably of the order of 50 000 times greater than that of fresh powder snow. Tensile strength, critical to the process of avalanche initiation, is affected by three main factors: it increases with density, decreases as temperature rises, and varies with grain size, reaching a maximum in the case of old fine-grained snow. The decrease of tensile strength with the formation of coarse crystals in the snow, as will be discussed below, is an important cause of instability. The viscosity of snow (**Figure 9.3**) becomes very sensitive to temperature as density increases; it also increases with grain size. Strictly, snow exhibits a range of behaviour that includes both viscous and elastic responses. Under stress it may deform elastically, flow as a viscous liquid, or both. Snow of lower viscosity can flow more easily and will not therefore accumulate stresses. Many experimental tests have been carried out on snow samples to support these generalizations. For example, Roch (1966) has studied the shear strength of different layers in the snow cover, showing not only that shear strength varies in a similar way to tensile strength as explained above, but that successive measures vary over time. The way in which the physical characteristics of a snow pack change over time is one of the most fundamental factors in the initiation of avalanches.

Temperature has been mentioned as a principal control on the main characteristics of snow. Temperatures within a snow pack are in turn strongly affected by the low thermal conductivity of snow (see p. 308), by the high heat of fusion of ice (about 80 cal g^{-1}), and by the amount of water or water vapour in the snow. The potential supply of heat to the base of a snow pack is limited to geothermal sources (capable of melting only about 1 cm a year) and to any heat stored in the ground from the previous summer (if the ground was then exposed). In temperate zones the base of a snow pack is nearly always at 0°C, but in arctic areas, permafrost may extend beneath thin snow (see p. 310). Heat supply at the snow surface is the most important variable, and it is this that primarily controls the temperature gradients in the snow. Thus weather conditions play a fundamental part in the changes of snow properties—density, viscosity, shear strength and so on—over time.

Processes of change in snow characteristics are referred to as *metamorphism* (Figure 9.2), and may be caused by temperature variations, by percolating snow melt, and by changes in pressure with snowfall. Added snow layers cause compaction and increase in density; with suitable temperature conditions, water vapour is transferred from snow-crystal points to the central parts of the crystals, giving more rounded crystal forms and reducing internal cohesion (destructive metamorphism). Steep temperature gradients in the snow and high air permeability also favour vertical transfer of water vapour, forming new crystals (constructive metamorphism) with a fragile structure known as 'depth hoar'. The formation of depth hoar reduces the mechanical strength of the snow, forming an unstable base for further increments.

Types of avalanche

Table 9.3 gives a general classification of avalanches and Figure 9.4 portrays some of the characteristic features of the main types. If there is little internal cohesion in the snow, a *loose-snow avalanche* will result. Loose snow becomes unstable on a given slope when one or more of the following changes occur: (i) addition of further light fluffy snow; (ii) metamorphism reduces internal cohesion to a critical level; (iii) lubrication by percolating meltwater. Loose-snow avalanches move as a formless mass and often show no clear sliding surface. They range in size from minor sloughs during a snowstorm, to the relatively rare large slips accompanied by destructive wind-blast. Clean powder-snow avalanches have little geomorphological effect. However, in spring or summer, wet snow may be involved, adding to the weight and destructive power of the avalanche.

The *slab avalanche* exhibits varying degrees of internal cohesion, ranging from a tendency for the snow crystals to adhere, to hard snow slabs that do not easily break. The beginning of a slab avalanche is often marked by a clearcut fracture, and there is usually a well marked sliding surface. Sometimes a subdivision into soft slab and hard slab is made, the latter retaining its slab form during travel even though broken into blocks. Slab formation is attributable to several factors, notably wind drifting and metamorphism of the snow, but the onset of slab avalanching depends not just on the existence of snow slab but on its anchorage to either

Table 9.3   Avalanche classification (de Quervain, 1966)

Criterion	Characteristics	
1. Type of rupture	*starting from a line:* slab avalanche (soft slab av., hard slab av.)	*starting from a point:* loose snow avalanche
2. Position of sliding surface	*within snow cover:* surface layer avalanche (new snow fracture, old snow fracture)	*on the ground:* entire snow cover av.
3. State of humidity	*dry snow:* dry snow avalanche	*wet snow:* wet snow avalanche
4. Form of track	*open flat track:* unconfined avalanche	*channelled track:* channelled avalanche
5. Form of movement	*whirling through air:* airborne powder avalanche	*flowing along ground:* (sliding av., flowing av.)
6. Triggering action	*internal release:* spontaneous avalanche	*external trigger:* (natural, artificial)

the ground or other snow layers beneath it, and on its tensile strength. Figure 9.5 shows the arrangement of forces at the moment of initiation of a deep slab avalanche. Here, depth hoar provides a weak layer beneath the slab. The shear stress τ that tends to break the bond of the slab to the underlying layer is given approximately by

$$\tau = \rho g h \sin \alpha \qquad (9.1)$$

where α is the angle of slope of the potential sliding surface, h is the thickness of the slab and ρ is the density of the snow. When τ exceeds the shear strength of the bond, sliding will begin. However, this simple relationship is complicated by the fact that the slab is also anchored at the top and sides, involving tension and shearing, and by the fact that snow on a slope will be undergoing constant slow downhill creep (see p. 314), adding stresses that contribute to instability. Slab avalanches are more likely to travel in contact with the ground and hence erode and transport debris. Thus, geomorphologically, there is an important distinction (Table 9.3) between airborne avalanches and those that slide or flow over the ground, though in practice, many avalanches fall into a mixed category. The term 'ground avalanche' (*Grundlawine*) is not an ideal one since it can be taken to include both avalanches that slide over the ground (and are therefore potential erosive as well as transporting agents) and those where the sliding surface is within the snowpack. Rapp (1960) suggests 'surface avalanche' (*Oberlawine*) for the latter type, but this term is also ambiguous. Here, 'ground avalanche' will be understood to mean any avalanche travelling mainly in contact with the ground, whether the latter is bedrock, snow or glacier ice.

Shreve (1968a) has discussed the possibility that some ground avalanches may move on a thin layer of compressed air, which would help to explain the high velocities sometimes observed, although evidence of erosion in the form of bedrock grooves and scars is difficult to reconcile with the air-cushion theory (Hoyer, 1971). Whereas ground avalanches are relatively dense (possibly up to $400\,\mathrm{kg\,m^{-3}}$, depending on the amount of included rock debris), airborne avalanches which lack significant contact with the ground consist of a snow/air suspension whose density is usually less than $10\,\mathrm{kg\,m^{-3}}$. The suspension behaves like a heavy gas, attaining high velocities because of the virtual absence of ground friction. Damage

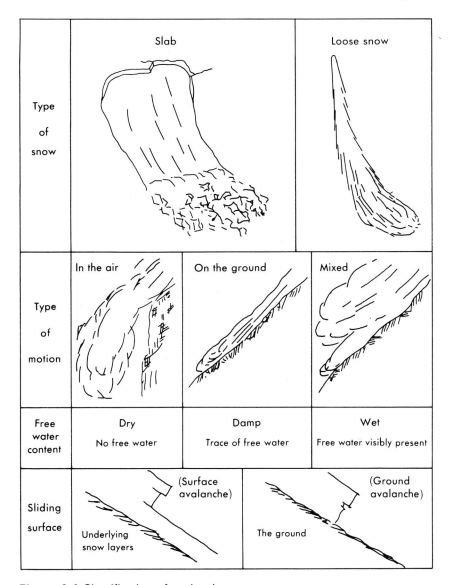

Figure 9.4 Classification of avalanches.

from such avalanches is principally caused by wind-blast (it has been shown that turbulent currents within the suspension may be moving at up to twice the maximum speed of the mass) and by pressure differences, capable of destroying forests and buildings. Wind-blast from the Huascaràn avalanche of 1962 in Peru defoliated trees and stripped branches in side-valleys not actually touched by the avalanche (Morales, 1966).

The distinction between clean and dirty types of avalanche is also important. The former move only snow or ice and are not significant geomorphological agents. Dirty avalanches, more common in the latter part of the snow-melt season when patches of bare ground are

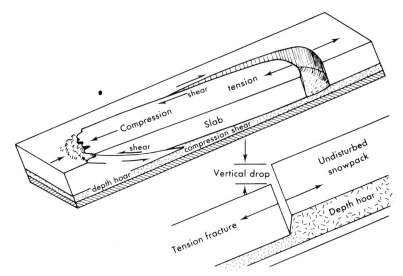

Figure 9.5 Diagram of forces at the moment of initiation of a deep slab avalanche (Bradley, 1966).

appearing and the ground itself is thawing, are now recognized as having very significant functions, eroding their tracks, capable of carrying a great deal of debris in a few minutes or seconds, and depositing distinctive and sometimes massive accumulations at their toe (see p. 323). Rapp (1960) in a detailed study of mass movements in Kärkevagge, Lappland, noted that, of 225 avalanches falling in 6 weeks in 1953, about 75 were dirty and therefore geomorphologically active.

Velocities and friction

Figure 9.6 shows the theoretical velocity profiles of a mixed type of ground and airborne avalanche (*left*) and of a slower moving ground avalanche (*right*). V_f is the speed that is observed and measured (mainly now by photogrammetric techniques); V_m is the maximum core velocity that is much more difficult to estimate but is nevertheless of great significance since it is the source of the greatest destructive forces (de Quervain, 1966). Actual speeds attained by avalanches are impressive but sometimes exaggerated. Experimental studies of artificially triggered avalanches on a 35-degree slope in Japan (Shoda, 1966) using photographic recording gave 57 km h^{-1} for the airborne type and 50 km h^{-1} for the ground type. Eye-witness descriptions of the Puget Peak avalanche in Alaska (1964) suggest a mean speed over 40 km h^{-1}, but possibly reaching 165 km h^{-1} in places (Hoyer, 1971). The Huascarán avalanche in Peru (1962) may have attained 100 km h^{-1} in a zig-zag course along ravines, showing super-elevation effects on the outsides of curves amounting to 150 m, but its average speed over the whole length was about 60 km h^{-1} (Morales, 1966). Loose-snow avalanches may achieve speeds of 200–300 km h^{-1} under favourable conditions. Friction absorbs a high but variable proportion of the energy available. For ground avalanches, values cover an enormous range depending on both snow type and ground type and on whether an air cushion is present. Within the moving mass there are internal losses of energy. Internal motion may be either laminar or turbulent. The former is less common and occurs when

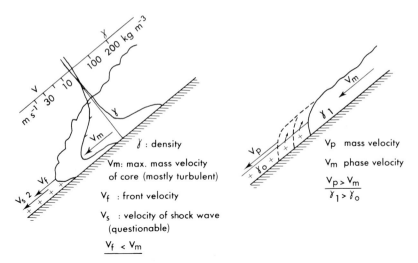

γ : density

Vm: max. mass velocity
     of core (mostly turbulent)

V$_f$: front velocity

V$_s$: velocity of shock wave
     (questionable)

$\underline{V_f < V_m}$

V$_p$ mass velocity

V$_m$ phase velocity

$V_p > V_m$
$\gamma_1 > \gamma_0$

Figure 9.6 Velocities in an avalanche. *Left*: mixed airborne and flowing avalanche; *Right*: slow-moving dry or wet avalanche (de Quervain, 1966).

sliding is within the snow layer or when an air cushion develops, but, depending on velocity, amounts of rock debris involved and track configurations and roughness, turbulence soon develops. If an avalanche is sliding rapidly over bedrock or rock debris, turbulence may be extreme, and not only will more material be carried in suspension but erosive effects on the ground surface will be considerable.

Terrain controls

There is no precise limiting slope angle for avalanching (Figure 9.7). It is uncommon on slopes of less than 15 degrees but much depends on other factors—snow depth and condition, slope roughness and vegetation. Under the most favourable conditions, avalanches have been known on slopes as low as 6 degrees. Large avalanches are most frequent in the range 25–50 degrees, above which they become rare because snow will be unable to adhere to the slope in any quantity. Slope configuration will be a major determinant of the tracks of ground avalanches: steep gullies and steep open slopes are natural avalanche paths. Orientation of a slope will affect the frequency and timing of avalanching. In the northern hemisphere, south-facing slopes receive a greater supply of radiant energy, and in spring such slopes may experience melting and wet-snow avalanching. In winter, on the other hand, snow temperatures may be higher than on other slopes, aiding destructive metamorphism and settlement of the snow and encouraging snow stability. North-facing slopes will be significantly colder; depth-hoar formation will be likely and dry-snow avalanches may be frequent and persistent. Lee slopes with respect to the prevailing snow-bearing winds will also be favoured avalanche areas, as the snow will lie deeper and overhanging cornices may build up.

Ground roughness and vegetation play an important role. Roughness controls the amount of snow needed to fill in irregularities and present a smooth sliding surface. Smooth grass slopes are conducive to avalanching, whereas a slope of large boulders will need a lot of accumulation before a sliding surface can be established and will discourage snow creep. Forest and even scrub will prevent avalanche initiation within it, but will not be immune to avalanches if these originate higher up the slope above the trees.

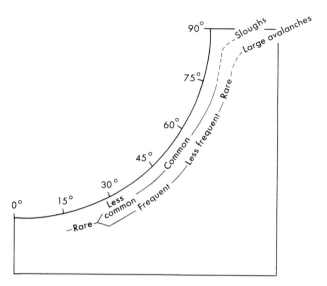

Figure 9.7 Relation of slope angle to probability of avalanche release (US Department of Agriculture, Forest Service, Handbook No. 194 (1968), p. 29).

Geomorphological activity of avalanches

Although much has been written on the prediction and control of avalanches, relatively little has been published until recently on the geomorphological role of avalanches. The most comprehensive and important study of the last 20 years is that of Rapp (1960), in which earlier work is also summarized. Although the valley walls in Kärkevagge, northern Scandinavia, are not particularly favourable to large-scale avalanching, being too steep and little dissected, Rapp shows that avalanching is the third most important process of vertical transfer of debris in the area, ranking after transport of dissolved salts and earth-slides/mudflows.

Table 9.4   Denudation of slopes by avalanches in Kärkevagge, 1952–60 (Rapp, 1960)

Process	t km^{-2} year^{-1}	Average movement (m)	Average gradient	Vertical transport (t–m year^{-1})
Small avalanches	1·4	100	30°	1 050
Large slush avalanches	14	200	30°	20 800

Table 9.4 shows some data based on field observation for two categories of avalanches. The second of these, slush avalanches, will be considered separately below. The average amount of debris transferred by each of the 'small' dirty avalanches was 5–10 m³.

The geomorphological role of avalanches can be considered under four headings:

1 Removal of loose debris from gullies, ledges and cliffs. Weathering processes, especially freeze–thaw action, produce broken rock which is periodically cleared out by avalanches.

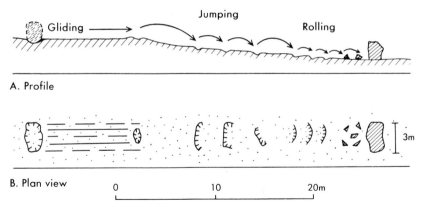

Figure 9.8 Transport of boulders by avalanches and associated scars on bedrock (Rapp, 1960).

2 Erosion of gullies and scarring of bedrock (Figure 9.8). Matthes (1938) showed how 'avalanche chutes' in the Sierra Nevada, California, were enlarged and given U-shaped profiles even in resistant bedrock. Chutes may begin life as minor stream-eroded gullies or lines of weathering and weaker rock etched by mass movement. Rock fragments borne at high speeds by avalanches can crack off new boulders from bedrock by impact. Measurements by the Swiss Institute on the Weissflujoch above Davos have shown forces in excess of $100\,t\,m^{-2}$. The avalanche chutes play an important role in concentrating avalanche energy. Bedrock scars, grooves, and cracks are common features (Hoyer, 1971; Peev, 1966; Rapp, 1960). At the same time, it should be remembered that many avalanches, perhaps the majority, have little or no erosive effect because they consist only of clean snow, are sliding over snow, or are of the airborne type.

3 Erosion and redistribution of talus. The effectiveness of avalanches in this respect is strongly dependent on whether the talus slopes have a protective snow cover (as is likely in winter or early spring) (Gray, 1973). Rapp (1959), Gardner (1970b) and others have observed that talus affected by avalanching tends to have lower average-slope angles than those of other types of talus slope, and that the avalanche talus slopes tend to show concavity in long profile. The same applies to avalanche boulder tongues (Rapp, 1959, 1960), the products of individual avalanches following defined tracks. On the talus slopes, debris tails below prominent boulders are characteristic.

4 Deposition. The process is essentially the dumping of debris and snow in varying proportions. The deposits of major dirty avalanches are of impressive size and testify to the great transporting power of the avalanches. Snow may survive in avalanche deposits for a year or more before finally melting. The following are examples of the huge debris volumes associated with catastrophic events:

Puget Peak avalanche (1964): 460 000 m³ rock (Hoyer, 1971)
1 360 000 m³ snow
Huascaràn ice avalanche (1962): 13 000 000 m³ of debris and ice fragments (proportion of ice, *c.* 25 per cent) including individual blocks up to 6000 t (Morales, 1966)

Table 9.5 shows measurements of debris accretion by small avalanches in Kärkevagge over a 6-week period. The rock deposits are characteristically unsorted, angular and unstable due to the gradual settling process as the snow content melts.

Table 9.5   Approximate quantities of avalanche debris in Kärkevagge, 8 May–22 June, 1953 (Rapp, 1960)

Section recorded:	A	B	D	F	G	H	J	K	L	Total
No. of dirty avalanches	11	13	18	8	11	2	8	1	1	73
Debris volume (m³)	3	5–6	2–3	1	0·5	0·5	0·5	10	0·1	c. 25

Slush avalanches

Slush flows have been defined by Washburn and Goldthwait (1958) as the 'mudlike flowage of water-saturated snow along stream courses'. Rapp (1960) considers that slush avalanches in northern Scandinavia and in other high-latitude areas are essentially the same phenomena, though the wet dirty ground avalanches of the Alps, released as snow slabs on steep slopes outside stream courses, are probably different. The slush avalanches of northern Greenland (Nobles, 1966) occur characteristically in the early part of the melt season, on both land and on glacier surfaces. Most are released as meltwater saturates thick snow resting on ice or ice-veneered bedrock, though a few start with the thawing of frozen waterfalls. They may entrain quantities of debris and thus become important geomorphological agents. Caine (1969) has shown how they affect talus-slope development and contribute to basal concavity. Their transporting power is impressive: Rapp (1960) describes three major ones in Kärkevagge that together moved 700 m³ of rock and earth in the period 1952–60 and showed that their effect was ten times greater than that of all other avalanches in the area (Table 9.4). One of these slush avalanches moved a boulder weighing 75 t for 120 m on a 5-degree slope.

Niveo-aeolian processes

In Table 9.1, three aspects of niveo-aeolian action are listed. The term niveo-aeolian is applied to phenomena associated with wind-blown snow and any included sediment transported and deposited by the wind. Corrasion by wind-blown sediment alone is considered in Chapter 10, but there is also evidence that, at low temperatures, corrasion by wind-driven snow grains can occur (Fristrup, 1953; Teichert, 1939); the hardness of individual grains can reach 6 on the Moh scale at $-50°C$ and etching of soft rock surfaces is possible. The most conspicuous features associated with niveo-aeolian action are, however, depositional. Cailleux (1972) describes how mixed accumulations of snow and wind-blown sediment build up to thicknesses of a metre or more over periods of years in the McMurdo Sound region of Antarctica. In lower latitudes, where summer temperatures are higher, the snow content melts out each year and only the aeolian sediment remains. Rhythmically stratified sequences can build up over time by this process, as Jahn (1972) has described in southwest Poland.

Quantitatively, niveo-aeolian processes are unimportant when ranked alongside other forms of nival activity, especially avalanching; on the other hand, few studies have yet been undertaken, and it may be that some deposits of loess and coversand are strictly niveo-aeolian, the melting-out of included snow playing a significant role in the textural and stratigraphic development of the deposits.

10

Aeolian processes

Andrew Warren

Winds are a result of the redistribution of solar energy from receiving to losing sites. On a global scale, the difference between net loss of energy near the poles and net income in the equatorial regions itself produces a complex pattern of surface winds. This is made even more complex by seasonal changes and local, usually diurnal, effects, as from the sea to the land, are superimposed.

These winds perform geomorphological work either indirectly through waves (Chapter 11), water or directly on the land surface. Since a cover of plants very considerably reduces the wind speed near the ground (Figure 10.1), and since much of the land is vegetated, the wind only manages to perform significant work on beaches swept clear by the waves or tides, on outwash plains of very mobile braided rivers, on bare fields, or in arid deserts. The forms that the wind creates in these places result from the interaction of five main factors (and some minor ones): the velocity and turbulence of the wind on the one hand, and the inherent roughness, cohesion and grain size of the bed on the other. In some places sparse vegetation adds an important complicating factor. Erosion, transport and deposition are the fundamental aeolian processes, although they merge very considerably with each other in practice.

Erosion

Small- or experimental-scale observations

Winds are retarded at the surface by friction. When the wind is light and the surface smooth, flow is laminar, and little or no geomorphological work is accomplished. Effective winds are faster and turbulent, and have a velocity profile as in Figure 10.1A, but with a laminar sub-layer about 0·05 mm thick at the surface. Most of the energy that is transferred to the surface is lost as heat, but some is used to lift and transport particles.

The lifting or 'entrainment' of particles is a function of six kinds of force: lift, shear and ballistic impact must overcome gravity, friction and cohesion.

Lift could happen by the Bernoulli effect, which is the effect that lifts an aircraft wing: air at a higher velocity has a lower pressure. Although not demonstrated experimentally, differences in velocity around grains at rest on a bed could happen in two ways: by the acceleration of the flow over the obstacle itself, or by the difference in pressure between the very slowly moving air in the voids beneath the particle, and in the steeply rising velocity gradient on its projecting parts (Chepil, 1945a). Both would induce low pressure above particles and help to lift them vertically into the airstream.

Shear on the bed is the major driving force in transport by the wind (e.g. Bagnold, 1941; Chepil and Woodruff, 1963). In detail, shear induces a difference in pressure between the upstream and downstream sides of a projection, so encouraging forward motion. Small

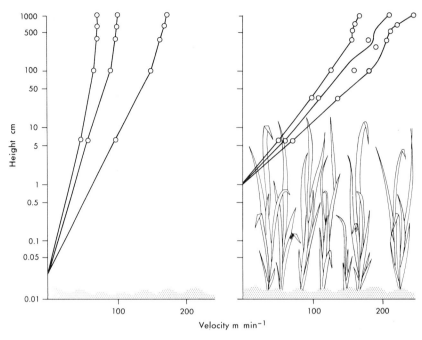

Figure 10.1 Wind-speed variation with height over a flat sandy surface and over a grassy field, showing the low velocities on vegetation-covered surfaces (data from Chepil and Woodruff, 1963).

particles which do not protrude above the laminar sub-layer will not be sheared off, but this can only happen on very smooth surfaces (Chepil, 1945b). If a particle were rolled up and over a larger one, shearing might actually lead to lift, but more commonly it simply rolls the particle forward on the surface. Bagnold (1941, p. 32) noticed such movement before lift in his laboratory experiments.

Both lift and shear are probably greatly affected by turbulence. Winds over rough granular beds develop shallow, turbulent boundary layers in which there is intense eddying. Occasional 'jets' of violent activity induce wild fluctuations of velocity and pressure on the bed. Grains on the bed can be seen to oscillate wildly, though somewhat less intensely than the air pressure, and then suddenly to depart, probably in a rare pocket of very low pressure (Lyles and Krauss, 1971). The fluctuations can greatly increase momentary values of velocity and, since shear varies as the second power of velocity, the effect on entrainment is marked. Such high values of shear therefore overcome static friction and start general movement.

Whatever the process that entrains the first few particles, most subsequent entrainment is by ballistic impact or bombardment by grains already lifted. One or two grains, once lifted, start a zone of movement that fans out downwind, a process known as 'avalanching'. On highly erodible soils, the maximum rate of movement (with all possible loose grains saltating) is reached over about 60 m (Chepil, 1959). If there are loose clods of soil, abrasion by bombardment will break them up and further increase the amount of soil released to the wind (Chepil, 1946).

For particles of the same specific gravity, in air of the same specific gravity, the force of gravity resists movement by way of the size of particles. Most sands are quartzose with a

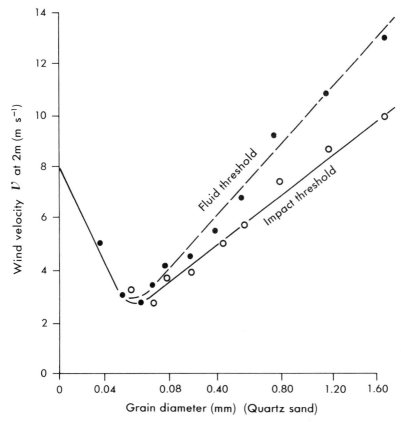

Figure 10.2 The relation between particle size and the threshold velocity for movement. The diagram has been constructed with data from Chepil (1945a) and Hsu (1973). For particles finer than 0·06 mm in diameter, the threshold velocity for movement *increases* as size *decreases*.

specific gravity of $2 \cdot 65$ g cm^{-3}; there are occasional admixtures of heavier mineral sands, such as magnetite, and lighter grains such as those of gypsum or shell. Silts and clays are often of lower specific gravity than sand, and aggregates that include organic matter are even lighter. Organic matter itself is the lightest of the eroded materials.

If, for the moment, we ignore turbulence, or assume it to be directly correlated with mean wind velocity, then for particles of the same density and shape, there should be a direct relation between the size of a particle and the windspeed that will begin to move it (shear increasing with velocity). For particles greater than $0 \cdot 1$ mm in diameter this indeed occurs (Figure 10·2). In reality, mineral grains greater than about 1 mm in diameter are only moved in very high winds, which occur for example where winds are funnelled around or between obstacles.

Detailed observations show that there is a wide range between wind velocities that will move only a grain or two, and those that will move a whole bed of sand (Lyles and Krauss, 1971). If the bed is of mixed grain-size this range is even greater (Chepil, 1945b). The wind needed to start movement by lift or shear is much faster than the one needed to maintain it by ballistic bombardment. Bagnold (1941) distinguished the first as the *fluid threshold velocity* and the second as the *impact threshold velocity* for grains of a particular size.

Cohesion is the second main force resisting entrainment. Even when particles are spherical, the magnitude of the cohesive force between particles less than 0·1 mm in diameter is greater than the weight of the particle. Small particles, moreover, are usually more irregular in shape, or indeed platy (see below), and this further encourages contact between them and so more cohesion; it also means that friction plays a greater role. It has been suggested (Smalley, 1970) that the force (F) needed to rupture a mixture of particles varies inversely as the third power of their diameter (D)

$$F \propto 1/D^3 \qquad (10.1)$$

In fact, where sands are virtually cohesionless, clays are inherently highly cohesive. These considerations alone explain the shape of the curve for small particles in Figure 10.2, but if we introduce moisture the effect is magnified. Sands drain rapidly, and since they hold water at low suction in large pores, they also dry rapidly in a wind. Clays, on the other hand, hold water tightly in small pores, from whence it can hardly be removed, even by evaporation, and where it helps to bind the particles tightly together. One would expect, therefore, an even sharper increase in threshold velocities on Figure 10.2 for wet, fine soils. Experiments show that when water is held at 15 atmospheres (1500 k Pa) or less (the permanent wilting point and wetter), the soil is almost totally resistant to wind erosion (Chepil, 1956).

Since cohesion is so important between particles smaller than 0·1 mm, silts and clays should be resistant to wind erosion. This is true if surfaces are smooth, if the soils are uniformly fine, and if no other material is blown onto them. These are stringent limitations. Because large grains can be either dislodged from a fine soil surface by the wind as aggregates or be blown into it from elsewhere, fine soils, especially silts, can be very erodible when dry. The wide occurrence of wind-blown loess is testimony enough to their mobility.

Clays and some silts can, and often do, resist erosion by their aggregation, either with salts or with organic matter (on eroding agricultural fields) but can travel both as single particles and as small aggregates. Most aggregates, however, are very susceptible to breakdown, and unless trapped in a short distance, soon disperse to their primary particles. The most erodible soils are organic peats which are both fine, light and irregularly shaped. Such soils erode easily when exposed in agricultural fields, as in Florida or eastern England (Robinson, 1969).

Where interparticle cements are really strong, as in old sedimentary, metamorphic or igneous rocks, entrainment can only occur when pieces are actually chipped off the hard surfaces, a process known as abrasion.

Medium- and large-scale evidence

The glacial outwash plains, dry river beds and beaches that are the main sources of wind-blown sediment are too frequently reworked by waves or running water to preserve signs of wind erosion. Only on ancient tablelands, high river terraces, between the dunes in very arid dune-fields or in ancient lake beds does the wind leave any permanent evidence from which to deduce something about the way in which it erodes.

Medium-scale (hand specimen) evidence for wind erosion occurs in distinctively sorted sediments and aerodynamically formed hard-rock surfaces. Desert pavements (stony mantles overlying finer soils) are probably more the result of the washing-away of fine material (Cooke and Warren, 1973) and the sorting of coarse from fine material under the continual disturbance of wetting and drying, heating and cooling or freezing and thawing, than of selective wind erosion, although wind-winnowing does play some part in the process in some areas.

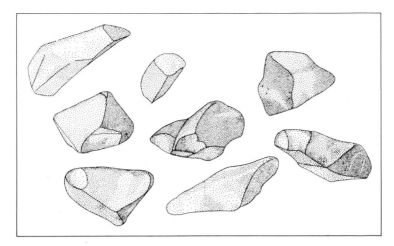

Figure 10.3 Ventifacts (wind-faceted stones) from Antarctica (drawn from a photograph in Lindsay, 1973).

Coarse-particle concentration at the surface also occurs, however, in loose sands, and here wind action is much more important. The concentration of coarse particles on the surface of eroding sands has often been observed in the laboratory. In the field, in the Ténéré Desert, for example, there are large expanses of relatively low relief whose surfaces are composed of a thin layer of coarse sand mixed with a small amount of very fine sand. One explanation of this bimodal 'lag' sand is that the medium-sized grains (about 0·18 mm) bounce quickly downwind off the coarse grains, while the fine sands (about 0·06 mm) are protected between the coarse particles (about 0·50 mm) and so remain to form the fine mode. The coarse sands are moved only by occasional strong winds (Warren, 1972). Surfaces with such sands are very extensive in all of the world's larger dune fields, often covering a much greater area than more accentuated dunes of finer, more uniform sand.

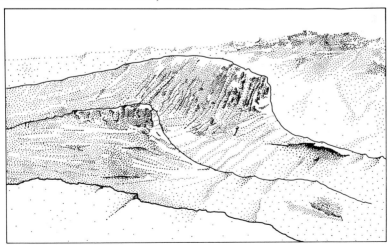

Figure 10.4 Yardangs (wind-eroded mounds) in California (drawn from a photograph in Black-welder, 1934).

On stable surfaces, cohesive pebbles are fluted or moulded by abrasion into aerodynamic shapes known as ventifacts (Figure 10.3). The winds produce smooth shapes around which they flow with minimum resistance. The rate at which abrasion occurs clearly depends on the intensity of the wind and the availability of particles for it to hurl at the rocks. In some dry, windy interdune corridors in the Sahara, in the Antarctic, Arctic and perhaps even

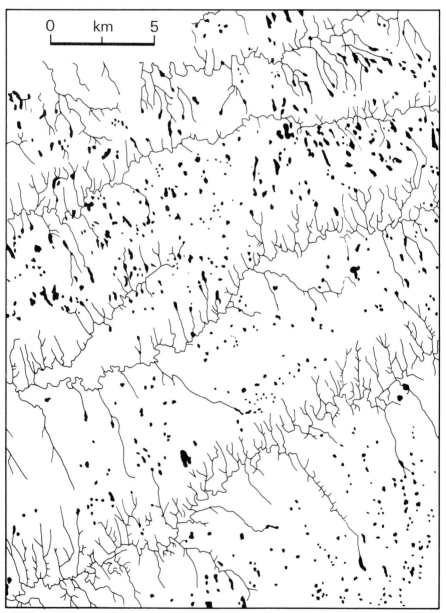

Figure 10.5 Wind-eroded grooves (as shown by the alignment of minor stream valleys) and hollows in central Kansas (after Frye, 1950).

between dunes behind temperate-zone beaches, ventifacts of a sort may appear in tens rather than hundreds of years. In the dry Coachella Valley of Southern California, bricks and lucite rods were distinctly eroded over a period of 11 years (Sharp, 1964). Sharp was able to throw light on the old problem of how three-sided ventifacts (*dreikanter*) are formed; he found that sand beneath the bricks used in his experiments was undermined by the wind, so that the block toppled slowly forward to reveal a new face to the wind. Other explanations of multiple-faceting are discussed in Cooke and Warren (1973).

At the large scale the wind can carve aerodynamically shaped mounds or yardangs (Figure 10.4) and can also excavate rounded hollows and long parallel grooves from incohesive and even cohesive deposits. Many such hollows, widespread on the North American High Plains (Figure 10.5) and possibly in northwestern Europe, are thought to have been eroded in dry late-Pleistocene times (Frye, 1950). They were probably formed in the same way as similar features, sometimes called pans, which occur in semi-arid climates. In these, it is thought that intermittent flooding and drying prepared loose sediment by disturbing and aggregating it with salt crystals, and that the wind then blew the aggregates onto nearby dunes, which are composed of redispersed clays and silts, particularly gypsum. A similar process occurs in some dry coastal environments such as the Gulf Coast of Texas and the Senegal delta (Bowler, 1973).

Grooving, sometimes on a vast scale, is the most spectacular of the effects of wind erosion. It was probably first noticed in the Great Plains in the Dakotas and Montana (Russell, 1929) where minor streams are oriented in the same direction over a vast area (Figure 10.5). Paha ridges, or long parallel mounds cut in silty material nearer to the Wisconsinan glacial front are probably related to the grooves (Flemal *et al.*, 1972) as are similar features in the Rocky Mountain States and Oregon (Lewis, 1960). In modern desert environments, similar or even more spectacular grooving has been noted in Iran (Weise, 1974), around the Tibesti Massif in the Sahara and in Peru. The reason for the concentration of wind erosion into lines parallel with the wind will be discussed below.

Transport

Entrainment and transport are really a continuum, but once in motion, particles take part in new processes. Aeolian transport makes a distinction between coarse particles that roll or hop slowly along the surface, and fine particles that move rapidly away in suspension.

Fine particles

The upward velocities in turbulence are easily able to lift silt and clay particles, and once lifted, these fall only slowly back to earth. If it were not for rain, they might, and indeed do, travel enormous distances. The platy shape of some fine particles is an additional factor in keeping them in motion. They meet other particles rarely, and in collision do not have either the momentum to induce abrasion in each other, nor the strength to confine damage to mere chipping. They thus remain platy, unlike sands, which become rounded in transport.

Saharan dust is frequently collected in northwestern Europe (e.g. Stevenson, 1969). On satellite imagery, clouds of dust have been seen travelling westward in the trade-winds from the southern Sahara towards the West Indies (Morales, 1979) where they have accumulated in soils. Dust reaching the Middle East from the northern Sahara appears to travel up to a height of 1·5 km in the atmosphere and stay there for a few days. It can be seen, on satellite imagery, to hang around the fronts of eastward-moving depressions. Several million tons of dust can be carried in a single heavy storm (Yaalon and Ganor, 1974). In the Pleistocene, some loess was carried far beyond the edges of the European ice sheets and from other sources such as braided streams and wind-eroding sediments in China, northern Pakistan and North America. Beyond the deep and obvious loess deposits, appreciable fractions of aeolian dust have been added to almost every soil on earth.

Most loess, however, does not travel far from its source. The eastern sides of the Mississippi and Missouri Rivers are followed by a high 'loess lip' where much of the dust must have lodged only a few kilometres from its source. Beyond the loess lip the thickness declines on a rapidly falling logarithmic curve. The proximate source of loessic dust seems often to have been the exposed banks of braided streams.

Airborne dusts are sorted in travel but not appreciably abraded (loess is created before its travel in the wind, which merely sorts it; see Smalley, 1977). In both the Great Plains and Hungary, the fine-sand fractions of Pleistocene loess decrease downwind (Swineford and Frye, 1951) and modern dust fall-out in Israel becomes finer northward away from its source in Sinai (Yaalon and Ganor, 1974).

Sands—small-scale and experimental observations

Sands are moved close to the surface. In 'creep', sands too coarse to be lifted roll along the surface under the force of bombardment by saltating grains. 'Saltation' (Latin *saltare*, to leap) is a hopping motion in which particles are ejected almost vertically into a faster and faster wind which drags their paths round to be momentarily parallel with the flow when the upward acceleration is balanced by gravity (Figure 10.6). The subsequent downward direction of motion is a resolution of the gravitational and drag forces which produces a straight line at between 6° and 12° to the horizontal.

Few saltating grains reach higher than the well defined upper edge of a cloud of bouncing grains. This is seldom higher than a metre, though this height depends on particle sizes, wind speeds, and the character of the surface. Of the few particles that do leave the cloud, a surprisingly large proportion are quite coarse, and field measurements show that the average grain size of particles in high winds actually increases with height in the lower part of the cloud (Figure 10.7). This is probably because of the better bouncing and greater momentum of large particles (Sharp, 1964).

The cloud of sand above an eroding sand patch removes energy from the boundary layer and slows it down. By contrast, saltation over a hard stony bed is faster (and higher) because of more frequent good bounces. Thus, if a sand stream, driven by a strong enough wind to move sand on the pebble surface, encounters a sand patch, it would be slowed down and might drop some of its sand load. A light wind, unable to move sand from between the pebbles of the pavement, might still be able to move it on a sand patch. These observations led Bagnold (1941) to the seemingly paradoxical conclusion that light winds erode, or elongate a sand patch or dune, while strong winds build it up. Surfaces that release large amounts of dust to the wind also slow it down.

Most experimental work has shown that the rate of sand transport is proportional to the third power of wind speed (Cooke and Warren, 1973). Other factors influencing the rate are the grain size and density, the air density (varying with altitude and temperature), the character of the surface, and the height at which the wind measurements are taken. A simplified formula, useful for comparative work, can be adapted from Hsu (1973):

$$Q = (4{\cdot}97D - 0{\cdot}47)\left(\frac{0{\cdot}4(\overline{V}_z - 275)}{\ln z(gD)^{1/2}}\right)^3 \qquad (10.2)$$

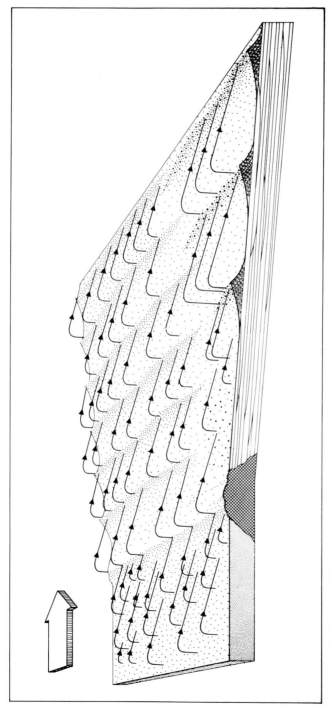

Figure 10.6 A simplification of the process of ripple formation by saltating sand grains. In the foreground, coarser sands, travelling as creep, collect at ripple crests, so helping to create longer-wavelength ripples.

in which Q=weight of sand moved in tonnes per metre width per annum; z is the height of wind-speed measurement in m; D is the mean grain diameter in mm, and $\overline{V}_z$ in cm s^{-1} is the hourly averaged annual wind velocity from a particular direction.

The power relation between weight moved and wind velocity means that much more sand is moved by strong than by gentle winds. But, as in most of the climatically-controlled geomorphological processes, there is a Poisson-type distribution of magnitude: few calms, a large number of gentle breezes and fewer and fewer stronger winds. In Figure 10.8, a hypothetical, but not unrealistic, distribution of windspeeds is compared to the relative amounts of sand that they would move. It can be seen that the moderately fast and moderately common winds do most of the work (as in most geomorphological processes—see Wolman and Miller, 1960).

In the case of dusts, the rate of wind erosion is influenced both by wind velocity and the absence of water. The recurrence interval of a synthetic measure (average annual wind

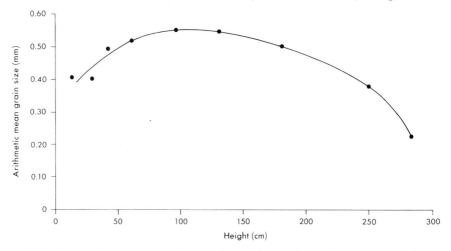

Figure 10.7 The relation between height and particle size in a saltating cloud of sand. The particles actually become, on average, coarser with increasing height in the lower part of the cloud. (after Sharp, 1964).

velocity divided by the Thornthwaite precipitation–evaporation ratio) shows that intense wind erosion occurs on average every five years on the American Great Plains (Chepil and Woodruff, 1963).

Like dusts, sands are sorted as they travel along in the wind. One aspect of this has been described above (the separation of a coarse or bimodal 'lag' sand from a medium-sized saltating sand). In fact, a whole range of sorting mechanisms fractionates sands by size: on ripples, on the windward and lee-side slopes of dunes, and between different parts of a dune field. An example of sorting occurs in the coastal dunes of the Texas Gulf Coast (Figure 10.9) (Mason and Folk, 1958). Beach sands are blown inland onto dunes and eventually form a low 'aeolian flat' behind these. As they leave the beach, a proportion of the coarsest sand is left behind and in transport more and more of this is dropped. On the aeolian flat a finer fraction of dust, perhaps lost in earlier transport, is added again.

Sands are sorted not only by size, but also by shape. Although irregularly shaped particles (such as long thin ones) may stand up to be lifted more readily than spherical ones, it seems that the spherical ones travel better, having less resistance to movement through the air

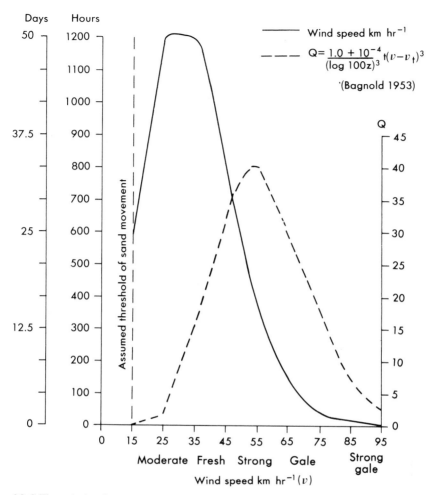

Figure 10.8 The relation between wind velocity and sand movement. The wind data are based on a hypothetical, but realistic, poisson distribution through a yearly cycle. Q refers to the rate of sand movement in tonnes/m-width/year. t is the period of time that winds of velocity v blow during the year and v_t is the threshold of movement. Neither 'moderate' nor 'gale'-force winds blow nearly as much as 'strong' winds blowing for no more than about 25 days in the year.

(Williams, 1964), and indeed they form the greatest proportion of dune sands. Transport, of course, further rounds the grains as they chip pieces off each other at each impact.

Sands—medium- and large-scale evidence

Although sand travels by creep, saltation and slip-face movements alone, some of it is temporarily stored during transport in bedforms (ripples and dunes). This has been termed 'bulk transport'. The sand particles are behaving in dunes much as cars in a traffic jam. A car (or grain) enters the jam (or dune) at a fair speed, slows down (or in a dune, stops) and, after a period leaves it again at a fair speed. The jam (or dune) moves forward at a speed of its own, though each of its constituents is stationary. Some grains merely leave one side of a dune and are re-incorporated in the other.

These temporary stores are created in two ways: ballistically or aerodynamically. Ballistic bedforms occur only at a small scale, namely as ripples (features between 0·5 cm and 2 m apart, and between 0·1 and 5 cm high). Most ripples were first recognized to be ballistic by Bagnold (1941). If saltating sand encounters a slight irregularity in the surface, more sand will be thrown up from the slope facing the wind than from the one in the lee which is protected from bombardment (Figure 10.10). If all the sand grains were of about the same size, they would all reach the surface at about the same distance downwind of the irregularity where they would build up a mound until its upwind slope provided again for the same amount of sand to leave as is received. This new mound would build up another to its lee and the process would fan out downwind until a large surface was ridged transverse to the wind.

If there were some coarse grains, these would travel as creep, but only on the bombarded upwind slopes. They would accumulate as a coarse deposit on the ripple crest from which they would be moved only slowly (Figure 10.10B). They would thus increase the ripple altitude. As they accumulate, the coarse grains provide a better bouncing surface for the saltating grains, allowing them longer saltation paths which, in turn, leads to a wider ripple spacing. The most extreme result of this process is the 'mega-ripple' found where there is a large proportion of coarse sand and winds strong enough to move it. These may be 5 or 6 m in wavelength and 50 cm in height (Sharp, 1964; Ellwood, Evans and Wilson, 1975).

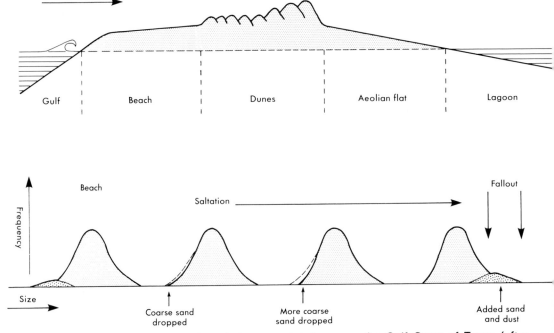

Figure 10.9 Sand sorting by the wind on coastal dunes on the Gulf Coast of Texas (after Mason and Folk, 1958). Sand blown from the beach has fewer coarse grains than that laid down by the waves. As it is moved inland more of the coarse load is left behind. In the 'aeolian flat' behind the dunes, the dune sand is joined by a finer fraction.

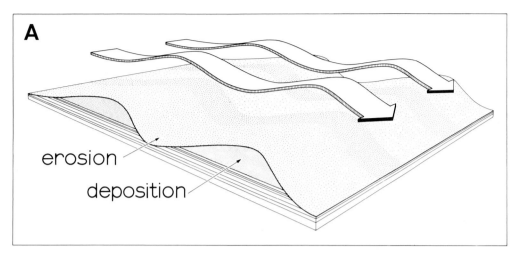

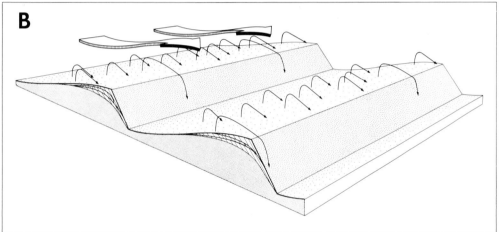

Figure 10.10 The hypothetical first two stages of transverse dune formation. **A**: a wave-like pattern in the wind erodes and deposits alternate ridges and hollows; **B**: slip-faces develop.

Aerodynamic or 'fluid drag' bedforms occur in at least three evidently distinct size groupings. The smallest are seldom seen, being of the ripple scale and usually overridden by the more pronounced ballistic process. Aerodynamic ripples have been created in the laboratory in fine, uniformly graded sands (Bagnold, 1941) and signs that aerodynamic motions at this scale occur can be inferred from the parallel sand streamers that can be observed in almost any blowing sand. Occasionally these are fixed into longitudinal ripples on the ground.

Aerodynamic bedforms, which occur clearly at two large scales (dune and mega-dune), are forms that result from an interplay between pre-existing secondary motions in the wind and the bed. They can occur on cohesive and particulate (clastic) beds. In cohesive beds they leave erosional forms that are almost always longitudinal (small ripple-sized flutes or huge grooves spaced at a kilometre or more). The more common aeolian bedforms are in loose sands. They seldom occur, if ever, in silts or clays.

Over a water surface the secondary motions of the wind (Chapter 3) generate moving waves or constantly shifting drift-lines. This shifting must also occur over a flat desert plain, but when the wind carries a large amount of loose sand, the motions are trapped by accumulations of sand and begin to interact with the bed, generating new secondary motions. The process can be complicated, especially where the wind direction and turbulence vary through a seasonal cycle. Here only some simple cases are outlined, partly by analogy with subaqueous ripple formation as described by Allen (1968).

Suppose that a flat plain is underlain by loose sand to a considerable depth and that blowing across it is a wind of constant direction, whose velocity exceeds the threshold speed for the sand. Wave-like motion develops transverse to the wind in some manner, perhaps during a rather rare temperature inversion, perhaps in the lee of an upwind escarpment. Where the waves descend near to the ground there is more sand movement than where they do not (Figure 10.10A). The sand eroded from the high-velocity strips moves into the low-velocity strips and builds up there until the velocity at the surface of the new mound is sufficient to remove as much as comes in (as with ripple formation). The process might have started with only a slight disturbance of the surface, but quite different, perhaps un-wavy, winds would tend to perpetuate it because they would be slowed down and accelerated according to the pre-existing pattern. This, then, is a process that, once begun, generates itself.

The process of bulk transport concerns the movement of these bedforms. In a regular current, the upwind slope of the dune is subject to more erosion than the lee. The eroded sand is deposited just beyond the crest where the current falls, and eventually reaches a slope just greater than the angle of rest for loose dry sand ($33° \pm 1°$). It then slides down to form a slip-face (Bagnold, 1941, pp. 201–3) (Figure 10.10B). At the downwind edge of a dispersed dune field in Mauretania, it has been calculated that $93\,000\,m^3\,year^{-1}$ are being rolled over or stored in this way whereas some 7 to 13 million m^3 are moved in saltation over the desert floor (Sarnthein and Walger, 1974). Small dunes travel more quickly than large ones, so that they often collide and merge with larger ones, and thus disappear more quickly (Bagnold 1941, pp. 203–5). In Peru, dispersed transverse dunes (barchans—see below) move at between 17 and 47 m per year (Hastenrath, 1967).

The pattern of straight parallel dunes, transverse to the wind, as derived above, is rare in nature. Further mechanisms have to be invoked to set in train more interactions to produce a more realistic picture. Suppose, now, that vortices appear, either weakly, or, as before, in a rare, brief event. These sweep sand from the part of the ridge affected by the diverging vortices and towards the zones of convergence. Because lower dunes travel more quickly, the eroded sections move ahead of the enlarged ones and the ridge becomes sinuous. Any wind, with or without vortices, that flows over these sinuous ridges, encounters oblique steps at the crests of the dunes, and generates new vortices in their lee (Figure 10.11). In a regularly repeating system, vortices meet at points, such as A in Figure 10.12, to produce vortex pairs which can then deform the next ridge downwind in the same way as the first. This is another example of a self-generating process. The pattern resulting from this hypothetical process is more regular than most found in nature, but is topologically similar. Such patterns are known as 'networks' or 'aklé'.

Most dune patterns can be seen as special cases of this model. Barchans (Figure 10.13) are a case where the sand is so sparse that the diverging vortices sweep it entirely on to the convergent areas. Barchans at In Salah in Algeria have been seen to generate distinct and strong vortices at their tips (Figure 10.13) which sweep sand downwind in well-defined streamers (Knott, pers. commun.). Barchans are more widely spaced than in network dunes, but have the same *en échelon* arrangement (Lettau and Lettau, 1969).

A more common, indeed widespread, pattern develops when longitudinal vortices dominate the flow to form long ridges (Figure 10.14). These frequently merge where rising vortices are replaced from either side. The transverse dunes become mere crenulations on the long ridge crests. The grooves eroded in cohesive rock which were mentioned above must be due to very persistent vortex flow of this type. In both these cases, the vortices must have been generated by intense heating from below.

The term 'seif' is usually applied to shorter, less regular features (Figure 10.15). They are often the elongated arms of barchans, as can be seen in Figure 10.16, and, as such, seem to be oblique to the main barchan-forming wind. There are probably many different ways in which seifs develop (Cooke and Warren, 1973, pp. 296–306), but many, if not most, are associated with seasonally or diurnally alternating wind régimes. In the late-Pleistocene Sand

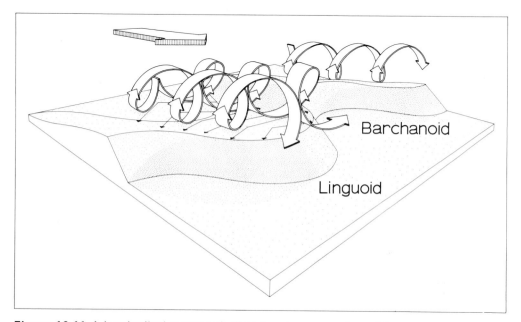

Linguoid

Barchanoid

Figure 10.11 A longitudinal vortex pair distorts a transverse dune ridge.

Hills of Nebraska, for example, a number of long ridges, up to 30 m high, are associated with barchans of a similar size (Figure 10.16). Several pieces of evidence (such as the orientation of former slip-faces) point to there having been two principal wind directions when the dunes were formed in a dry late-Wisconsinan period (Warren, 1976a). Strong winter northwesterlies evidently created mega-barchans with roughly south- and east-pointing wings. Weaker, probably summer, southwesterlies encountered the northern slopes of the east-pointing wings as oblique steps, steepened them into slip-faces, and formed eastward-moving vortices on their far sides. As the bedform developed, both sets of winds induced eastward-moving vortices in between which high, long, seif-like, sand ridges were formed. This reconstruction differs little in essence from the pattern of seif development proposed originally by Bagnold (1941, pp. 222–4). However, the Nebraskan ridges probably diverged considerably from the resultant direction of sand transport, which was thought to coincide with seif orientation by some workers (e.g. McKee and Tibbitts, 1964). The down-wind

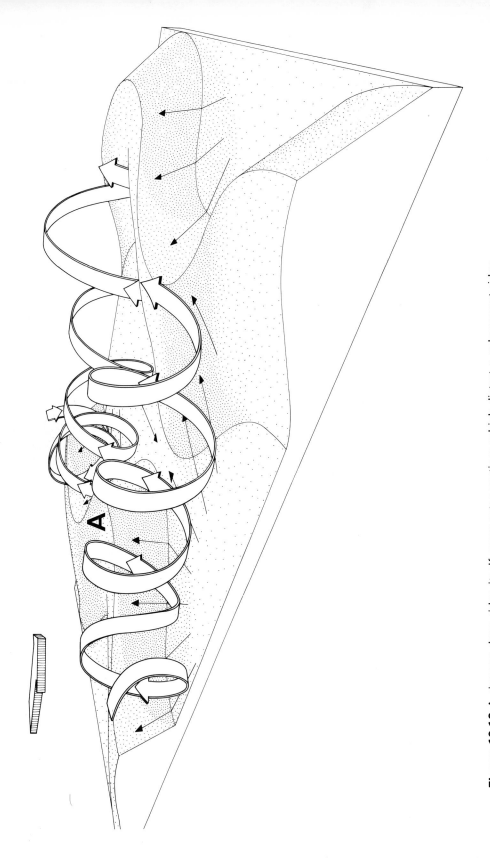

Figure 10.12 A sinuous dune ridge itself propagates vortices which distort a subsequent ridge.

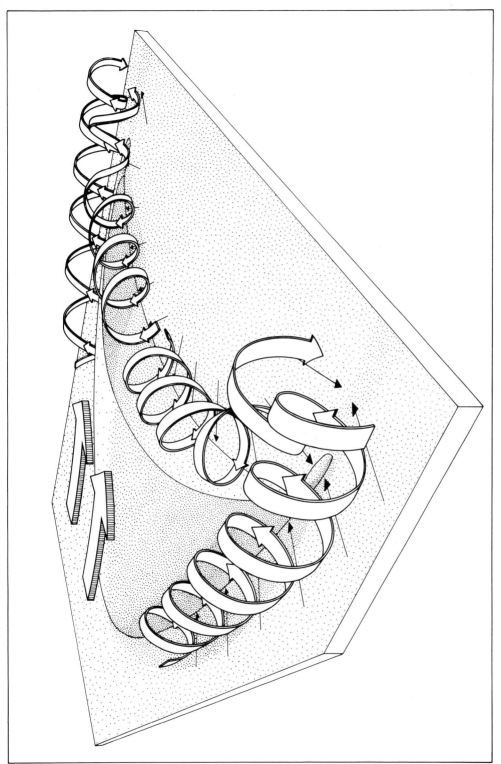

Figure 10.13 The pattern of flow around an isolated barchan dune. This is reconstructed from field data collected near In Salah in Algeria by P. Knott (Pers. comm.).

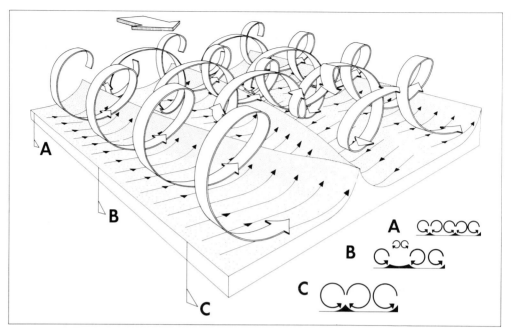

Figure 10.14 Longitudinal dune ridges and a hypothetical reconstruction of a vortex system (formed mainly by surface heating) that might form them. The ridges are similar to those found in central Australia. The central vortex pair is rising buoyantly and being replaced from either side to leave a Y-junction similar to those found in Australia.

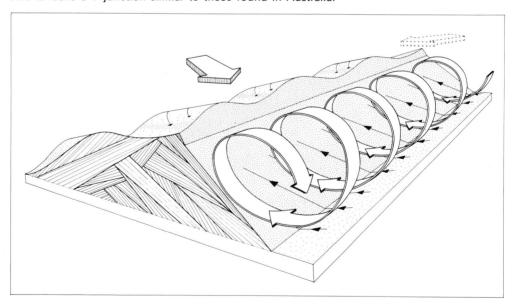

Figure 10.15 The formation of a seif dune by winds blowing from two principal directions, either at different seasons or at different times of the day. The slip face and shallower beds dip in two directions.

extensions of zones of dense sand in Nebraska (i.e. the resultant direction of sand movement) cut across the lines of the seifs (Figure 10.16).

In any desert landscape there are numerous scarps and hills that divert sand flow and allow dunes to grow in calmer pockets. These dunes are known generally as 'obstacle dunes'. Some of the highest dunes in the world grow in zones parallel to escarpments where winds approaching the scarp create a huge standing vortex that sweeps sand from an intervening corridor together with sand brought in by the wind (Figure 10.17). Because they mirror the form of a sinuous scarp, these are known as 'echo-dunes'. The highest recorded dune in the world (400 m) in the Issauane-n-Tifernine in Algeria occurs in such a position.

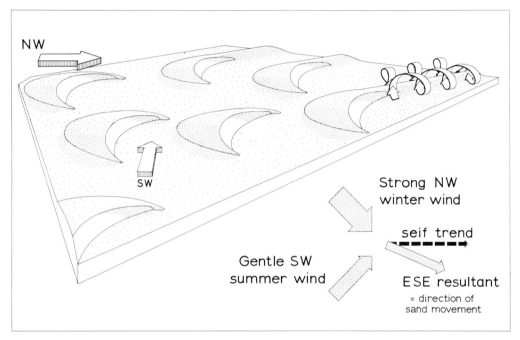

Figure 10.16 A hypothesis for the formation of mega-dune barchans and seifs in Nebraska in late Pleistocene times. Strong northwesterlies probably formed the 100-metre-high barchans, while weaker southwesterlies helped to form the east–west seif-like ridges (after Warren, 1976b).

The coarse (or bimodal), slow-moving, and medium-sized, fast-moving fractions that are sorted by the wind into different parts of a sand sea (see above) form different dune types. Coarse sands can be moved only by high winds, and usually form transverse features with long wave-lengths, perhaps because the faster winds have longer wave-length eddies. These dunes, with firm surfaces and no slip-faces, are known in Arabia as *zibar*. The medium-sized fractions that bounce off downwind accumulate in softer, higher dunes with slip faces (Warren, 1972).

The effect of the interaction of roughness, pressure gradient and Coriolis force (discussed on p. 65) on dune trends is most clearly seen on coasts where an onshore wind leaves the smooth surface of the sea and encounters the rough surface of the land covered by coastal dunes (Warren, 1976b). Figure 10.18 shows some examples. In coastal dunes, as in deserts,

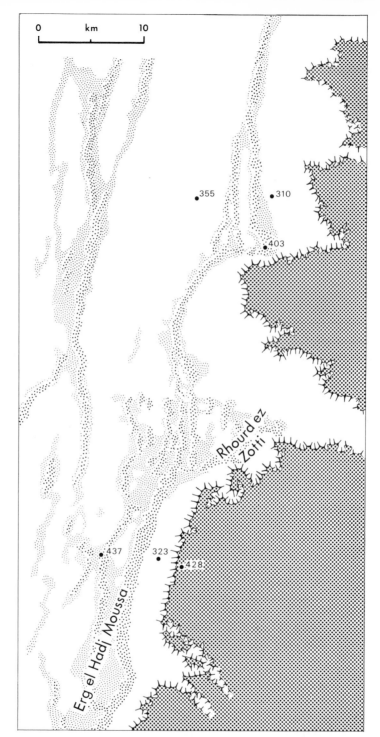

Figure 10.17 Echo dune near El Golea in Algeria. The mega-dune, on the left, shown with a coarse stipple, rests on lower dunes with a closer, finer stipple. The scarp in the centre fringes a stony plateau to the right (heavy stipple). The dune follows the line of the scarp, the intervening corridor being kept free of sand by a 'bolster' of air as shown in Figure 3.17. The dominant summer wind is from the north-east. Heights in metres (from IGN 1:200 000 series, sheet NH-31-XVI, Hassi Djafou).

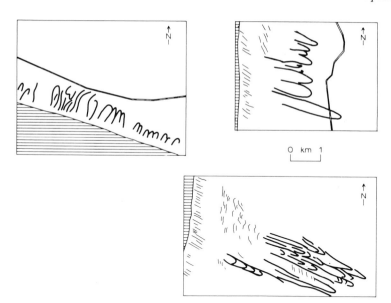

Figure 10.18 Coastal longitudinal, transverse and parabolic (heavy line) dunes in California (after Cooper, 1967). The transverse and longitudinal dunes near to the shore are not at right-angles, because of the surface roughness effect.

the rougher surfaces of transverse dunes evidently deflect the wind more than those of longitudinal ones, as can be seen on the Figure.

A common, but not universal feature of dunes is the coexistence of two distinct size-groupings. Dunes (0·1 m to about 15 m high and 3 m to 600 m apart) may occur beside and on top of 'mega-dunes' (20 m to 400 m high and 300 m to 3 km apart). The separation, which is also found in bedforms beneath the sea and large rivers (Jackson, 1975), may be associated with two scales of atmospheric turbulence. An overlapping system of dunes and mega-dunes, each with different grain sizes, and the interactions of a complex wind régime, can produce some very complicated patterns (Plate XXVII).

Deposition

As between erosion and transport, there is no clear demarcation between transport and deposition except with fine particles. The biggest mega-dunes, especially if they are longitudinal or echo-dunes, are turning over and moving very slowly indeed. If dunes ride up over each other, if their lower strata are cemented by salts such as gypsum, or if the tectonic or climatic environment changes, parts of them can be preserved as geological strata.

Fine particles

Fine clay-sized particles would remain permanently in suspension in the air, were they not washed down by rain, for which they themselves may act as condensation nuclei. Others may be precipitated by electrical effects. For example Beavers (1957), observing that dust storms often exhibit strong electrical activity, probably because of the intense movements of particles in them, speculated that the Pleistocene quartzitic silts blown only a short distance from the Missouri and Mississippi floodplains might have one type of electrical charge, and

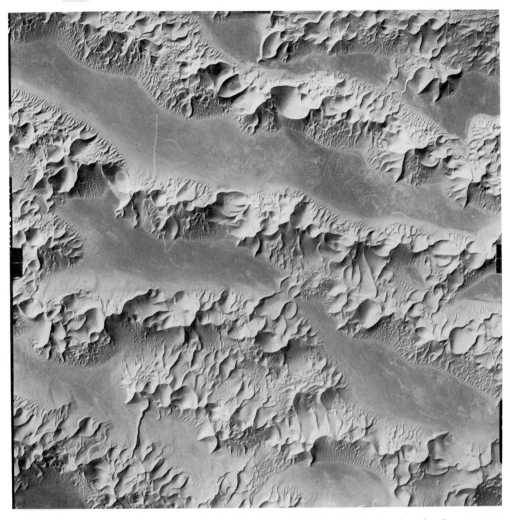

Plate XXVII A complex dune pattern in the Grand Erg Occidental, Algeria, showing long mega-dune ridges with superimposed dunes. (Photo NH31–XI–271, IGN, Paris)

would be precipitated by far-travelled clays with the opposite charge. Precipitation by rain or by electrical effects would explain the extraordinarily uniform, unbedded nature of loess deposits, and the frequent presence of montmorillonitic clay that is not found in the silt source areas. Calcium carbonate is another common constituent of some loesses and, like the clay, is probably much farther-travelled than the silt.

Loessic silt and clay deposits do not exhibit any bedforms, except when there is a considerable mixture of sand, but they are nonetheless not uniform. The 'loess-lips' of the Missouri and Mississippi were mentioned above. In addition loess tends to cloak the windward sides of valleys more than the leeward (Yaalon, 1974) and long loess-free 'shadows' occur to the lee of higher hills.

Loess is an inherently more cohesive material than sand, because of inter-particle chemical

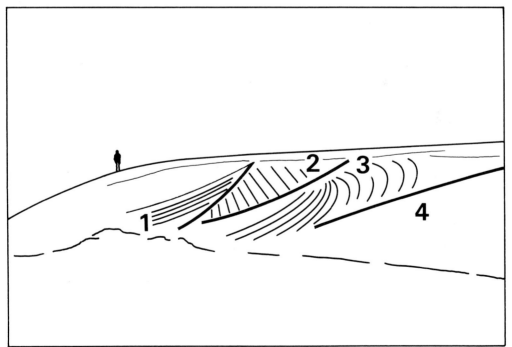

Plate XXVIII A seif dune wetted in winter and eroded by winter winds to reveal successive summer cycles of deposition, labelled 1, 2, 3 and 4. (A.W.)

bonds, because of greater moisture retention, and because of the irregular shape of fine particles. The clays precipitated with it add to its cohesion, as does the calcium carbonate. Moreover, loess deposits are much less permeable than sandy ones so that not only are they eroded
by streams, but support steep cliffs on stream banks.

Sands

By far the most widely preserved sand structures are parallel beds of slip-faces at 30–40°
(Figure 10.19). These are created either by avalanching of dry sand, or by the slumping
of blocks weakly cemented by dew or rain. Avalanche deposits produce thick beds with
coarse grains on top, which thicken downslope. Mass-movement produces minor faulting
and contortions, especially towards the foot of the slip face (McKee *et al.*, 1971). Slip-face
beds are very loosely packed, and provide a soft surface where they appear on the windward
slopes of migrating dunes. In buried deposits it is often found that steep-angle beds have
a bimodal distribution of directions. These are the directions of the two wings on either side
of the crescentic dunes.

   Between dunes, and in limited areas on the crests, low-angle fore-set beds are laid down,
usually from migrating ripples or from suspension. These are more regular and thinner than
the beds on slip faces. The virtual absence of preserved ripple marks distinguishes aeolian
from fluvial sand deposits.

   In dunes such as seifs, in which the wind direction alters seasonally or diurnally, each
new wind erodes a surface across the old beds and deposits material on new parts of the
dune. The eroded minor unconformities are characteristic of seif-bedding, as are beds laid
down at angles between 12° and 25° on the dune flanks (Figure 10.15). Repeated regular
cycles of erosion and deposition can sometimes be clearly seen on seif flanks (Plate XXVIII).

The effects of sparse vegetation

Numerous stems or branches of grasses and low bushes considerably retard the wind at
ground level. At a larger scale the arrangement of scattered trees and bushes in a climax
desert-vegetation community seems also to reduce most winds to well below the damaging
thresholds (Marshall, 1971). In the relatively untouched arid scrublands of Australia, where
Marshall worked, there seems to be little likelihood of entrainment or transport, except on
dry wastes and saline lake beds.

   Vegetation interacts with aeolian sedimentation only on the margins of really arid or
vegetation-free areas, for example where sand is swept into an oasis in which bushes are
sustained by shallow groundwater, or where sand sweeps off a beach kept clear by waves
and tides, and into a vegetated hinterland.

   The principles of the process are similar in each case. The wind velocity up to tens of
centimetres above the surface is sharply retarded by the plants (Plate XXVIII), and sand
drops out around the first few stems. In the desert, where hardy, woody shrubs are the important survivors, this produces hummocky, isolated 'shrub-coppice dunes' (Gile, 1966). On
temperate coasts (but not evidently on the majority of tropical ones), most of the plants that
tolerate these situations are grasses, and the dunes that build up form a more continuous
ridge. In both cases some plant species have adapted to being buried by growing upward
through the accumulating sand and the dune grows with them. The best known of the
temperate plants adapted to this growth is 'marram' (*Amophila arenaria* in Europe) which
cannot survive in a fully active state unless new sand and its nutrients are constantly being
added from the beach. Marram dunes reach enormous heights (300 m) on the Aquitanian
coast of France and appear on beaches in many parts of the northern hemisphere. Where

the rainfall is sufficient these dunes hold large groundwater reservoirs, and the ultimate height of the dunes must be controlled by the wind velocities, which are greatly increased at the summit (Olson, 1958). Lower on the beach than the marram, more salt-tolerant (but less sand-tolerant) grasses and herbs trap lower 'fore-dunes'.

In arid climates, sand-trapping plants often cement the dunes they grow in with root-linings of salts as they filter off water from the saline groundwater. In time the plants may die to leave cemented mounds which are then eroded like yardangs by the wind.

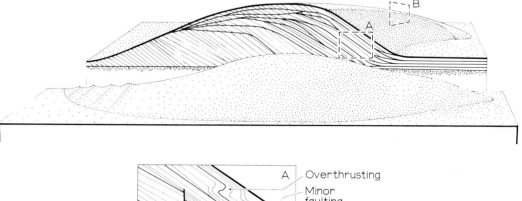

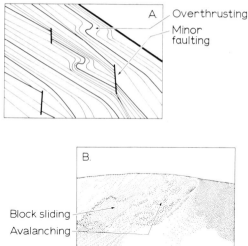

Figure 10.19 Bedding on a barchan (data from McKee, 1966 and from field photographs).

In temperate climates, salts such as calcium carbonate are progressively lost from the fixed sands and they become less, not more cohesive, but their subsequent life is controlled by a different set of processes. If the sand-colonizers are destroyed, either artificially or by rare storms, or if they are replaced by more fragile, acidophilic plants, and these are in turn destroyed, the wind, funnelled between remaining vegetation, attacks the sand surface again and erodes a blow-out or deep hollow. This can happen with increasing speed as the wind is funnelled more and more effectively and as eroded sand blasts the surrounding vegetation.

On many low and not very quickly growing dunes, the blow-out, deepened to a pebbly dune base or to the water-table, is the end of the process. On coastal dunes with a strong

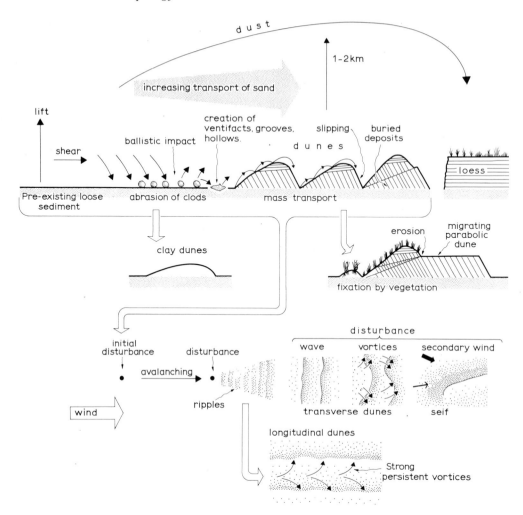

Figure 10.20 A summary of aeolian geomorphological processes.

sand supply, especially on the west coast of North America, the sand from the blow-out buries vegetation downwind, and new sand, both from erosion and straight from the beach, funnels in and feeds an advancing dune, on either side of which slowly colonizing trailing arms of sand are left. These are known as 'parabolic' dunes (Figure 10.18). They are also found in semi-deserts where sand may advance over vegetation, as at White Sands, New Mexico.

Conclusion

The processes of aeolian geomorphology are summarized in Figure 10.20. Loose sediments are entrained by a variety of processes such as shear, lift and ballistic impact. The bed is left with a characteristic mixture of coarse and fine particles, and cohesive materials are blasted into aerodynamic shapes. Dust is lifted high into the air, travelling long distances downwind before being deposited as loess. Sand moves nearer the ground and, when there is enough of it, accumulates as dunes which themselves move slowly forward as wave-like forms. If vegetation intrudes, these dunes have a distinctive form, but otherwise they are subject to wave or vortex-like motions, developing transverse or longitudinal forms except in complex wind régimes.

11

Marine processes

Malcolm W. Clark

Since the mid-60s the momentum of research into nearshore processes has increased, and much of this work is beginning to bear fruit. Like many contemporary advances the origin of this research can be traced to the Second World War, when it became important to determine (for example) beach slope for planned amphibious landings. Today, many parties participate in research in this area—mathematicians, engineers, geologists, biologists and even a few geomorphologists. It is usually sandy beaches that have attracted the attention of the field worker. Shingle beaches have a few advocates, but apparently lack the mass appeal of sand, while rocky coasts, although not ignored, are even less popular. Lest this be seen as a manifestation of the 'bucket and spade' syndrome, it is worth pointing out that sandy beaches do have some properties which make them particularly suitable for research. Access is usually fairly easy, especially compared to rocky coasts; many sandy beaches are particularly extensive, especially in the United States where much research is carried on; the rate at which processes operate is quite fast—it is possible to see something happen with every wave reaching the beach. More recently the recreational value of beaches has encouraged workers to increase their understanding of ways to maintain beaches in a pleasing form. These factors also give a clue to the tendency for most nearshore research to be carried on in temperate areas, mirroring the distribution of the developed nations. Despite this apparent accent on field studies, it must be appreciated that for field study to be effective there must also be sound theoretical and laboratory work.

The single most remarkable feature of the nearshore is the periodic nature of the processes. No other geomorphological process has the same range of periodic phenomena—for example, waves, surf beat and tides are present. A different part of the nearshore is affected from moment to moment. The line between sea and land is never static. On a time-scale of minutes, this gives an interaction zone typically measured in metres (with successive wave swash and backwash), but in hours the zone may be measured in kilometres (with tidal cycles). In no other geomorphological study area are there such characteristic fluctuations. Although outside the scope of this book, there are easily observed periodicities of form in the nearshore, but it must be stressed that the link between 'periodic' process and 'periodic' form is not necessarily through the periodicity, tempting though the inference may be.

The nearshore may be defined fairly closely as the area landward of the breakers. This is not a zone that remains constant in width, since the point at which breaking occurs depends on the characteristics of the waves. Confining attention rigidly to this zone disregards the events and processes that lead up to breaking, and can at best permit an incomplete understanding. Some waves that approach the coast never break, and yet the notion of a nearshore zone, with processes and events quite distinct from the 'deep' sea, remains. While a precise definition may be attempted, a looser notion may be more useful.

The nearshore requires understanding of the interplay of one solid and two fluid media. It is a complex zone compressed into a narrow strip. The complexity of the interactions at the coast is reflected in our imperfect understanding of those interactions. Observation can be difficult in the coastal area (if exhilarating), especially when observations must be made over a full tidal cycle, below the sea's surface, or in the winter (when the beach can change very rapidly, and is arguably at its most interesting). Any recording instruments must be particularly robust, and if they are not to be moved by the sea, must be well anchored.

Waves and wave transformations

The major agencies that power processes in the nearshore area are waves. They provide the energy to move sediment and they are responsible for the generation of currents. Waves can be generated over any stretch of water, and represent a transfer of energy from air (wind) to the water over which it is passing. Waves generated by wind action have periods between 300 and 0·01 s. Those of period greater than 0·1 s are termed 'gravity' waves. On the surface of the sea a wide range of wave sizes is present. There is sufficient fluctuation in wind velocities that the spread of values for individual waves gives a more mixed pattern. This diversity

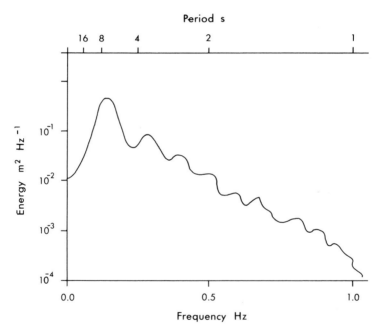

Figure 11.1 Typical wave spectrum with major 'swell' peak at about 7 s period, and lesser 'storm' peak at about 3·8 s period.

is best represented by a spectrum. The waves that reach the shore are often the result of more than one generating event, so the spectrum may have more than one peak (Figure 11.1). Incident waves frequently show a long-wave component from swell, and a shorter-wave component from 'local' storms. The magnitude of waves depends on the characteristics of the generating event. The wave length (λ), height (h) and wave velocity (or celerity) (C) are sufficient to characterize the wave (Figure 11.2).

Since waves play such an important part in nearshore processes it is valuable to mention some of the main wave theories and their range of application to the 'real' world. In general terms, the more awkward and difficult to use, the 'better' the expression of the theory fits the observed waves. The theory with widest application is the 'cnoidal wave theory'. Komar (1976), a mathematician as well as geologist, describes the mathematics of the cnoidal theory as 'difficult'. However, it turns out that two other wave theories are limiting cases of the cnoidal wave, namely the Airy wave and the 'solitary' wave. Airy wave theory finds wide application, largely because of its comparative simplicity. In essence, the Airy wave is a sinusoid, but it should be applied only to deep-water waves of small amplitude. Some modifications have been made to the theory to permit application to shallow water. Within this context, shallow implies that $d/\lambda_\infty > 0.05$, where d is still-water depth, and λ_∞ is the 'deep'-water wave-length. This will give an error of up to about 5 per cent in predicted height, which may be acceptable for many applications. The solitary wave is particularly important for the examination of processes in the nearshore, since waves in this zone are similar in

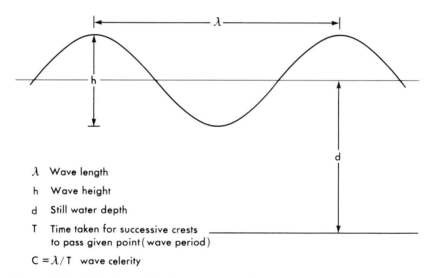

λ   Wave length

h   Wave height

d   Still water depth

T   Time taken for successive crests
     to pass given point (wave period)

$C = \lambda/T$ wave celerity

Figure 11.2 Definition of basic wave parameters.

form to the solitary wave. Unlike other waves, the solitary wave is not oscillatory—that is to say, there is no wave period associated with it (it is indeed solitary). At first sight it seems singularly inappropriate, since waves in the nearshore are periodic, but they do take a form very close to the 'classical' solitary wave—quite peaked, with long flat troughs between. A solitary wave is also progressive, that is to say, there is mass transport of the fluid in the wave (this is also termed a wave of translation). Stokes waves are also used to model waves, but they are appropriate to fairly high waves in shallow to deep water. The Stokes and Airy waves are quite similar over much of their range, given the same amplitude and period, but the Stokes wave has a flatter trough and slightly steeper crest. In general terms, Airy theory and solitary-wave theory can between them provide quite a good approximation in the nearshore, and as a consequence they tend to be the main theories used. Since there are always complications involved, because of interaction between waves, bottom topography and even local winds, it is unclear how far more 'precise' theories will benefit understanding of the processes in the nearshore. In addition it must be remembered that all the

Plate XXIX Three Cliffs Bay, Gower, South Wales: the interaction of sea, sand and rock. Wave fronts enter the bay well separated, but the swash-zone is considerably more complex. (M.C.)

wave theories have been developed explicitly with reference to a flat, not to a sloping bottom. Applying them outside their domain will introduce errors.

As waves approach the coastline, their characteristics change in two important ways (Plate XXIX). Refraction of the wave fronts occurs, and the waves break. Refraction of water obeys the same law as the refraction of light (Snell's law). For the change in phase velocity,

$$\sin \alpha_1 / \sin \alpha_2 = C_1 / C_2 = \text{constant} \qquad (11.1)$$

where α_1 and α_2 are the angles between wave crests and the bottom contours, and C_1 and C_2 the respective phase velocities. A useful conceptualization of wave refraction is to draw 'wave rays' which begin in deep water normal to the wave crests, and are therefore parallel. As refraction occurs, these wave rays lose their parallelism; if the wave rays were drawn initially at equally spaced intervals, the distance between adjacent rays after refraction can help to indicate the increase or decrease in the distribution of wave energy due to refraction. The procedure is not as simple as it may seem since wave rays can cross one another, but in simple topography this is useful and easily understood (Figure 11.3). This gives the following relationships:

$$\cos \alpha_1 / \cos \alpha_2 = L_1 / L_2 = \text{constant} \qquad (11.2)$$

where L is the distance between adjacent wave rays.

Since the relationship is through Snell's Law, it is obvious that waves of different celerities will be refracted differently. Long waves (e.g. swell) are affected much farther off-shore,

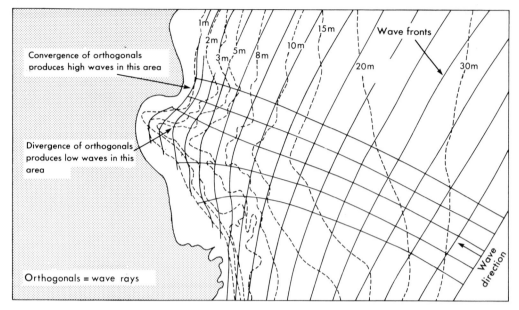

Figure 11.3 Idealized relationship between monochromatic waves and bottom topography (after Goldsmith, 1976).

and it is possible to observe a variety of directions in the wave fronts on the beach, as the whole spectrum of waves reacts to the bottom topography.

For shallow water, the energy flux (P) is given by an expression of the form

$$P = \kappa_1 Ch^2 \tag{11.3}$$

where C is phase velocity, h is wave height and κ_1 a coefficient varying between $0.125\,\rho g$ in shallow water and $0.0625\,\rho g$ in deep water; ρ is the fluid density and g the acceleration due to gravity. This quantity has the dimensions of power (watts, or joules per second). Energy flux on the shore is altered by refraction, since the energy 'concentrated' in the unit length of wave is either expended over a greater length, or over a shorter one. Because of the bottom topography, headlands are generally areas where refraction leads to the concentration of energy (and hence are zones of increased erosive potential). In bays, refraction leads to a reduction in the energy incident on the beach, making them areas of reduced erosion. The effect of refraction is commonly to make the wave crests more nearly parallel to the shore.

The type of wave presents an important variable. In general, three main breaker types (Figure 11.4) are recognized. These are not distinct types, but merely points along a spectrum of possible types. The wave type can be predicted from the relationship $h_\infty / \lambda_\infty \tan^2\beta$ (Galvin, 1968), where β is beach slope and the subscript ∞ refers to deep water. A value of over 4·8 suggests a 'spilling' breaker where the front face of the wave is covered by foam, bubbles and turbulent water; values between 0·09 and 4·8 suggest a plunging breaker—perhaps the most spectacular—where the front face of the wave steepens until vertical and the crest curls and falls into the base of the wave, often rebounding. Values lower than 0·09 are typical of surging breakers, where the base of the wave rushes up the slope of the beach,

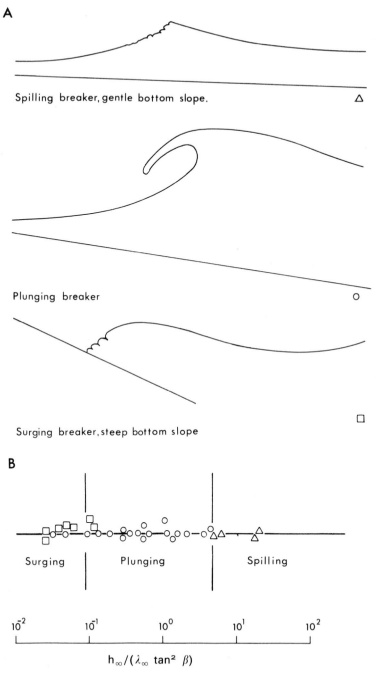

A

Spilling breaker, gentle bottom slope. △

Plunging breaker ○

Surging breaker, steep bottom slope □

B

Surging | Plunging | Spilling

10^{-2} 10^{-1} 10^{0} 10^{1} 10^{2}

$h_{\infty}/(\lambda_{\infty} \tan^2 \beta)$

Figure 11.4A: Main types of breaking wave; **B**: graph to show spectrum of breaking wave types related to wave height (h_{∞}), beach slope (β) and wave length (λ_{∞}) (see text: after Galvin, 1968).

with only minor production of foam and bubbles. This type of wave resembles a standing wave as it moves up and down the beach slope; it does not really break at all.

Both spilling and surging breakers retain their shape fairly well, and changes occur only slowly as they move through the nearshore, while plunging breakers deform very dramatically and in a short space of time. Surging breakers are most commonly associated with steep beaches (as implied by the $\tan^2\beta$ term in the equation). Waves break when the velocity of the water particles in the crest exceeds the rate at which the crest itself is moving. In shallow water the character of the waves changes. From a rather chaotic jumble, the waves appear to separate and approach the shore as fairly short-crested waves, separated by troughs which are long and flat (solitary waves). This transformation begins, approximately, when the water depth is about half the deep-water wave-length. Wave-length and velocity decrease and the height increases. The wave period, however, remains constant. The height of the wave before it breaks is much greater than the deep-water wave from which it originated (Figure 11.5).

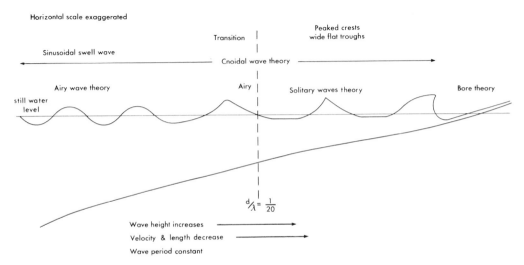

Horizontal scale exaggerated

Transition

Peaked crests
wide flat troughs

Sinusoidal swell wave

Cnoidal wave theory

Airy wave theory

Airy

Solitary waves theory

Bore theory

still water
level

$$\frac{d}{\lambda} = \frac{1}{20}$$

Wave height increases

Velocity & length decrease

Wave period constant

Figure 11.5 The change in characteristics of waves, and the applicability of major wave theories.

The energy of the wave is dissipated partly by bottom friction (although this is now held to be a minor factor), while much more is dissipated by the turbulence in the breaking wave, as the water is mixed. Some notion of the energy loss can be gained by observing the waves on ridge-and-runnel beaches, where waves break over the submerged bars, re-form in the runnel, and finally break on the coast. This type of beach often forms a reasonable sea defence, since the maximum wave energy is seldom concentrated on the uppermost parts of the beach. Losses of energy do not end with the breaking of the wave. Work is done in pushing water up the beach on the uprush. The backwash often interferes with the next breaking wave, reducing the amount of 'free' energy that it has for its uprush.

Motion in a fluid is often described in terms of either a Lagrangian or Eulerian frame of reference. Essentially the Lagrangian frame is one where an individual fluid 'particle' is monitored, using techniques that may depend on particles of neutral buoyancy (such as drifters), or on sand particles or bubbles within the flow. The Eulerian approach is one where the history of a point in space is examined; fixed current meters provide information in the Eulerian frame. The theories of wave form cannot cope with the discontinuity of breaking.

Up to this point, it is usual to consider that the wave energy remains reasonably constant, with little loss from friction or internal damping. When the wave breaks there is considerable energy loss. The mechanics of the breaking wave are poorly understood, but Longuet–Higgins (1976; Cokelet, 1977) has introduced a 'mixed Eulerian–Lagrangian' approach to wave breaking which has permitted wave shape to be predicted beyond the point of breaking. In addition, Miller's (1976) laboratory work has given a clearer qualitative understanding of wave breaking, especially in the formation of breaker vortices which have direct implications for the interaction of breaker and sediment.

Radiation stress and edge waves

It is possible to examine the behaviour of waves in the nearshore in terms of energy, mass and momentum, since all three must be conserved. When applied (Collins, 1976) to longshore currents it is simplest to use momentum. Only a small proportion of wave energy is used in driving the currents, and the derivation of an energy budget is particularly difficult. The volume of water brought to the shore in mass transport is also hard to predict, making an analysis based on conservation of mass difficult. An approach through the conservation of momentum is more easily applied. A longshore current is generated parallel to the shore when waves approach the coast at an angle. Outside the breaker zone there is no longshore current. The breaking waves provide the energy for the current. Longuet–Higgins (1972) points out that the driving force for the longshore current is proportional to the energy dissipation, and that without some dissipation of energy by breaking, no current would be generated. It is only recently that an adequate theory of current generation in the nearshore has appeared, through the concept of radiation stress. Radiation stress may be defined as the excess flow of momentum associated with water waves. It can be shown, using these concepts, that there is a difference between the still water level in the surf zone, and the mean water level (Figure 11.6). Theory indicates that mean water level will be lower where the wave

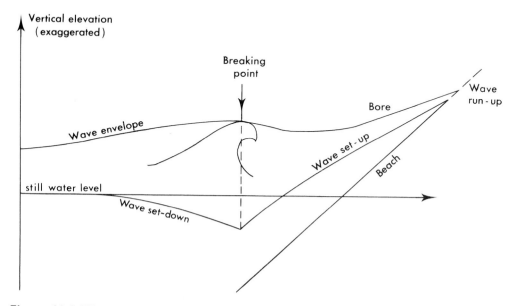

Figure 11.6 Effects of radiation stresses in the surf zone (Collins, 1976).

height is a maximum, while shoreward the mean water level increases, to a value in excess of the still water level. This variation is termed wave set-up and set-down. Observation confirms the theory, although there is some suggestion that the type of breaking wave may alter the magnitude of the set-up and set-down. The extent of wave set-up has been suggested as a way in which a head of water may be provided to contribute to the longshore current. Since larger waves break earlier than small ones, they begin their set-up earlier and the actual water rise associated with them is greater. The longshore current is a combination of the current due to longshore variation in wave height and the oblique wave approach. These currents need not be complementary, but may be opposed (in a net sense), so that there is no continuous longshore current apparent along the whole beach. Even when there is no oblique wave approach there may still be longshore currents. However, these may produce no 'net' current along the beach, but instead a series of cells. Although this discussion indicates that the longshore current is restricted to the area landward of the breakers, the horizontal eddy mixing effect allows driving forces in the surf zone to produce a current beyond the breakers. The mid-surf current may be estimated by

$$v = \kappa_2 v'_m \sin \alpha' . \cos \alpha' \tag{11.4}$$

where v'_m is the maximum horizontal orbital velocity in the breakers and α is as in equation 11.1 (the prime indicates the 'breaker' value). The whole distribution across the surf zone is more complex, but has been given by Komar (see Komar, 1977b, and Figure 11.7).

The observed variations in wave height may be the result of four factors (Wood, 1976): (i) bottom geometry, operating through refraction; (ii) interactions when wave trains of

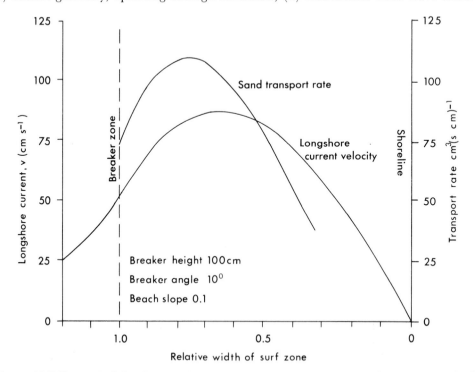

Figure 11.7 Theoretical distribution of longshore sand transport and longshore current velocity across the surf zone (after Komar, 1977b).

similar frequency arrive from two distinct directions; (iii) from local storm-generated waves which interact with the swell and generate a variation along the shore; and (iv) edge waves.

The presence of edge waves on beaches was predicted by theory. The idea behind such a wave is that an incident wave may be partly reflected and may be radiated back out into the ocean, or may be refracted back into the nearshore so that the wave becomes 'trapped' (Figure 11.8). Edge waves have become very popular, and can be used to explain surf beat and the spacing of rip currents (as well as a number of nearshore forms). Observations in the laboratory and on natural beaches have confirmed the presence of such waves, but the links with other processes and forms are not unequivocally defined. The edge wave is confined to the nearshore, and is negligible more than one wave-length from the shoreline. In form it resembles a standing wave normal to the shore. This carries with it the implication that the edge wave may interact with the incident waves; there should be 'nodes' where the edge

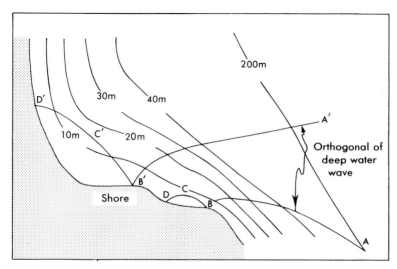

Figure 11.8 Waves trapped in the nearshore zone (after Isaacs, Williams and Eckart, 1951). Deep-water waves are trapped by reflection at the beach (B and B') and in deeper water (C and C').

wave does not contribute at all to run-up, while at the anti-nodes the full extent of the edge-wave amplitudes should be experienced (see Komar, 1976). This standing wave may also affect breaker height, producing the observable variations (required by the theory of radiation stress), and this is turn will encourage the presence of rip currents where the breaker height is smallest. Since the rips are a response to the edge wave, they should occur at regular intervals. It has been suggested earlier that other mechanisms could lead to the variations in wave height, so that edge waves, attractive though they are, are not essential to explain the presence of rips.

Rip currents (Figure 11.9) do not seem to be invariably present, but there must always be some mechanism leading to the return of water seawards. Longuet–Higgins (1953) has suggested theoretically that there might be some return flow at intermediate depths, but this has not been described in the field, although Schiffman (1965) suggests an intermittent, low velocity return near the bed.

Munk (1949) observed rather long-period waves on beaches with a period typically about

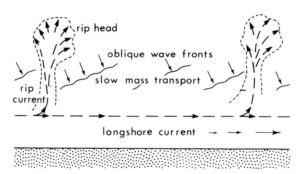

Figure 11.9 Nearshore circulation system (after Komar and Inman, 1970).

six times that of the incident waves. He termed this wave 'surf beat'. It is a low amplitude wave. Explanation for the phenomenon depends on the incidence of two swell trains which interfere to produce a regular variation in wave height. Inevitably, radiation stress may be employed to provide an explanation (Longuet–Higgins and Stewart, 1962), while Gallagher (1971) suggested that interaction among wind-generated waves could excite edge waves in the surf-beat range. Guza and Inman (1978) also relate edge waves to surf beat. Actual longshore velocities calculated from the theory of radiation stress correspond reasonably well with observation in the field and in model studies. Keeley and Bowen (1977) and Meadows (1976) observe that the longshore current may fluctuate markedly over a short period of time. Some rip currents have also been noted to 'pulse'.

Tides

A key variable in the examination of coastal processes is the action of tides. The magnitude of tides determines the extent of the coast which is commonly affected by the sea. Tides are mainly caused by the rotation of the moon around the Earth, and represent the balance of gravitational forces in the sun, moon and Earth system, as they affect a mobile surface level—the oceans and seas. The size and shape of the ocean basins help to determine the tidal characteristics. They have natural periods of oscillation which may correspond with the tidal cycles, or may bear no obvious relationship. The Atlantic basins tend to have an oscillation period of about 12 hours, corresponding to the 12-hour (actually 12 hours 25 minutes) lunar tides, so that the semi-diurnal tides are each about the same height. The Pacific basins may produce oscillations which are not multiples of 12 hours, so that the sequence of tides, although regular, is not simple. Such tides are termed mixed. Where the period of oscillation is close to 24 hours (as the Gulf of Mexico), diurnal tides are found—that is, one high tide per day. The general form of tide height is a sinusoid, so that mixed tides represent a summation of two series with different frequencies—hence the apparent awkwardness (but underlying regularity).

The magnitude of tides varies greatly. Enclosed seas (Mediterranean, Baltic, Great Lakes, etc.) have very limited tidal ranges, and variation in water level is mainly the product of local meteorological conditions. It should not be assumed that oceanic locations will automatically have large tidal ranges. The tidal range in some parts of the Pacific basin is very small—in mid-ocean about 50 cm. Where there are extensive continental shelves, or where local conditions lead to funnelling (as in some bays) tidal range may be enhanced. Tidal ranges for the North Sea (a fairly extensive shallow sea) are often somewhere over 4 m, while the Severn Estuary amplifies the tidal range to something over 7 m. The Rance estuary in Brittany has an even greater tidal range. The Bay of Fundy produces the greatest tidal range, at about 15 m maximum. Within the Bay there is marked variation, with a figure of less than half towards the mouth of the bay.

Tidal range is of course not constant through the lunar month, but depends on the configuration of the sun, Earth and moon system. When all three bodies are in line the tidal range is at a maximum—a spring tide. When they are out of phase by 90 degrees (moon in 'quadrature') the tides are at a minimum—a neap tide. This is a fortnightly cycle. This lunar cycle superimposes a further sinusoidal cycle on the model suggested earlier, but with a frequency of about 2 cycles per month. Further examination of all the tide-producing components reveals more sinusoidal components, with frequencies that vary between (approximately) 2 cycles a day, and something of the order of 1 cycle in 1600 years. Regular variation can also occur in basins when a resonant standing wave is set up—a seiche. This can add another component to the tidal height. The Bay of Fundy has a seiche with a period close to the tidal period, further amplifying the range. Seiches are also set up on continental shelves.

Even using all these elements to predict tide heights does not, though, lead to a completely accurate picture. Local meteorological conditions can contribute markedly to local variations in tide height. In extreme circumstances these are 'storm surges', and while it is convenient to discuss these changes in the context of 'exceptional' events, the general principles remain valid for less dramatic events. There are two main (but related) causes for storm surges. Onshore winds help to pile up water in the nearshore zone, augmenting the existing water level. Extreme low pressure also tends to encourage the accumulation of water. These phenomena are related; intense low pressure always has high winds associated with it, but the direction of the winds need not be onshore. Conversely, high pressure or off-shore winds will tend to diminish water levels at the coast. When storm surges occur simultaneously with high and spring tides they can result in extensive inundation of the land behind the immediate coastal strip (and consequent loss of life and damage to property). Even when these events are not simultaneous, damage can be catastrophic—the 1953 storm surge in the North Sea (Steers, 1953) fortunately occurred nearer a neap tide, and not at the highest point of that tide. Nevertheless, damage was widespread in the low-lying areas around the North Sea basin. The interaction of less dramatic meteorological conditions and nearshore processes has received some attention, and Fox and Davis (1976) have shown that the characteristics of waves reaching the coast are modified. In particular, wave height is changed, and also the longshore current. They developed empirical models for a number of locations around the United States, relating the changes observed to prevailing weather patterns. Sonu *et al.* (1973) have examined the interaction of local diurnal winds (on- and off-shore sea breezes) and the nearshore. In particular, they found that this periodic phenomenon introduced a high frequency peak in the wave spectrum which dominated the background swell in the afternoon and evening, and also that these breezes had systematic interaction with the nearshore currents (with significant diurnal variations).

Indirectly this helps to point to an important distinction that can be made between coasts that are dominated by swell-generated waves, and those dominated by storm waves (Davies, 1973). Broadly speaking, much of northwestern Europe falls in the zone dominated by storm waves, while much of the United States coastline is characterized by a swell-wave régime. Swell-generated waves originate in the temperate storm belts, but by the time they have reached the coast they are long low waves of reasonably constant direction. They tend to be fairly consistent throughout the year. Storm-wave coasts tend to have shorter waves of greatly varying direction throughout the year. Energy levels are higher, superimposed on a background of swell. Fox and Davis (1976) help to put the influence of local conditions into perspective by their observations that, on the Oregon coast (dominated by storm waves in the winter), the spring tide is about 4 m, and normal (i.e. swell background) waves are about 2·5 m high, but winter storms may produce waves of up to 8 m. This may be contrasted with the relatively protected Texas coast at Mustang Island with a spring tide of 1 m and a maximum breaker height of about 1·2 m. While there must be substantial differences imposed between these two locations, a site such as Lake Michigan, with a 10 cm tide but a breaker height of up to 1·8 m, represents yet another range of possibilities.

In the northern hemisphere, tidal streams rotate in an anticlockwise direction. This imparts a residual velocity which may average out at about 40 cm s^{-1}. This is a gross feature observable over the whole tidal basin; at a smaller scale, over a tidal cycle, an interesting feature of tidal currents is the asymmetry often present. The velocity of the ebb and flow currents is often different and in some areas this is enhanced by the 'mutual evasion' of ebb and flow channels (*cf.* King, 1972). Off Gibralter Point, there is a series of longitudinal sand ridges, parallel to the coast. These separate 'channels' are dominated by either ebb or flow. Not only do these channels contain tidal streams of differing velocities, but an ebb channel may maintain ebb flow even after the turn of the tide. Similar observations of ebb and flow channels have been made in estuaries. It is particularly noticeable that these systems are best developed in areas where sediment is abundant and easily transported.

Swash zone

The swash zone of the beach is potentially the most dynamic part of the nearshore. Much of the energy of the breaking wave is expended in this area, but wave motion does not translate directly into swash motion. Research in the swash zone is reasonably well supported by field, laboratory and theoretical work. Laboratory and theoretical studies have, however, tended to direct attention to monochromatic waves, and often to impermeable slopes. This does not reflect naturally occurring beaches with their spectrum of waves and a permeable (and deformable) run-up slope. The behaviour of spilling breakers in the nearshore can be approximated to that of a bore. Bores can be classified by reference to the ratio h/d, where h is wave height and d is water depth. When $h/d < 0.28$, the bore is described as undular. An undular bore has the property that, when the energy flux through the bore is too great to be dissipated by turbulence at the bore face, it is radiated as 'wavelets' behind the bore face. At values over 0·75, the bore is fully developed and large-scale turbulent breaking develops. Between these two values some turbulent breaking may take place and also some wavelets may be present. Note that the incoming wave may interact with backwash to increase the water depth before it, so that an undular bore develops, radiating some of the energy as wavelets. Suhayda and Pettigrew (1977) have documented the systematic changes that take place as a wave breaks and crosses the surf zone, and show that, while the behaviour does deviate from that predicted by bore theory, some bore-like characteristics do develop.

The run-up on the beach provides another significant variable since it determines the amount of the beach that will be affected by waves. Huntley, Guza and Bowen (1977) have compared the run-up spectra of four beaches, and demonstrate that all four have a similar form. An implication of their analysis is that wave energy at the shoreline is determined by conditions at the shoreline, rather than those at the breakers. A further implication is that the run-up spectrum is controlled not by the incident waves but by the nearshore standing wave. Such a wave may be the result of the partial reflection of the shoreward incident wave and would have a period greater than that of the incident wave. In one case they acknowledge that edge waves might also be involved. The period of the incident waves and the swash period are not in a perfect one-to-one relationship. There are fewer swash cycles than incident waves. Kemp (1975; Kemp and Plinston, 1968) formulates this into a relationship between the time taken to uprush, t, and the wave period T. The ratio t/T he terms 'phase difference': it may be employed to distinguish three characteristic types of interaction in the swash. Where the value of $t/T < 0.3$, the broken wave is able to 'surge' up the beach and return as backwash to the breaker point before the succeeding wave breaks. As the time of uprush increases, a point is reached when the backwash is not completed before the next wave breaks. Thus there is interference between backwash and uprush. When $t/T > 1.0$, this 'transition' stage is replaced by what Kemp terms 'surf' conditions, as the breakers continually spill water into the swash zone. These three categories correspond closely to the wave type. The surge condition is associated with waves surging or breaking on steep beaches; the transition condition with plunging breakers and the surf condition with spilling breakers on beaches of low slope. In transition and surf conditions the swash period will be longer than the period of the incident waves.

Groundwater in the beach

The swash and backwash are modified by percolation, since some of the water may be lost to the beach water-table; similarly, augmentation of the backwash can occur when the water table is high. The relationship between the beach water-table and the position of the shoreline is quite critical to the movement of sediment. This has been recognized at the level of tidal cycles by Grant (1948), Emery and Foster (1948) and Duncan (1964). The main points of the interaction are that on a rising tide the beach water-table is low, much of the water contained in the swash percolates into the sediment and, as a result, sediment contained in the swash is likely to be deposited. The backwash is correspondingly decreased; the reduced backwash is less capable of transporting material back downslope, so that the net effect is deposition at the swash margins. When the foreshore is saturated, as on a falling tide, the backwash is accelerated by addition of water to the backwash from the beach water-table. This escaping water not only adds to the flow volume but also dilates the sand, making it more erodible. Waddell (1976) has refined this model considerably. Notably, he has shown that the beach water-table responds to shorter period variation. He found that the beach water-table responds not to the incident-wave period but to the period of the standing wave produced by reflection, and also to the swash period. The incident-wave spectrum had peaks at about 0·16 Hz and 0·025 Hz; the swash had a peak at 0·1 Hz; the groundwater spectrum was dominated by a major peak at 0·025 Hz, corresponding to the standing-wave component, but there was also a minor peak at 0·1 Hz, corresponding to the swash period. In this context it is significant that Huntley and Bowen (1975) found a correspondence between swash period and the dominant edge wave of a steep ($\beta = 0.13$) beach. (Similar observations are made by Huntley *et al.* (1977) and Guza and Inman (1978), relating swash period and edge-wave frequency.) Standing waves and edge waves depend on some degree of reflection from the shoreline, and are most likely to be produced by conditions of low phase difference. When the inshore region is truly surf (high phase difference) the reflection is reduced and the presence of such waves reduced. Even when edge or standing waves are absent, the lower-amplitude surf beat may provide low frequency fluctuations which would be suitable driving mechanisms for Waddell's model. He suggested that there is transmission of water in the saturated portion of the beach matrix, which responds to the position of the nearshore standing wave. The changes in groundwater level cause a shift in the location of the boundary between water-table and the beach slope. The model outlined for tidal cycles can therefore be applied at the level of the period of oscillation of the standing wave. Because of fluctuations of the water-table, there can be a portion of the beach face which is alternately saturated and unsaturated, and which would therefore alternately experience conditions favourable to deposition and then unfavourable to deposition. Waddell demonstrates that virtually all sand-level variations occurred at periods greater than 0·025 Hz, suggesting that they were more likely to be linked to the groundwater fluctuations than the swash frequency. The characteristics of the swash are also important. Miller (1976) relates the flow and the sedimentation, suggesting that low-energy bores tend to transport material upslope (which may remain there), while high-energy high-volume bores have large backwash volumes that move the sediment back downslope. Waddell points out that the low-energy bores are likely to be those generated by waves of high phase difference, while high-energy bores are likely to be those that are relatively undisturbed.

Sediment interactions

Sediment motion can be conveniently divided into two categories—suspension and bed load. Although Yalin (1977) points out that the separation into suspended and bed load is an idealization of a continuous natural transition, there is one category of suspension which Bagnold (1963) has shown to be quite distinct; material remains in suspension when its settling velocity (ω) is below a critical value determined by the local bed slope (β) and the horizontal fluid velocity (v):

$$\omega < v \tan \beta \qquad (11.5)$$

Fine material below a certain size is not deposited, being uniformly distributed through the fluid, allowing it to diffuse through the surf zone and out into deeper water, or to be swept out by rip currents. Inserting some reasonable values into the expression, Bagnold demonstrates that with a beach slope of 2·86° (tan β = 0·05), and a value of 25 cm s^{-1} for the mean fluid velocity, ω is about 1·25 cm s^{-1}. For quartz, this represents grains with a diameter of about 0·15 mm. It is perhaps significant that this is a commonly observed lower limit to beach sand sizes; material finer than this is rare.

Allowing that suspended and bed load are not necessarily distinct, there is an important difference to be made between the nearshore zone and rivers. The principal fluid movements in the nearshore zone are unsteady and oscillatory, not unidirectional—although there are currents such as tidal streams and longshore currents that are almost unidirectional. It is also an area that is influenced by the added turbulence of breaking waves. The oscillatory motion of the waves generates a velocity field which can be established from one of the wave theories. Airy wave theory is easy to manipulate in this respect, and the expression

$$v_m = \pi h / (T \sinh[2\pi d / \lambda]) \qquad (11.6)$$

where h is wave height, T is wave period, d is water depth, λ is wave length and π is the constant (3·14159...), may be used to give the bottom orbital velocities, although this may provide an underestimate.

Komar and Miller (1973) have discussed the threshold of sediment motion under oscillatory waves, and developed a relationship that allows a threshold to be calculated for a given grain size and density from wave period and orbital velocity. The longer the wave period, the deeper the sediment that may be set in motion. An implication of the relationship is that waves can affect sediment in quite deep water, well beyond the usual notion of the nearshore. Draper (1967) found wave activity at the sea bed around northwestern Europe to at least 170 m. A response of this movement of sediment is the formation of ripples, but the presence of ripples is no proof of wave action, since ripples can also be formed from unidirectional flows (*cf*. Allen, 1968). An examination of the ripple characteristics can usually allow the generating process to be established. In the nearshore area, Dingler and Inman (1976) show that the relationship between bedforms and the wave conditions is rather close. In the surf zone, ripples may disappear because of the high shear stresses present, sheet flow takes place, and the bed becomes plane (Figure 11.10). Ripples were found to move onshore, but this need not imply that there is onshore movement of the material composing the ripples at the same rate (Inman and Bagnold, 1963). If ripples are replaced by a plane bed it takes only a few waves exerting a lower shear stress before they reappear. Because the lag time between process and bedform is so short, it is difficult to discuss one without the other. Kennedy and Locher (1972) have described the sequence of events which takes place when waves pass over ripples. During the horizontal passage of fluid, a burst of sediment is thrown upward

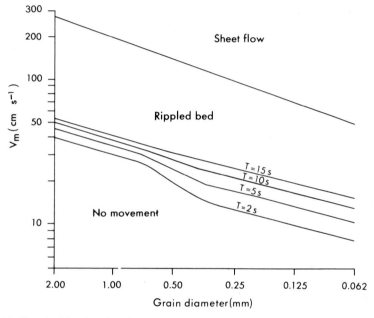

Figure 11.10 Threshold velocities for initiation of quartz grain movement in water, (lower set of lines, according to wave period) and conversion from rippled to flat bed (upper line) (after Clifton, 1976).

into the fluid, from the region just downstream of each ripple crest, by the eddy that forms in the lee of the ripples, and from the areas immediately upstream from the ripple crests by the high velocities and shear stresses in these regions. The sediment cloud entrained near each crest will be swept along a distance depending on the wave-length in the fluid and the wave-length of the ripple; over this period the sediment cloud disperses and settles out. When the velocity reverses, the clouds entrained may not have settled back to the bed and may be swept back in the opposite direction. Inman and Tunstall (1972) show how the shape of the ripple is an important control on the movement of the entrained sediment. They relate this to the intensity of the vortex in the lee of the ripple crest. Large vortices are generated on the lee side which is steeper. Net sediment motion is in the direction of the orbital velocity which is out of phase with the maximum vortex formation (Figure 11.11).

Bed load is particularly evident in the swash and backwash where the shear stresses may be very high. Describing the backwash transport simply as bed load hardly does it justice,

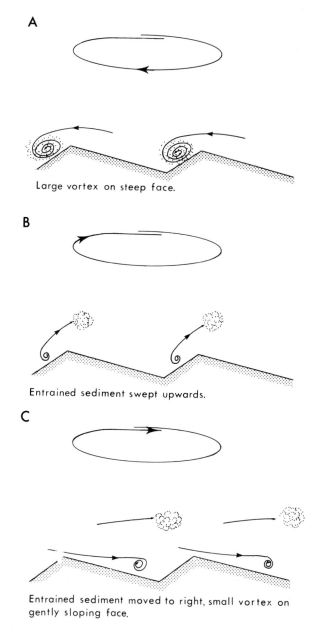

Figure 11.11 Net sediment movement related to bedforms (after Inman and Tunstall, 1972). **A**: large vortex on steep face; **B**: entrained sediment swept upwards; **C**: entrained sediment moved to right, small vortex on gently sloping face.

since the whole bed may be moving as a traction carpet, where the support for the load is provided by the intergranular collisions; moreover, flow out of the bed will reduce the effective particle weight.

A consequence of the oscillatory flow provided by waves is the asymmetry of the flow (Figure 11.12). As the figure shows, the onshore velocity is higher, but of shorter duration than the return offshore velocity. From the motion-threshold values introduced earlier, it is clear that coarser grains tend to move onshore, while finer grains move offshore. The onshore movement may be balanced by gravity which will tend to exert an offshore trend. There may, therefore, be a point at which grains of a given size remain, oscillating only a little. This is termed the 'null point'. The null-point concept has had a rather chequered career, and much contradictory evidence has been accumulated. Since the null point will vary according to wave conditions and the size of the bed material, laboratory experiments with monochromatic waves and a narrow range of grain sizes may not be representative of field conditions. The presence of bedforms may also affect the model.

With unidirectional flows, the amount of material in suspension increases as the shear stresses increase. In the nearshore, the presence of waves helps to place extra material into

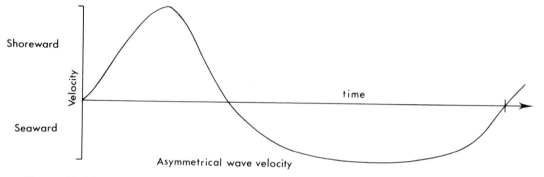

Figure 11.12 Asymmetrical wave velocity in the nearshore zone.

suspension for slightly different reasons. At the breakers themselves, Madsen (1974) reports from laboratory work that the passage of breaking or near-breaking waves can induce a horizontal flow within the bed itself, producing a momentary instability. This instability need not throw material into suspension, but grains near the bed–water interface would be unable to resist any force due to the turbulent flow of water above the bed. Miller (1976) has shown that the vortices associated with breakers may entrain sediment: as a wave breaks it forms a series of vortices that contribute to the decay of the wave. This vortex system is more pronounced in plunging breakers than spilling, where they are confined near the free surface. The vortices impart a transfer of momentum to the bottom sediment. This again lends support to the idea that breaker type is one of the fundamental variables in the nearshore. Brenninkmeyer (1976) also notes a very particular form of suspension—sand fountains. These are short spurts of suspension which occur under the influence of progressive waves. The fountains are quite variable in characteristics, but of short duration. Within the swash he found that material was likely to be suspended (i) by the horizontal and vertical fluid velocities, and their fluctuations, which he observed were capable of suspending even coarse sand; (ii) by the air trapped in the breaking wave which could have an entraining effect when the bubbles penetrated the bottom sediment; (iii) by a hydraulic jump formed when an incoming bore was unable to progress against the backwash, or (iv) because of bedforms.

The actual depth of disturbance of the beach sediments provides some kind of clue to the amount of material that is moving in the nearshore. Continuous measurement of the bed is difficult, especially if the bed is below water. In such a situation the most readily made measurement is over a tidal cycle, and so it may include the addition and subtraction of many movements caused by the breaking waves, nearshore currents, swash and backwash. King (1951) has made such measurements in the surf zone (with spilling waves), and found depths of disturbance of up to about 5 cm. Waddell (1976), measuring the bed continuously in the swash zone noted that up to 5 cm of sediment may be removed in a single 'swash event'. Accretion was less dramatic. Brenninkmeyer (1976) describes changes in elevation of 6 cm per wave as 'not uncommon'. Williams (1971), on the other hand, found much greater depths of disturbance, but he was dealing with plunging breakers. Besides accounting for these differences by breaker type, Madsen's work (1974) also helps to reconcile the observations to some extent: the momentary instability that occurs when a breaking wave passes over the bed may lead to some movement of the sediment at depth, providing some mixing, which would be distinct from either bed-load movement or the passage of bedforms. Again, the type of breaker is significant in determining the pressure gradient which provides the mechanism for the instability.

Dean (1973) presents a model of sediment motion in the nearshore which depends upon breaking waves placing sand into suspension over part of the water column. After this suspension phase, as the particles return to the bottom, the direction of horizontal displacement depends on whether the particle is acted on by an on- or off-shore velocity. If the individual particles require a short time to settle, relative to the wave period, then the particle will tend to move onshore. If the fall time is long, the net direction will be offshore. Defining a parameter

$$D = f/T \tag{11.7}$$

where f is the fall time of a sediment particle, when $D < 0.5$, onshore movement occurs; when $D > 0.5$, offshore movement. It is apparent from this model that finer particles, by virtue of their greater fall times, would tend to be moved offshore, while coarse material would accumulate in the nearshore. The very coarse material that could not be moved into the breakers would tend to accumulate at the breakers. This is a fair representation of the observed pattern. However, the model has some shortcomings; it does not consider bedload, and it also omits any mention of the asymmetry of wave velocities; the actual fall times may be difficult to calculate, since the settling velocities of particles in the surf zone are likely to be modified from their single-particle still-water values. Single-particle still-water settling velocities can be calculated fairly easily and fairly accurately. In the turbulent water of the nearshore, settling velocities will be reduced; both turbulence and sediment concentration are known to reduce the settling velocity. It is possible that breaking waves impart an 'eddy viscosity' larger than molecular viscosity; sediment concentrations of 0.1 by volume may reduce settling velocities by up to 40 per cent. Taken together, some retardation is likely to take place. In order to put fall times into some kind of perspective, the single-particle still-water settling velocity for a quartz grain 0.20 mm in diameter is just over 2 cm s^{-1}.

In the breaker zone, Miller and Ziegler (1958) propose a similar model where coarse material accumulates at the breakers, while finer material is found both seaward and landward; again, they envisage that in the turbulent region of the breakers, fine material does not settle out as rapidly as the coarse material. They do, however, suggest that some fines may be present, trapped by the coarser material as it settles out. Miller and Ziegler's model is more comprehensive than this, since it encompasses the whole nearshore zone. They divide the nearshore into three parts—breakers, swash and the shoaling zone. By shoaling zone they mean the area outside the breakers where waves are being modified appreciably by the bottom. This distinction is quite useful, since the processes are likely to be quite different in these three areas. Their model for the shoaling zone is based on the null-point concept. In the swash zone they conclude that the material would be graded with the finest deposited highest up the beach, resulting in a swash zone with fine material coarsening to the breakers. This echoes a model for swash transport proposed by Evans (1939), recently supported in fieldwork by Komar (1977a).

These qualitative descriptions of sediment motion in the nearshore do not permit estimation of the actual amounts being moved, or the principal type of transport. Traditionally, it has been assumed that the bulk of sediment movement is as suspended material. This view is supported by the large amounts of suspended sediment commonly observed in the nearshore, but this is hardly quantitative evidence. Thornton (1968, 1972) has used bedload traps in order to arrive at an estimate of the amount of material moving as bedload. Unfortunately bedload traps are not highly regarded and are probably not particularly accurate indicators. Thornton estimates the trap efficiency as between 40 and 100 per cent, and comments that in some cases it may be more than 100 per cent efficient. He compares the amount being moved alongshore and caught in the traps with the theoretical total longshore movement of sediment. This analysis indicates that bedload is the minor component compared to suspended sediment, with the proportion of bedload to suspended load being somewhere in the region of 10–40 per cent. The ratio I_s/I_t (immersed weight of suspended load/immersed weight of total load) is then between 0·7 and 0·9.

This approach requires some clarification. There are two implicit assumptions present: first, the total longshore sediment load is equated with the total amount in motion. It is possible for the net amount passing a point on the beach to be different from the total amount in motion. Consider the possibility that the longshore current is unsteady, with occasional reversals; the net amount in motion might then be zero, but the total amount is non-zero. The second assumption revolves around the expression for total sediment movement. There are several formulae available for this calculation but they are not completely accurate, and can be in error by more than an order of magnitude.

Komar (1976) has also made a similar type of analysis, but considers the total amount of material in suspension. If the product of the longshore current and the suspended-sediment concentration is calculated, it will provide an estimate of the total amount being moved in suspension, and may be compared with the total amount being moved alongshore. This leads to a relationship of the form

$$I_s/I_t = 15c/\tan \beta \qquad (11.8)$$

for quartz grains. Values of c, the volume concentration, are commonly between $3·2 \times 10^{-4}$ and $4·6 \times 10^{-4}$. On a beach of slope 2°, this would result in a value of about 0·2, suggesting that suspension transport is rather minor.

Brenninkmeyer (1976), from his work on sand fountains, suggests that the total quantity of material thrown into suspension is of the order of two to five times greater than the amount moved by longshore currents, and he infers a major role for suspension. There is an obvious paradox here, which may be resolved to some extent. Brenninkmeyer does not distinguish between movement normal to the shoreline and movement parallel to it; he does suggest that more than two-thirds of the sand in suspension is not involved in longshore transport, but moves normal to the shoreline. Inman and Bagnold (1963) implied a figure of the order of 0·1 for the ratio of the onshore–offshore movement to the longshore movement. This latter figure would relegate suspension transport to much the same figure as that determined by Komar.

All of these analyses ignore explicit mention of the type of wave, and hence the phase difference, which might be expected to contribute to substantial differences between locations.

Komar (1971) distinguishes between sediment moved by the longshore current and that moved in the swash. The swash movement is a zig-zag (or saw tooth) where material is moved upslope obliquely, and then returns downslope (usually) normal to the beach. This gives a net travel along the beach, although the actual distance travelled is greatly in excess of the longshore movement. He derives an expression of the form

$$I_t = \kappa_3 P' \sin \alpha' \cos \alpha' \tag{11.9}$$

where κ_3 has a value <0.77 and P' is the energy flux, for the longshore movement of the sediment in the swash. Komar and Inman (1970) show that the same expression will predict the longshore movement due to the longshore current (oblique wave approach) and the variation in breaker height. This is rather encouraging since it lends support to the validity and generality of the expression. They also argue that this expression is equivalent in form to a model suggested by Bagnold (1963). He considered that the wave activity would be sufficient to support the sediment, although it would not impart any net transport. Any superimposed unidirectional current, however, would be sufficient to move the sediment. Using this argument, an expression of the form

$$I_t = \kappa_4 P' \cos \alpha' <v> / v'_m \tag{11.10}$$

where $\kappa_4 = 0.28$, v is the current velocity and v'_m is the maximum orbital velocity in the breakers, is derived. The cosine term in both equations will usually be about unity, and can generally be omitted. κ_3, the dimensionless proportionality coefficient in 11.9 is a kind of efficiency parameter, relating input power to the work done.

Indirectly the value of κ_3 gives a notion of the kind of energy losses present in the nearshore. The expression gives only a maximum figure to the sediment load, which applies to times when conditions are fully and most favourably developed. Individual observations may show less material being moved (by as much as an order of magnitude). Suhayda and Pettigrew (1977) point out that most of the energy lost in breaking is lost near the break-point as turbulence rather than bottom friction. Unless both of these processes are equally effective in producing movement of sand and water, correlation of sand-transport rates with total wave energy would produce a proportionality which is not constant. These equations give a gross figure, but Komar is prepared to go further and, with the aid of some convenient assumptions, predict the transport rate right across the breaker zone (Figure 11.7). The distribution has a maximum at a position just shoreward of the breaker zone, but seaward of the maximum in the longshore current distribution. This is because the sand transport is proportional to the product of the longshore-current velocity and the stress exerted by the waves and longshore current. The wave stress increases to a maximum at the breakers. Komar (1977b) has used the actual value of κ_3 as yet another way of demonstrating that suspension is of minor importance in the nearshore. If suspension had some importance, it would be anticipated that some relationship would appear between the mean settling velocity and κ_3. No such relationship can be discerned.

Tracers

Much work has been carried out using tracers (usually fluorescent, but also radioactive) to tag grains which then enable the motion of sand, and sometimes shingle, to be followed (*cf.* Ingle, 1966). Such techniques are not without their problems, both from a practical point of view, and from the point of view of interpretation. The released grains can become difficult to find, either becoming buried, or passing outside the area examined. The rate of recovery is always low, certainly below one per cent, which throws some doubt on whether the results are representative. Even if there is no net movement of sand, there will be an apparent movement towards areas of maximum diffusion. This implies that material (wherever emplaced), will tend to appear to migrate towards the breakers. Some studies (e.g. Murray, 1967) have used grains which were 'colour coded', in the sense that different size grades were given different colours. Since there is a fairly limited range of identifiable fluorescent colours, this has some limitations, but the approach can provide some information on the differential motion of grain sizes. Komar (1977a) extended this by coating the whole size range, then sieving the recovered samples and counting the number of tagged grains in each size range. This is a rather time-consuming procedure, but it has allowed Komar to verify that coarser material tends to move more swiftly along the beach. This is a reflection of the mechanics of swash transport. The coarse grains tend to move less far than finer grains up the swash, and are therefore more frequently subject to swash movement.

Cliff erosion

Shepard and Grant (1947) identified a number of factors which they considered important in examining the effectiveness of erosion on rocky coasts: the hardness of the rock; the presence of structural weaknesses; the configuration of the coast (straight, crenulated, etc.); the solubility of the rock or the rock cement; the height of the cliffs (arguing that higher cliffs are inherently more unstable); and the nature of the wave attack (noting the effect of refraction, concentrating wave energy on headlands). To some degree these factors are echoed by Davies (1973) who identified six processes important in the erosion of rocks:

1 Wave quarrying, where the waves pull away already prepared material, is well known. Jointing and cleavage make rocks more susceptible to quarrying, stressing the importance of structure and lithology. Assisting in preparing the rock surface, by analogy with the pressure-release or unloading phenomena found in glaciated areas, may be the repeated application and release of a substantial weight of water above the bed. Certainly as erosion takes place and material is removed, some pressure release must take place. Even the oscillatory behaviour of the waves may be sufficient to encourage the breakdown of rock in place.

2 Wave abrasion takes place when the waves can move sand or pebbles across the sea bed. It has been noted that some seaweeds attach themselves to pebbles (Emery and Tschudy, 1941; Kudrass, 1974); this makes it easier to move the pebbles, making it possible for larger pebbles to abrade the cliffs or the sea bed. In addition, some attrition takes place by pot-holing (Plate XXX). The swirling action of the incoming tide helps pebbles erode the bed in a roughly circular hole. Bradley (1958) suggests that significant abrasion will not take place more than 10 m below sea level, although this seems a little conservative. In any case he feels that material beyond the surf zone is too fine to abrade the bed. Again this may be slightly overstating the case. The breaking wave itself can exert a substantial force, especially when it traps a pocket of air. Such forces have been shown to be quite powerful, exerting extremely localized, short-lived 'bursts' which may weaken rock, extending existing fracture patterns. Between them, these two processes can produce abrasion platforms. These platforms can be quite extensive around the shores of Britain, and provide some clue to the erosive (and transporting) power of the sea. It is unwise to regard these impressions uncritically. Contemporary erosion rates are rather slow, and these platforms were probably created

Plate XXX Culver Hole, Gower, South Wales: pot-hole formation in limestone, probably assisted by solution. The rocks above high tide level are noticeably less smooth. Boot size 8 for scale. (M.C.)

during other periods when sea level was similar to today's. Sea level has been at approximately its present position for only about 4000 years. The shore platform provides a good example of negative feedback: as the platform is widened, the waves reaching the back of the platform have dissipated more energy in reaching it, and are therefore less powerful, reducing their erosive and transporting power. This is an oversimplification, since other forms of erosion need not be limited by widening the platform.

3 Water-layer weathering is the result of successive wetting and drying of the rock surface, and requires semi-isolated bodies of water. Platforms developed mainly by this process are consequently usually near the base level of low-water mark.

4 Sea-water solution is also isolated by Davies, although this must be related to water-layer weathering. Sea water may be heavily charged with dissolved carbon dioxide. This is obviously a much more important process with certain rock types, notably calcareous rocks. The coincidence of carbonate rocks and tropical climates in certain areas gives the impression of greater solutional activity in the tropics, but this may be fortuitous. Solutional features have been noted in the Antarctic, but in general other processes tend to be more active outside the tropics, so that the forms associated with solution are less evident.

5 Frost weathering is also recognized as a coastal process. Three factors must be considered: the availability of fresh water (in many cases from snow), the susceptibility of the rock to frost riving, and the removal of the debris, so that it does not form a blanketing layer. Frost weathering on the coast need be no different from frost weathering elsewhere, except perhaps

to note that the presence of salt may help depress freezing temperatures, so that it may not be quite as prevalent as at a comparable inland site.

6 There is, finally, erosion in the nearshore zone from living organisms. Bio-erosion has been widely noted, but much more is known about carbonate rocks in tropical environments than any other rock types or climates. Molluscs have been shown by Ahr and Stanton (1973) and Vita-Finzi and Cornelius (1973) to promote cliff sapping. Many organisms bore into rock to obtain food or shelter. Neumann (1966) noted that biological erosion increased progressively upward towards the water surface, but terminated abruptly at the first level of periodic prolonged exposure to the air. He concluded that browsing organisms smoothed surfaces of the rock, and that above the spray zone the rock surface was too rough for chitons. Browsing organisms include fish. The spray pits that occur in many carbonate rocks may be the result of algae, and not of actual spray solution, which is rather difficult to visualize. Blue-green algae are known to dissolve calcium carbonate by their excretions, and Neumann noted that algal cells may be present a few millimetres into the rock. Some erosion rates for common marine intertidal snails have been estimated (North, 1954). These snails scrape algae and fine detritus from the rocks, also removing particles from the rock itself. Erosion rates of 1 cm every 16–40 years were estimated, a figure which was within the range estimated for other processes in the area. A notable feature of bio-erosion is the zonation present. This may even manifest itself in different colours on the shore. Above high-tide level, orange lichens may be found; these do not tolerate flooding but prefer spray wetting, and also benefit from bird droppings; below these, black lichens may colonize the rocks, tolerating immersion, as long as they are dried out periodically. North's snails showed a similar zonation; *Littorina scutulata* tended to be found in a zone about 1 m either side of high-tide mark, while *L. planaxis* preferred the zone 2–3 m above spring high-tide mark. Similar zonations can be found for many marine organisms.

Integrating all these processes can only be achieved with reference to particular cases. Wilson (1952) discussed the influence of rock structure on coastal cliff development in Cornwall. The rock types included slates, grits, phyllites, lavas and intrusives. The shape of the cliffs has been influenced by bedding, jointing, and to a lesser extent, rock type. Massive joint-bounded blocks had been wrenched away from their settings, but the debris of ancient rock falls protected the bases of many cliffs from further erosion. Trenhaile (1972) examined the shore platforms of part of Glamorgan, and noted that the detailed platform geometry was determined by structure and lithology. The removal of limestone joint blocks was made easier by the erosion of interbedded shales. Solution was also present, but of secondary importance. Similarly, pot-holing, abrasion and other processes rarely played an important part in the development of the platform. Nevertheless, Trenhaile found recession rates of up to 10 cm year^{-1} over the last 50 years or so, but also suggested that more rapid retreat may have taken place in the last 2500 years. The reduction in erosion rate may be a result of the wider platform which the waves must cross.

Benthos

Biotic activity may also be found on sand beaches. This particular aspect has received rather piecemeal attention, but Howard and Dorjes (1972) and Hill and Hunter (1976) demonstrate the zonation of species across the beach from backshore to the offshore (Figure 11.13). Howard and Dorjes found over fifty species present on two tidal flats (with crustaceans the most common species). On the flatter of the two tidal flats, they noted that the fine fraction had two sources: the primary source was from suspended matter, rich in organic nutrients, which was the main food source for many of the organisms. The secondary source of fine material was the fecal matter excreted by the organisms. Each ripple trough acted as a collection zone for fecal matter. Two main habitats were identified—the drier, cleaner, steeper sandy backshore, and the muddy flats. They observed complete reworking of the sand making up the flats at least as deep as 15 cm. The ghost crab (Hill and Hunter, 1973) has been observed to burrow up to 70 cm into the beach. Various polychaetes (worms) can often be seen on fairly flat sandy beaches, where their tubes, covered by sand or incorporated shells, shell fragments and plant debris, poke up above the surface. Other species such as cockles, sand fleas and crabs are also familiar in the temperate areas.

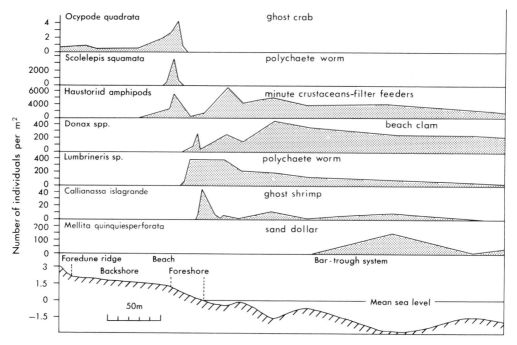

Figure 11.13 Distribution of major species, north Padre Island, Texas (after Hill and Hunter, 1976)

Some beaches owe their origin entirely to biological activity. Shell fragments can make up the greatest portion of the sediments on some beaches; indeed there are few beaches which do not contain some shell fragments, usually from species, such as the cockle, which live within the beach. Where there are offshore beds of mussels, oysters, or other molluscs, these can form the main constituent of the beach, especially when there is little other sediment available. Such beaches can be found in western Ireland and Scotland. It is usually argued that tropical beaches have much higher carbonate productivity, and therefore provide a higher proportion of the beach sediments, but the productivity of temperate areas should not be underestimated. The sandy beaches of northwestern Europe owe their origin more to fluvioglacial debris deposited in the Pleistocene than to any inherent production of sand-sized materials by processes operating today. In tropical areas, beaches may be composed of a wide variety of carbonate debris—foraminiferans, echinoids, sponge spicules, calcareous algae, coral fragments and oolites, among others.

Conclusion

It is clear that one of the most significant variables in the nearshore zone is breaker type. It is directly related to a whole series of other variables—phase difference, energy dissipation through 'instability', through sediment motion and (perhaps) nearshore currents. Since breaker type and beach slope are intimately related, surging breakers, and hence steep beaches, will be associated with partial reflection of the wave from the beach, leading to the formation of edge waves, beach water-table fluctuations, surf beat and rip currents. The short-term water-table fluctuations affect the history of deposition and erosion on the beach (and through this the local slope). But it should not be assumed that the breaker type is constant for a given beach. Komar (1971) writes this larger by stating that 'the processes and features we see at low tide or within the inner surf zone may not necessarily be most important at high tide or within the breaker zone and outer surf zone.' This helps to stress the fact that breaker type can vary through the tidal cycle (*cf*. Orford, 1977). Beaches are seldom planar, and commonly have a steeper backshore, so that the breaker type may change as the tide comes in, from spilling, through plunging to surging. (Although this is predicated on the assumption of some tidal range, storm surges in areas with little tidal range will produce much the same conditions.) The different breaker types are only convenient ways of referring to fairly well-defined sections of a continuous spectrum. Many beaches have a ridge-and-runnel topography, where waves may break over the ridge at one point in the tide, but at higher stages will not be modified substantially by the ridge. So the observed sequence of breaker types may be quite complex.

Since wave height and length also help to determine breaker type, different ocean or sea conditions can provide another source of temporal change. Longer swell waves behave differently from locally generated swell waves—they break earlier, and they refract earlier. The pattern of wave fronts in the nearshore is not as regular as the bald statement of the refraction equation might suggest (this also points to a slight problem with all the sediment-transport equations which use breaker angle—which breaker?).

The sediment in the nearshore zone does not form a smooth inactive floor for the breaking waves, but interacts with the fluid, and forms a variety of structures. This interaction must be taken into account for an understanding of the transformation within the fluid, and the movement of the sediment.

Added to these complications, the nearshore zone forms a habitat (or part of a habitat) for a variety of organisms, from large mammals to algae. They have their own interactions which can produce rather intriguing variations to the themes already present. Ignoring them would be foolish, especially on rocky coasts where they may contribute markedly to the erosion, in the tropics, where they can form corals, but also on sandy temperate beaches where they rework so much of the sediment, perhaps producing much of the fine material which gives the nearshore zone its turbid appearance.

It is encouraging to look back over the last twenty years of work in the nearshore zone and the increasingly intimate relationship between theoretical development and field observations. However, Miller and Barcilon (1976) caution this euphoria by pointing out that 'it would be premature to think that a general model encompassing all possible situations in the littoral environment is imminent'.

12

Processes and interrelationships, rates and changes

J. B. Thornes

In this book we have mainly been concerned with the mechanics of processes. In the wider subject of geomorphology this represents the starting point for investigating the variation of landforms through space and time. The landform variations are responses to spatial and temporal variations in processes, and to spatial variations in the materials and forms upon which the processes operate. In this last chapter some attention is paid to the nature of the interaction between processes in both a spatial and temporal sense and an attempt is made to link the rather mechanical approach to processes with the wider field of study. As Chorley (1978) argues, 'Geomorphology can only continue to make a unique contribution to Earth science if, in the study of processes, physical truth is sufficiently coarsened in both space and time [*from the mechanical approach*] as to accord with the scales on which it is profitable to study geomorphologically-viable landform objects.' The physical truth is here extended spatially in terms of the concept of process domains and temporally in terms of the notions of stability and instability.

Process domains

By a domain we mean the zone in which a particular process operates when plotted on a graph. *Existence domains* define the presence or absence of an object, event or process according to the values of the axes on which the domain is mapped. In Figure 12.1 a phenomenon or process could be defined as occurring when $Y \leqslant Y_1$ and $X \leqslant X_1$, so that the shaded area is the domain of the process or phenomenon.

An example of this idea is shown in Figure 12.2. This deals with a two-layered soil system, with hydraulic conductivities K_1 and K_2 for the upper and lower layers respectively; i is the rainfall intensity, measured in the same units as the hydraulic conductivity. If i/K_1 is less than one, then all the rain is absorbed, so there can be no infiltration overland flow. If the conductivity of the lower horizon is higher than that of the upper horizon, then there will be vertical percolation and no saturated overland flow (K_1/K_2 is less than one). The existence domains on the graph therefore indicate where and under what conditions the two types of overland flow would occur (Whipkey and Kirkby, 1978). Such existence domains are quite familiar in geomorphology; other examples include the existence domains of various flow types according to the Reynolds and Froude numbers (Figure 3.1, p. 41) and the existence of a mineral in an equilibrium chemical solution (Figure 6.7, p. 197).

At one time it was thought that the processes discussed in this book had existence domains which were quite rigidly defined. In deserts, for example, it was thought that there were situations in which moisture was physically absent, and therefore a spatial existence domain for fluvial processes could be defined. Likewise, an existence domain was postulated for 'limestone processes'. It is clear now that both water and calcium are ubiquitous so that the concept loses its utility.

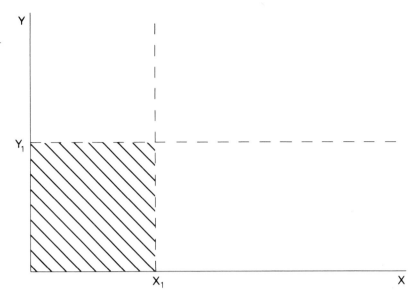

Figure 12.1 Definition of an existence domain on the basis of the critical values of two variables. The same idea may be expressed as a function of several variables in a higher order space.

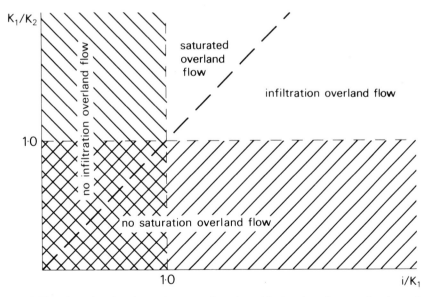

Figure 12.2 The domains for the existence of various kinds of surface and sub-surface flow (Whipkey and Kirkby, 1978).

Instead we have to think of *intensity domains* for processes, following the ideas of Peltier (1950), recently revitalized by Kirkby (1978). Processes are not mutually exclusive; rather one tends to dominate another because it is working more effectively in a particular set of circumstances. This idea is illustrated in Figure 12.3. In the range of values of an environmental factor (for example, temperature) one process is dominant in the zone A–B, whilst the other is dominant in the zone B–C. Peltier (1950) developed an intuitively based set of process-intensity graphs using annual rainfall and mean annual temperature which were then merged to produce a set of process domains for different types of weathering (*cf.* Figure 4.2). An example of an empirically defined process-domain diagram is that for zones of meandering and braiding described by Leopold and Wolman (1957). Here the shift from one process to another is thought to occur over a relatively sharp line according to channel slope and discharge conditions (see Figure 12.6, p. 385).

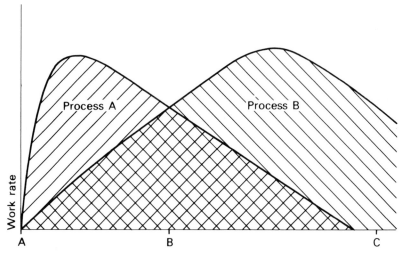

Figure 12.3 The concept of intensity domains. At the point B there is a shift in the relative dominance of process in terms of their intensity of process. This need not imply a change in form (after Kirkby, 1978).

When the graph axes are geographical coordinates, the domains are spatially defined, and as with existence and intensity domains, movement across the boundaries between processes or their relative dominance may be expected to produce a change in dominant form. The shift across the boundary between the domains of glacial- and non-glacial processes usually produces a strong contrast in form and represents the boundary of an existence domain. The shift from soil creep to wash, on the other hand, usually represents a shift between intensity domains, for example between the upper and lower part of a hillslope. The lines or zones between the relative dominance of one process over another are *transition thresholds*, which may be more or less sharp. These are somewhat different from *stability–instability thresholds* (Kirkby, 1978) which are discussed below. There have, as yet, been no studies that have defined spatial intensity domains, though there is a widespread assumption that they exist. This takes an implicit form in the study of morpho-climatic regions where landforms are accounted for in terms of climate through an assumed set of climate-process and process-form relationships. Unfortunately these links are still as yet poorly understood, as indicated by the collection of papers edited by Derbyshire (1976).

The properties of materials, such as their shearing resistance, are incorporated into the process-domain concept by considering *effective intensity*. This refers to the capacity for doing work when the threshold for change has been exceeded. For example, the effective rainfall is sometimes used to denote the rainfall available after interception, evaporation and infiltration losses have been accounted for. The application of stress to a stream bed only becomes geomorphologically effective when the critical value for entrainment has been passed. In this way, relative dominance of process depends on the material involved. Some processes involving a large application of stress may achieve very little effective work. Another process, in the same environmental and lithological conditions, may do a great deal of effective work for the application of relatively little stress. Thus, weathering processes may effect considerable geomorphological change compared with fluvial processes in granite terrains.

As the intensity of process shifts in relation to shifts in the controlling variables, such as climatic changes, the relative dominance of one process may give way to that of another. Process domains become superimposed in space through time. If the processes are associated with particular landforms, then superimposition of landforms may occur subject to lags in adjustment. This is evident in the post-glacial development of the British Isles. Although fluvial processes are now dominant, in the glacial areas there has been a lagged response in landforms due to the persistence of features produced in the glacial process domain. With a recurrence of cold conditions in the Little Ice Age there was a lagged but perceptible effect in landform terms which has been summarized by Goudie (1978). Again, the complex, lagged and superimposed responses of morphology to changes in climate, livestock density and incidental effects have been examined for the American Southwest by Cooke and Reeves (1977). In climato-genetic landform studies, form is used to infer past dominant processes and, through the intensity-domain concept, the changes that have occurred in the controlling environmental variables. The difficulties associated with this procedure are outlined by Thornes and Brunsden (1977).

Process rates

Attention has so far been given to spatial variations in the intensity of processes, and it has been assumed that values of the controlling variables are averaged over time so that short-term fluctuations are irrelevant. In the process domains identified by Peltier, mean annual temperature and mean annual rainfall were the criteria thought to be relevant, for example. Changes in the controlling variables, leading to changes in relative dominance of processes may also occur in the short term. This can be illustrated with reference to soil moisture. In a semi-arid environment, soil moisture controls both surface wash and creep according to the lithology; in winter, creep processes appear to dominate whereas in summer, wash is most important in terms of effective work done.

In addition to their importance in elucidating landform genesis, process rates are also relevant in applied studies, for example in determining road alignments, in the construction of flood defences, for ploughing policies on hillslopes and so on. They may also be used to establish crude 'geomorphological clocks' if their rate is fairly constant. For example, an average erosion rate in a catchment might be determined from deposits trapped within the catchment. Annually produced lake bottom deposits (varves) have been widely used in Fennoscandia to elucidate the chronology of deglaciation and land uplift. Finally, 'raw' rates are used to calibrate the parameters of mathematical models of geomorphological processes.

Most studies of process rates embody the concept of *magnitude and frequency* developed by Wolman and Miller (1960). This states that the total amount of work done by a spectrum of events depends not only on the magnitude of the events but also upon the frequency with which they occur. The frequency distribution ($f(s)$) of the duration of events of different magnitudes is obtained (Figure 12.4). By multiplying magnitude by the frequency at that magnitude, total work done, $T(s)$, is obtained. This product curve will have a peak which is to the right of that of the frequency curve, indicating that most of the work is done by events which are somewhat larger than those which occur most commonly. If there is a sharp, well-defined peak this implies that a particular magnitude of event does more work than other magnitudes. It was generally thought that in fluvial processes events of moderate magnitude and frequency performed most of the work. This view has found some support in studies of soil erosion by Hortonian overland flow (Pearce, 1976).

The curve $f(s)$ and hence $T(s)$ may be multimodal when there are two or more dominant frequencies. In semi-arid channels for instance there are many small flows and a few very large flows, with hardly any flows between. This situation gives rise to some problems in

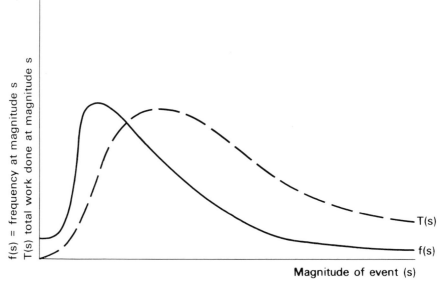

Figure 12.4 Frequency and magnitude distributions yield total work done.

process-response analysis when attempts are made to relate a landform property to a 'dominant' event, such as bankfull discharge. Moreover, the concept fails to identify the significance of differences of kind as well as differences of degree in events of different magnitudes. Some events have long-lasting effects which may be out of proportion to the work done in the total spectrum. It is worth noting, however, that the idea of a process domain based on intensity does not rely on the unimodality of the curve of total work performed. The problems also arise in terms of attempting to relate work performed to the sculpture of the landscape. Effectiveness should be thought of in terms of 'the ability of an event or combination of events to affect the shape or form of the landscape' (Wolman and Gerson, 1978). The magnitude and frequency concept tends to emphasize destructive effects, but those that tend to restore the surface of the landscape to the condition existing before the new landforms were created must also be considered. Recently there has been a revival of interest in extreme events, especially as analytical tools become available for investigating the succeeding transient as opposed to steady behaviour (Thornes, 1978). We should avoid the danger, however, of simply linking catastrophic response, as reflected by landforms, to catastrophic occurrences of low frequency, high magnitude events. Small events may, themselves, produce catastrophic behaviour depending on the stability or instability of the geomorphological system, just as small shifts in the ozone balance could lead to catastrophic changes in the global climate.

Before turning to this question of stability and instability it is worth mentioning that the establishment of rates of operation of processes is essentially a matter for field and laboratory investigations and even now, the available data are few and highly restricted spatially (Young, 1974; Douglas, 1976; Slaymaker, 1978). Field observations invariably suffer from great logistical problems during extreme events such as floods and landslides. Sampling problems are particularly acute when the cyclical components are small relative to irregular events. Laboratory data, on the other hand, are generated for highly specialized conditions and often suffer severe problems of scale reduction and are difficult to compare with field rates. Rates of operation may also be inferred from datable sites or extrapolated from rapidly evolving situations. These and other methods and problems associated with the establishment of process rates are discussed in Thornes and Brunsden (1977).

Stability and instability: sensitivity to change

Processes in operation may or may not result in landform change. The extent to which they do so depends on the relative stability of the landform concerned. Variables which can be used to characterize the condition of the system at any time are called *state variables*. Examples of state variables include the angle of a hillslope, the moisture content of soil, the temperature of glacier ice and the mean particle size of material at a location on a beach. It is important that the state variables observed are those particularly sensitive to change. For example, hillslope elevation and shape may have a low sensitivity to change compared with the soil characteristics on the hillside. The patterns taken on by state variables through time are sometimes used to infer the conditional stability of the system. A condition involving small fluctuations about a mean value through time is classified as 'steady-state' behaviour. This is attributed by some to the existence of self-regulatory mechanisms which tend to damp down any change. Inference concerning the likelihood of change from state variables and the classification of state-variable patterns in geomorphology is discussed by Chorley and Kennedy (1971) and Bennett and Chorley (1978) and is not pursued any further here.

Another approach is to enquire as to the conditions that determine the relative stability of a set of landforms. Figure 12.5 shows the analogy of a set of particles resting on a hill and is taken from Curtis (1976a). In this diagram three boulders are momentarily at rest on a hillside. A and C require only a small amount of energy to promote change, though A is more unstable in so far as it will change its state (the height above the reference plane) much more compared with C. Particle B has a potentially unstable state, relative to C, but must cross the threshold at D before its instability can be fully realized (it is said to be metastable). Thus, not only is the application of energy to be taken into account but also the overall instability of the system and any barriers to change.

In geomorphological literature the thresholds are usually thought of as barriers that must be overcome by an input of activation energy before change can occur. The critical shear stress required for entrainment is a good example, the shift from laminar to turbulent flow another. The idea that resistance to change depends on the ratio of stress to resistance is inherent in the 'structure' element of Davis's 'Structure, process and stage'. Schumm (1973) differentiated between those thresholds that are crossed as a result of the application of stresses external to the geomorphological system and those resulting from changes within the system.

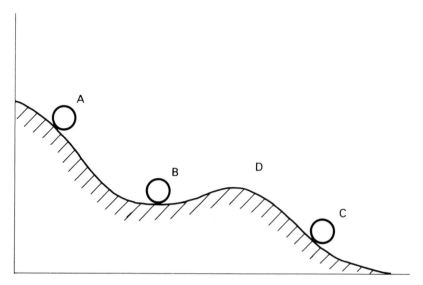

Figure 12.5 Relative stability exemplified by the height of a particle above some arbitrary datum.

The shift from channel aggradation to channel entrenchment in the floors of emphemeral valleys is an example of the latter type of process (Schumm and Beathard, 1976) and is comparable to the transitional thresholds between processes referred to earlier. The degree of stability in such systems depends on the height of the barrier to be overcome relative to the applied stress (e.g. the 'safety factor' in mass-movement). This can also be thought of as the position of the system in relation to the discriminating line between two processes. Figure 12.6 shows this interpretation with respect to the discriminating line between meandering and braided channels located empirically by Leopold and Wolman (1957). The point B, being close to the discriminant function, is relatively unstable when compared with points A or C.

Analytical techniques are now available to test the stability of dynamic systems when these systems can be expressed by differential equations. The mathematical model of the process-form relationship is perturbed (given a mathematical 'kick'). If the perturbation grows, the system is dynamically unstable, whereas if the perturbation damps, it is dynamically stable. Smith and Bretherton (1972) used this idea to examine the conditions under which a small hollow on a hillside would grow or fill up. If the hollow increases, the existence of the hollow tends to create flow convergence which in turn enlarges the hollow, provided that the increase in the rate of production of water to the hollow is greater than the increase in the rate of production of the sediment. Thus the situation is unstable. If the perturbation is filled up, minor changes through time occur, but otherwise the hillside is stable.

This leads to the recognition of a second type of threshold, that leading to a change from a stable condition in the system to an unstable condition. Following Kirkby (1978) this is shown graphically in Figure 12.7. He considers the hypothetical case of soil weathering as a function of soil depth. At shallow depths the weathering rate increases with depth because residence time is longer. If the soils are too deep, the residence times are too long. The peak rate occurs when the residence times are the same as the time required for chemical equilibrium (*cf.* Nortcliff, Waylen and Thornes, 1978). To the right of this point an increase in depth (A–A[1]) leads to a decrease in weathering rate, or a decrease in depth leads to an

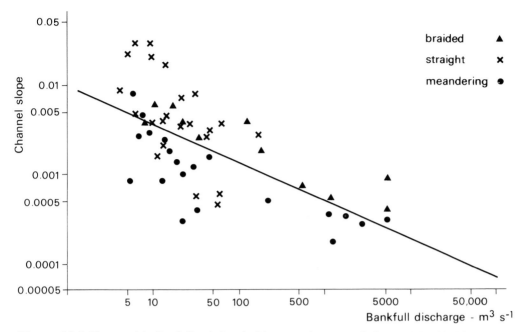

Figure 12.6 The empirically defined threshold separating meandering and braided flow.

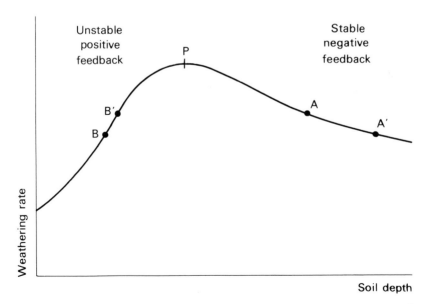

Figure 12.7 Stability (right-hand limb) and instability (left-hand limb) in terms of negative and positive feedback (after Kirkby, 1978).

increase in weathering rate. In either case, negative feedback ensures that the right-hand limb of the curve is stable. To the left of the maximum, a shift from B to B^1 increases the depth and increases the weathering rate. In this zone, where positive feedback occurs, a move in either direction will generate further change. Thus to the left of P the system is inherently unstable, whilst to the right the system is inherently stable. In this case the instability threshold (at P) separates stable and unstable behaviour. This particular argument can be generalized to any process curve with a maximum or minimum where there is uni-directional feedback from process to controlling factor.

Relaxation

Change in process type or rate may not be followed immediately by a landform response because the system has to have time to propagate these effects. The time required for the initiation of adjustment of morphology following a change in input is the *response time* or lag time and the time required to reach equilibrium with the changed conditions is the *relaxa-tion time*. Stable systems move towards the former steady-state conditions, as identified by the state vectors, following a perturbation. Unstable systems move progressively towards a more stable condition as indicated by Figures 12.5 and 12.7. The path followed in this process is the relaxation path. Such paths have been described for large floods (Schumm and Lichty, 1963), slopes on Mississippi bluffs (Brunsden and Kesel, 1973) and for deglaci-ated screes (Thornes, 1971). The most fundamental investigation of form response to change in process by following the evolution of form along a relaxation path is Allen's investigation of river-bed dune forms. In his study, Allen also provides theoretical arguments as to what form the relaxation path should take (Allen, 1976, 1977). The general evidence appears to indicate that for uninterrupted relaxation paths, the form taken is one of initial very rapid change followed by a progressive decrease in the rate of change. The complications of inter-related processes, varying degrees of stability and instability and the fact that we may be at any place along the relaxation path arising from process change, implies that the relation-ships between process and form need to be established primarily in this direction.

Process and form

The relationships between process and form have been at the core of geomorphology through-out the evolution of the subject. Two essentially contrasting approaches have been involved. In the first, the relationship between form and process is largely inferred from the investiga-tion of form properties. So, for example, continental drift was inferred from the morphology of the contents and the operation of glaciers was inferred from the distribution of till and erratic materials. Likewise the existence domains of periglaciation were inferred from ice-wedge forms and other characteristic features. It has to be admitted that this approach has had a high degree of success, particularly in certain cases, as the development of plate tec-tonics indicates. A logical step is to infer sequences of processes from sequences of forms and deposits, or spatially varying processes from spatially varying forms. In turn process, inferred from form, is used to infer the existence of controls on process, most notably climatic controls and, at the end of the chain, a sequence of forms may be used to infer a sequence of climates.

In the middle of the century the technical apparatus available to geomorphologists led to a much more sophisticated application of the basic inductive approach. In particular the description of form properties was greatly enhanced by the application of more carefully designed sampling schemes to the selection of information and by the application of correla-tion and regression techniques. It is difficult to point to real progress in process-form relationships which results directly from this development but it did greatly enhance our appreciation of the need for general statements about landforms. This has arisen in part from the recognition of the geometrical regularity in some, but not all, of the landforms that are studied. More important, description of dynamic aspects of geomorphological systems has led to the reappraisal of 'internal' mechanisms by which matter is stored in and conveyed through the system. It has also led to a questioning of the effect of dramatic changes of input or the internal operation of the system on the response which landforms represent.

The second approach seeks to establish form properties and their development on the basis of an understanding of the processes which is then applied to a set of initial conditions. Three steps are involved: (i) the identification of 'characteristic forms' on the basis of the environmental controls of the processes and the materials and forms on which they operate; (ii) specification of the conditions leading to change either from within the system or from the external environment and (iii) examination of the response to these conditions through time and space during the relaxation period. The latter step is based on the kinds of argument outlined earlier in this chapter. The two approaches are of course complementary, though the former may profitably be thought of as a body of information by which to examine the results of the latter. At the turn of the century it was argued that until the geomorphologists knew more about process there was little prospect of understanding long-term landform development and change. We should now be in a better position to address ourselves to these problems.

References

Ackers, P. and Charlton, F. G. 1970: The geometry of small meandering streams, *Proc. Instn civ. Engrs*, Paper 73285, 289–317

Adam, J. I. 1961: Tests on glacial tills, *Proc. 14th Can. Soil Mech. Conf., Nat. Res. Coun. Can., Ass. Commn Soil Snow Mech., Tech. Mem.* 69 (ed. E. Penner & J. Butler), 37–54

Addison, K. 1975: Aspects of the glaciation of Snowdonia, North Wales, Unpubl. D. Phil. thesis, Univ. of Oxford

Adie, A. J.: Figures quoted in Merrill (1897) from *Trans. R. Soc. Edinb.* 13, 366

Ahlmann, H. W. 1919: Geomorphological studies in Norway, *Geogr. Annlr* 1, 1–148 and 193–252

Ahlmann, H. W. 1935: Contribution to the physics of glaciers, *Geogrl J.* 86, 97–113

Ahnert, F. 1976: Brief description of a comprehensive three-dimensional process-response model of landform development, *Z. Geomorph., Suppl. Bd* 25, 29–49

Ahnert, F. 1977: Some comments on the quantitative formulation of geomorphological processes in a theoretical model, *Earth Surface Processes* 2, 191–202

Ahr, W. A. and Stanton, R. J. 1973: The sedimentological and paleoecological significance of *Lithotrya* a rock-boring barnacle, *J. sedim. Petrol.* 43, 20–3

Akimtzev, V. V. 1932: Historical soils of the Kamenetz-Podolsk Fortress, *Proc. 2nd int. Conf. Soil Sci.* 5, 132–40

Albritton, C. C. (ed.) 1963: *The fabric of geology*, Stanford, California, Freeman Cooper, 372 pp.

Alden, W. C. 1928: Landslide and flood at Gros Ventre Wyoming, *Tech. Publs Am. Inst. Min. metall. Engrs* 140, 14 pp. (also *Trans.* 76, 347–61)

Allen, J. R. L. 1968: *Current ripples, their relation to patterns of water and sediment motion*, North Holland, Amsterdam, 433 pp.

Allen, J. R. L. 1970a: *Physical processes of sedimentation*, Allen and Unwin, London, 248 pp.

Allen, J. R. L. 1970b: A quantitative model of grain size and sedimentary structures in lateral deposits, *Geol. J.* 7(1), 129–46

Allen, J. R. L. 1974: Reaction, relaxation and lag in natural sedimentary systems: General principles, examples and lessons, *Earth Sci. Rev.* 10, 263–342

Allen, J. R. L. 1976: Bed forms and unsteady processes: some concepts of classification and response illustrated by common one-way types, *Earth Surface Processes* 1, 361–74

Allen, J. R. L. 1977: Changeable rivers: some aspects of their mechanics and sedimentation, in *River channel changes* (ed. K. J. Gregory), Wiley, Chichester, 15–46

Allix A. 1924: Avalanches, *Geogrl Rev.* 14, 519–60

American Society for Testing Materials 1971a: Standard method of test for soundness of aggregates by use of sodium sulphate or magnesium sulphate, *A.S.T.M. Stand.* C88–71A, 117–18

American Society for Testing Materials 1971b: Specifications for facing brick (C216), *A.S.T.M. Stand.*, Footnote 1, 167–8

Anderson, A. G. 1967: On the development of stream meanders, *Proc. 12th Congr. int. Ass. Hydraul. Res.*, Ft Collins, 1, 370–8

Anderson, D. H. and Hawkes, H. E. 1958: Relative mobility of the common elements of weathering in some schist and granite areas, *Geochim. cosmochim. Acta* 14, 204–10

Anderson, D. M. and Morgenstern, N. R. 1973: Physics, chemistry and mechanics of frozen ground: a review, in *Permafrost: North American Contribution to the Second International Conference, Nat. Acad. Sci. Washington D.C.*, 257–88

Anderson, E. W. and Finlayson, B. 1975: Instruments for measuring soil creep, *Tech. Bull. Br. geomorph. Res. Grp* 16, 32 pp.

Anderson, M. G. and Burt, T. P. 1977: A laboratory model to investigate the soil moisture conditions on a draining slope, *J. Hydrol.* 33, 383–90

Anderson, M. G. and Calver, A. 1977: On the persistence of landscape features formed by a large flood, *Trans. Inst. Br. Geogr.*, N.S. 2, 243–54

Anderson, T. and Flett, J. S. 1903: Report on the eruption of the Saufrière in St Vincent in 1902 and on a visit to Montagne Pelée in Martinique, *Phil. Trans. R. Soc. Lond.* A, 200, 353–553

Andersson, J. G. 1906: Solifluction, a component of sub-aerial denudation, *J. Geol.* 14, 91–112

Andrews, J. T. 1970: A geomorphological study of post-glacial uplift in Arctic Canada, *Inst. Br. Geogr. Spec. Publ.* 2, 156 pp.

Andrews, J. T. 1972: Glacier power, mass balances, velocities and erosion potential, *Z. Geomorph., Suppl. Bd* 13, 1–17

Andrews, J. T. 1975: *Glacial systems*, Belmont, California, 191 pp.

Andrews, J. T. and Smithson, B. B. 1966: Till fabrics of the cross-valley moraines of north-central Baffin Island, N.W.T., Canada, *Bull. geol. Soc. Am.* 77, 271–90

Annersten, L. J. 1966: Interaction between surface cover and permafrost, *Biul. peryglac.* 15, 27–33

Anon. 1966: Ordered water molecular pressures offer new clues to rock failure, *Engng Min. J.* 167, 92–4

Apperly, L. W. 1968: Effect of turbulence on sediment entrainment, Unpubl. Ph.D. thesis, Univ. of Auckland, N.Z.

Armstrong, A. 1977: A three-dimensional simulation of slope forms, *Z. Geomorph., Suppl. Bd* 25, 20–8

Arni, T. H. 1966: Resistance to weathering, *A.S.T.M. Stand. Spec. Tech. Publ.* 169A, 261–74

Atkinson, T. C., 1978: Techniques for measuring subsurface flow on hillslopes, in *Hillslope hydrology* (ed. M. J. Kirkby), Wiley, Chichester, 73–120

Aubert, G. 1963: La classification des sols, *Cah. Pedol. ORSTOM* 3, 1–9

Bagnold, R. A. 1941: *The physics of blown sand and desert dunes*, Chapman and Hall, London, 265 pp.

Bagnold, R. A. 1953: Forme des dunes de sable et régime des vents, in *Actions éoliennes*, Centre nat. Rech. Sci., Paris, Coll. int., 35, 29–32

Bagnold, R. A. 1954a: Experiments on a gravity-free dispersion of large solid spheres in a Newtonian fluid under shear, *Proc. R.Soc.* A. 225, 49–63

Bagnold, R. A. 1954b: Some flume experiments on large grains but little denser than the transporting fluids, and their implications, *Proc. Instn civ. Engrs* 6041, 174–205

Bagnold, R. A. 1956: The flow of cohesionless grains in fluid, *Phil. Trans. R. Soc.* A, 249, 235–97

Bagnold, R. A. 1963: Mechanics of marine sedimentation, in *The sea* (ed. M. N. Hill), vol. 3, Interscience, N.Y., 507–28

Bagnold, R. A. 1966: An approach to the sediment transport from general physics, *U.S. geol. Surv. Prof. Pap.* 422–I, 37 pp.

Bagnold, R. A. 1973: The nature of saltation and of bed load transport in water, *Proc. R. Soc.* A, 332, 473–504

Bain, G. W. 1931: Spontaneous rock expansion, *J. Geol.* 39, 715–35

Bakker, J. P. 1960: Some observations in connection with recent Dutch investigations about granite weathering in different climates and climatic changes, *Z. Geomorph., Suppl. Bd* 1, 69–92

Bakker, J. P. 1965: A forgotten factor in the development of glacial stairways, *Z. Geomorph.* 9(1), 18–34

Bakker, J. P. and Le Heux, J. W. N. 1946: Projective-geometric treatment of O. Lehmann's theory of the transformation of steep mountain slopes, *Proc. K. ned. Akad. Wet.* B, 49, 533–47

Bakker, J. P. and Le Heux, J. W. N. 1947: Theory on central rectilinear recession of slopes, *Proc. K. ned. Akad. Wet.* B, 50, 959–66 and 1154–62

Bakker, J. P. and Le Heux, J. W. N. 1950: Theory on central rectilinear recession of slopes, *Proc. K. ned. Akad. Wet.* B, 53, 1073–84 and 1364–74

Bakker, J. P. and Le Heux, J. W. N. 1952: A remarkable new geomorphological law, *Proc. K. ned. Akad. Wet.* B, 55, 399–410 and 554–71

Bakker, J. P. and Levelt, Th. W. M. 1964: An enquiry into the problems of a polyclimatic development of peneplains and pediments (etchplains) in Europe during the Senonian and Tertiary Period, *Publs Serv. Carte géol. Luxemb.* 14, 27–75

Bakker, J. P. and Strahler, A. N. 1956: Report on quantitative treatment of slope recession problems, *International Geographical Union, First report on the study of slopes*, 30–2

Ball, D. F. 1966: Late-glacial scree in Wales, *Biul. peryglac.* 15, 151–63

Baltzer, R. A. and Lai, C. 1968: Computer simulation of unsteady flows in waterways, *J. Hydraul. Div. Am. Soc. civ. Engrs* 94 (HY4), 1083–117

Banham P. H. 1975: Glacitectonic structures: a general discussion with particular reference to the contorted drift of Norfolk, in *Ice Ages: ancient and modern* (ed. A. E. Wright and F. Moseley), Liverpool, 69–94

Baren, J. van 1931: Properties and constitution of a volcanic soil built in 50 years in the East-Indian archipelago, *Communs géol., Inst. Agric. Univ. Wageningen, Holland*, 17 pp.

Barnes, P., Tabor, D. and Walker, J. C. F. 1971: The friction and creep of polycrystalline ice, *Proc. R. Soc.* A, 324, 127–55

Barr, D. J. and Swanston, D. N. 1970: Measurement of creep in a shallow, slide-prone till soil, *Am. J. Sci.* 269, 467–80

Bartlett, W. H. 1832: Quoted in Merrill, 1897: *Am. J. Sci.* 22, 136

Barton, D. C. 1916: Notes on the disintegration of granite in Egypt, *J. Geol.* 24, 382–93

Bateman, A. M. 1950: *Economic mineral deposits*, Wiley, New York, 220 pp.

Bates, C. G. and Henry, A. J. 1928: Forest and stream-flow experiment at Wagon Wheel Gap, Colo. Final report on completion of the second phase of the experiment, *Mon. Weath. Rev., Suppl.* 30, 79 pp.

Battey, M. H. 1960: Geological factors in the development of Veslgjuv-botn and Vesl-Skautbotn, in *Norwegian cirque glaciers* (ed. W. V. Lewis), *R. geogrl Soc. Res. Ser.* 4, 5–10

Battle, W. R. B. 1960: Temperature observations in bergschrunds and their relationship to frost shattering, in *Norwegian cirque glaciers* (ed. W. V. Lewis), *R. geogrl Soc. Res. Ser.* 4, 83–95

Battle, W. R. B. and Lewis, W. V. 1951: Temperature observations in bergschrunds and their relationship to cirque erosion, *J. Geol.* 59, 537–45

Bauer, F. 1962: Karstformen in den Östereichischen Kalkhochalpen, *Act. 2nd int. Congr. Speleol.* (1958), 1, 299–329

Baulig, H. 1956a: *Vocabulaire franco-allemand de géomorphologie*, Paris

Baulig, H. 1956b: Pénéplaines et pédiplaines, *Soc. belge Études géogr.* 25, 25–58

Bazett, D. J., Adams, J. L. and Matyas, E. L. 1961: An investigation of a slide in a test trench excavated in fissured sensitive clay, *Proc. 5th int. Conf. Soil Mech.* (Paris), 1, 431–5

Beavers, A. H. 1957: Source and deposition of clay minerals in Peorian loess, *Science, N.Y.* 126, 1285

Behre, C. H. 1933: Talus behaviour above timber in the Rocky Mountains, *J. Geol.* 41, 622–35

Benedict, J. B. 1970: Frost cracking in the Colorado Front Range, *Geogr. Annlr* A, 52, 87–93

Bennett, J. P. 1974: Concepts of mathematical modelling of sediment yield, *Wat. Resour. Res.* 10(3), 485–92

Bennett, R. J. and Chorley, R. J. 1978: *Environmental systems*, Arnold, London, 624 pp.

Berg, T. E. 1969: Fossil sand wedges at Edmonton, Alberta, Canada, *Biul. peryglac.* 19, 325–33

Berry, L. and Ruxton, B. P. 1959: Notes on weathering zones and soils on granitic rocks in two tropical regions, *J. Soil Sci.* 10, 54–63

Beskow, G. 1935: Tjälbildningen och tjällyftningen med särskild hänsyn till vägar och järnvägar, *Sver. Geol. Unders. Afh.*, Ser. C, 375. English translation by J. O. Osterberg (1947), NWest. Univ. tech. Inst., 145 pp.

Birch, F. (ed.) 1942: *Handbook of physical constants (Revised ed.)*, *Geol. Soc. Am. Spec. Pap.* B6, 325 pp.

Birkeland, P. W. 1974: *Pedology, weathering and geomorphological research*, Oxford Univ. Press, 285 pp.

Birot, P. 1954: Désagrégation des roches cristallines sous l'action des sols, *C.R. Acad. Sci.* 238 (10), 1145–6

Birot, P. 1960: *Le cycle d'érosion sous les différents climats* (Centro de Pesquisas de geografia do Brazil. Univ. of Brazil, Rio de Janeiro), 137 pp. Trans. C. I. Jackson and K. M. Clayton, 1968: *The cycle of erosion in different climates* (Batsford, London), 144 pp.

Birot, P. 1962: *Contribution à l'étude de la désagrégation des roches*, Centre Docum. Univ. Paris, 232 pp.

Birot, P. 1968a, *The cycle of erosion in different climates*, Batsford, London, 144 pp. See Birot, P. (1960)

Birot, P. 1968b: Les développements récents des théories de l'érosion glaciaire, *Annls Géogr.* 77, 1–13

Bishop, A. W. 1967: Progressive failure, with special reference to the mechanism causing it, *Proc. geotech. Conf.* (Oslo), 2, 3–10

Bishop, A. W. 1973: The stability of tips and spoil heaps, *Q.J. Engng Geol.* 6, 335–76

Bishop, A. W. and Bjerrum, L. 1960: The relevance of the triaxial test to the solution of stability problems, *Proc. Am. Soc. civ. Engrs* (Conf. on the Shear Strength of Cohesive Soils), 437 pp.

Bishop, A. W. and Morgenstern, N. R. 1960: Stability coefficients for earth slopes, *Géotechnique* 10, 29–150

Bjerrum, L. 1954: Geotechnical properties of Norwegian marine clays, *Géotechnique* 4, 46–69

Bjerrum, L. 1955: Stability of natural slopes in quick clay, *Géotechnique* 5, 101–19

Bjerrum, L. 1971: Subaqueous slope failures in Norwegian fjords, *Publs Norw. geotech. Inst.* 88, 1–16

Bjerrum, L. and Jørstad, F. 1968: Stability of rock slopes in Norway, *Publs Norw. geotech. Inst.* 79, 1–11

Black, C. 1969: Slopes in S.W. Wisconsin, U.S.A., periglacial or temperate, *Biul. peryglac.* 18, 69–82

Black, R. F. 1963: Les coins de glace et le gel permanent dans le Nord de l'Alaska, *Annls Géogr.* 72, 257–71

Black, R. F. 1973: Cryomorphic processes and micro-relief features, Victoria Land, Antarctica, in *Research in polar and alpine geomorphology* (3rd Guelph Symposium on Geomorphology, ed. B. D. Fahey and R. D. Thompson), 11–24

Blackwelder, E. 1925: Exfoliation as a phase of rock weathering, *J. Geol.* 33, 793–806

Blackwelder, E. 1926: Fire as an agent in rock weathering, *J. Geol.* 35, 135–40

Blackwelder, E. 1933: The insolation hypothesis of rock weathering, *Am. J. Sci.* 26, 97–113

Blackwelder, E. 1934: Yardangs, *Bull. geol. Soc. Am.* 45, 159–66

Blake, D. H. and Ollier, C. D. 1971: Alluvial plains of the Fly River, Papua, *Z. Geomorph.* 12, 1–17

Blinco, P. H. and Simons, D. B. 1974: Characteristics of turbulent boundary shear stress, *J. Engng Mech. Div. Am. Soc. civ. Engrs* 100 (EM2), 203–20

Bloom, A. L. 1969: *The surface of the Earth*, Prentice-Hall, New Jersey, 152 pp.

Blong, R. J. 1973a: A numerical classification of selected landslides of the debris slide—avalanche-flow type, *Engng Geol.* 7, 99–114

Blong, R. J. 1973b: Relationships between morphometric attributes of landslides, *Z. Geomorph., Suppl. Bd* 18, 66–77

Bluck, B. J. 1974: Structure and directional properties of some valley sandur deposits in southern Iceland, *Sedimentology* 21, 533–54

Boch, S. G. 1946: Les névés et l'érosion par la neige dans la partie nord de l'Oural, *Bull. Soc. géogr. U.S.S.R.* 78, 207–34 (original in Russian; translated by C.E.D.P., Paris)

Bolt, G. H. and Groenevelt, P. H. 1972: Coupling between transport processes in porous media, *Proc. 2nd Symp. on Transport Phenomena in Porous Media* (Int. Ass. hydraul. Res.—Int. Soil Sci. Soc.), 630–52

Bolton Seed, H. 1968: Landslides during earthquakes due to soil liquefaction, *J. Soil Mech. Fdns Div. Am. Soc. civ. Engrs* 94, SM5, 1054–122

Bomford, G. 1971: *Geodesy*, Oxford, 3rd ed., 731 pp.

Boorman, L. A. and Woodell, S. R. J. 1966: The topograph, an instrument for measuring microtopography, *Ecology* 47, 869–70

Bott, M. H. P. 1971: *The interior of the Earth*, Arnold, London, 316 pp.

Boulton, G. S. 1967: The development of a complex supraglacial moraine at the margin of Sørbreen, Ny Friesland, Vestspitsbergen, *J. Glaciol.* 6, 717–35

Boulton, G. S. 1968: Flow tills and related deposits on some Vestspitsbergen glaciers, *J. Glaciol.* 7, 391–412

Boulton, G. S. 1970a: On the origin and transport of englacial debris in Svalbard glaciers, *J. Glaciol.* 9, 213–29

Boulton, G. S. 1972b: Modern Arctic glaciers ad depositional models for former ice sheets, *Q.J. Geol.* glaciers, *J. Glaciol.* 9, 231–45

Boulton, G. S. 1971: Till genesis and fabric in Svalbard, Spitzbergen, in *Till, a symposium* (ed. R. P. Goldthwait), Ohio State Univ. Press, 41–72

Boulton, G. S. 1972a: The role of the thermal régime in glacial sedimentation, *Inst. Br. Geogr. Spec. Publ.* 5, 1–19

Boulton, G. S. 1972b: Modern Artic glaciers as depositional models for former ice sheets, *Q.J. Geol. Soc. Lond.* 128, 361–93

Boulton, G. S. 1974: Processes and patterns of glacial erosion, in *Glacial geomorphology* (ed. D. R. Coates), State Univ. of New York, Binghamton, 41–87

Boulton, G. S. 1975: Process and patterns of subglacial sedimentation: a theoretical approach, in *Ice Ages: ancient and modern* (ed. A. E. Wright and F. Moseley), Liverpool, 7–42

Bowen, N. L. 1928: *The evolution of the igneous rocks*, Princeton Univ. Press, Princeton N.J., 288 pp.

Bowler, J. M. 1973: Clay dunes: their occurrence, formation and environmental significance, *Earth Sci. Rev.* 9, 315–38

Boyce, R. B. and Clark, W. A. V. 1964: The concept of shape in geography, *Geogrl Rev.* 54, 561–72

Boyé, M. 1950: *Glaciaire et périglaciaire de l' Ata Sund nord-oriental, Groenland,* Paris; Expéditions polaires francaises, 1

Boyé, M. 1952: Névés et érosion glaciaire, *Revue Géomorph. dyn.* 3, 20–36

Boyé, M. 1968: Défense et illustration de l'hypothèse du 'défonçage périglaciaire', *Biul. peryglac.* 17, 5–56

Bradley, C. C. 1966: The snow resistograph and slab avalanche investigations, *Proc. int. Symp. on scientific aspects of snow and ice avalanches* (Davos, 1965), *Int. Ass. scient. Hydrol. Publ.* 69, 251–60

Bradley, W. C. 1958: Submarine abrasion and wave-cut platforms, *Bull. geol. Soc. Am.* 69, 967–74

Bradley, W. C. 1963: Large-scale exfoliation in massive sandstone of the Colorado Plateau, *Bull. geol. Soc. Am.* 74, 519–28

Brahms, A. 1757: Elements of dam and hydraulic engineering, Aurich, Germany, 1, 105 pp.

Brakensiek, D. L. 1967: Kinematic flood routing, *Trans. Am. Soc. agric. Engng* 10(3), 340–3

Brenninkmeyer, B. 1976: Sand fountains in the surf zone, in *Beach and nearshore sedimentation*, ed. R. A. Davis and R. L. Ethington, *Spec. Publs Soc. econ. Palaeont. Miner.*, Tulsa, 24, 69–81

Brewer, R. 1964: *Fabric and mineral analysis of soils*, Wiley, New York, 470 pp.

Bridge, J. S. 1977: Flow, bed topography, grain size and sedimentary structure in open channel bends: a three-dimensional model, *Earth Surface Processes* 2, 401–16

Bridge, J. S. 1976: Bed topography and grain size in open channels, *Sedimentology* 23, 407–14

Bridge, J. S. and Jarvis, R. S. 1976: Flow and sedimentary processes in the meandering River South Esk, Glen Clova, Scotland, *Earth Surface Processes* 1, 303–36

Bridgman, P. W. 1912: Water in the liquid and five solid forms, under pressure, *Proc. Am. Acad. Arts Sci.* 47, 439–558

Bridgman, P. W. 1914: High pressures and five kinds of ice, *J. Franklin Inst.* 177, 315–32

Brown, C. B. 1924: On some effects of wind and sun in the desert of Tumbez, Peru, *Geol. Mag.* 61, 337–9

Brown, C. B. 1950: Sediment transportation, in *Engineering hydraulics* (ed. H. Rouse), Wiley, New York, Ch. 12

Brown, W. R. 1964: Grey billy and the age of tor topography in Monaro N.S.W., *Proc. Linn. Soc. N.S.W.* 89, 322–5

Brunsden, D. 1973: The application of systems theory to the study of mass movement, *Geologia appl. Idrogeol.* 8, 185–207

Brunsden, D. and Jones, D. K. C. 1971: The morphology of degraded landslide slopes in south west Dorset, *Q.J. Engng Geol.* 5, 205–22

Brunsden, D. and Jones, D. K. C. 1974: The evolution of landslide slopes in Dorset, *Phil. Trans. R. Soc.*, A, 283, 605–31

Brunsden, D. and Kesel, R. H. 1973: Slope development on a Mississippi River bluff in historic time, *J. Geol.* 91, 576–97

Bryan, K. 1934: Geomorphic processes at high altitudes, *Geogrl Rev.* 24, 655–6

Bryan, K. 1946: Cryopedology—the study of frozen ground and intensive frost action with suggestions on nomenclature, *Am. J. Sci.* 244, 622–42

Budd, W. F., Jenssen, D. and Radok, U. 1971: Re-interpretation of deep ice temperatures, *Nature, Lond. (Phys. Sci.)* 232, 84–5

Büdel, J. 1957: The Ice Age in the tropics, *Universitas* 1, 183–91

Büdel, J. 1963: Klima-genetische Geomorphologie, *Geogr. Rundsch.* 7, 269–86

Bunting, B. T. 1961: The role of seepage moisture in soil formation, slope development and stream initiation, *Am. J. Sci.* 259, 503–18

Buss, E. and Heim, A. 1881: *Der Bergsturz von Elm*, Wurster & Cie, Zürich, 163 pp.

Butler, B. E. 1959: Periodic phenomena in landscapes as a basis for soil studies, *Commonw. scient. ind. Res. Org. Aust., Soil. Publ.* 14, Canberra, 22 pp.

Butler, B. E. 1967: Soil periodicity in relation to landform development in south eastern Australia, in *Landform studies in Australia and New Guinea* (ed. J. N. Jennings and J. A. Mabbutt), Univ. Press, Cambridge

Cabrera, J. G. and Smalley, I. J. 1973: Quick clays as products of glacial action, a new approach to their nature, geology, distribution and geotechnical properties, *Engng Geol.* 7, 115–34

Cailleux, A. 1952: Polissage et surcreusement glaciaires dans l'hypothèse de Boyé, *Revue Géomorph. dyn.* 3, 247–57

Cailleux, A. 1972: Les formes et dépôts nivéo-éoliens en Antarctique et au Nouveau-Québec, *Cah. Géogr. Québ.* 16, 377–409

Cailleux, A. and Tricart, J. 1950: Un type de solifluction: les coulées boueuses, *Revue Géomorph. dyn.* 1, 4–46

Caine, T. N. 1963: Movement of low-angle scree slopes in the Lake District, northern England, *Revue Géomorph. dyn.* 14, 171–7

Caine, T. N. 1969: A model for alpine talus slope development by slush avalanching, *J. Geol.* 77, 92–100

Calkin, P. and Cailleux, A. 1962: A quantitative study of cavernous weathering (tafonis) and its application to glacial chronology in Victoria Valley, Antarctica, *Z. Geomorph.* 6, 317–24

Callendar, R. A. 1969: Instability and river channels, *J. Fluid Mech.* 36, 465–80

Capps, S. 1941: Observations of the rate of creep in Idaho, *Am. J. Sci.* 239, 25–32

Carlston, C. W. 1963: Drainage density and streamflow, *U.S. geol. Surv. Prof. Pap.* 422C, 8 pp.

Carlyle, W. J. 1965: Shek Pik dam, *Proc. Instn civ. Engrs* 30, 557–88

Carol, H. 1947: Formation of *roches moutonnées*, *J. Glaciol.* 1, 57–9

Carroll, D. 1962: Rainwater as a chemical agent in weathering—a review, *U.S. geol. Surv. Wat.-Supply Irrig. Pap.* 1535–9, 18 pp.

Carroll, D. 1970: *Rock weathering*, Plenum Publ., New York, 203 pp.

Carruthers, R. G. 1953: *Glacial drifts and the undermelt theory*, Harold Hill, Newcastle-upon-Tyne, 42 pp.

Carson, M. A. 1969: Models of hillslope development under mass fracture, *Geogrl Anal.* 1, 76–100

Carson, M. A. 1971a: *The mechanics of erosion*, Pion, London, 174 pp.

Carson, M. A. 1971b: Application of the concept of threshold slopes to the Laramie Mountains, Wyoming, *Inst. Br. Geogr. Spec. Publ.* 3, 31–48

Carson, M. A. 1977: Angles of repose, angles of shearing resistance and angles of talus slopes, *Earth Surface Processes* 2, 363–80

Carson, M. A. and Kirkby, M. J. 1972: *Hillslope form and process*, Univ. Press, Cambridge, 475 pp.

Carson, M. A. and Petley, D. J. 1970: The existence of threshold hillslopes in the denudation of the landscape, *Trans. Inst. Br. Geogr.* 49, 71–95

Cartledge, G. H. 1928: Studies in the periodic system, *J. Am. chem. Soc.* 50, 2855–72

Casagrande, A. 1932 in A. C. Benkelman and E. R. Olmstead: A new theory of frost heaving, *Natl. Res. Coun. Highw. Res. Bd, Proc. 11th Ann. Mtg* 1, 168–72

Casagrande, A. 1965: Role of 'calculated risk' in earthwork foundation engineering; Terzaghi Lecture, *J. Soil Mech. Fdns Div. Am. Soc. civ. Engrs* 91, SM4, 140 pp.

Casagrande, A. 1971: On liquefaction phenomena, Report of lecture, November 1970, prepared by Green, P. A. and Ferguson, P. A. S., *Géotechnique* 21, 197–202

Casagrande, A. and MacIver, B. N. 1970: Design and construction of tailing dams, *Symposium on the stability of open pit mining* (Vancouver), 55 pp.

Chamberlin, T. C. 1894: Proposed genetic classification of Pleistocene glacial formations, *J. Geol.* 2, 517–38

Chandler, R. J. 1969: The effect of weathering on the shear strength properties of Keuper Marl, *Géotechnique* 22, 403–31

Chandler, R. J. 1970a: The degradation of Lias Clay slopes in an area of the East Midlands, *Q.J. Engng Geol.* 2, 161–81

Chandler, R. J. 1970b: A shallow slab slide in the Lias clay near Uppingham, Rutland, *Géotechnique* 20, 253–60

Chandler, R. J. 1972: Periglacial mudslides in Vestspitzbergen and their bearing on the origin of fossil solifluction shears in low-angled clay slopes, *Q.J. Engng Geol.* 5, 223–41

Chandler, R. J. and Pook, N. 1971: Creep movements in low gradient clay slopes since the Late Glacial, *Nature, Lond.* U229, 399–400

Chapman, C. A. 1958: The control of jointing by topography, *J. Geol.* 66, 552–8

Chen, Y. H., Holly, F. M., Mahmood, K. and Simons, D. B. 1975: Transport of materials by unsteady flow, in *Unsteady flow in open channels* (ed. K. Mahmood and V. Yevjevitch), 1, 313–63

Chepil, W. S., 1945a: Dynamics of wind erosion. I. Nature of movement of soil by wind, *Soil Sci.* 60, 305–20

Chepil, W. S. 1945b: Dynamics of wind erosion. II. Initiation of soil movement, *Soil Sci.* 60, 397–411

Chepil, W. S. 1945c: Dynamics of wind erosion. III. The transport capacity of the wind, *Soil Sci.* 60, 475–80

Chepil, W. S. 1946a: Dynamics of wind erosion. IV. The translocating and abrasive action of the wind, *Soil Sci.* 61, 167–77

Chepil, W. S. 1946b: Dynamics of wind erosion. V. Cumulative intensity of soil drifting across eroding fields, *Soil Sci.* 61, 257–63

Chepil, W. S. 1959: Equilibrium of soil grains at the threshold of movement by wind, *Proc. Soil Sci. Soc. Am.* 23, 422–8

Chepil, W. S. and Woodruff, N. P. 1963: The physics of wind erosion and its control, *Adv. Agron.* 15, 211–302

Chorley, R. J. 1969: The role of water in rock disintegration, in *Water, Earth and Man*, Methuen, London, 53–73

Chorley, R. J. 1978: Bases for theory in geomorphology, in *Geomorphology: present problems and future prospects* (ed. C. Embleton *et al.*), Oxford Univ. Press, 1–13

Chorley, R. J. and Kennedy, B. A. 1971: *Physical geography: a systems approach*, Prentice-Hall, 370 pp.

Chow, ven T. 1959: *Open-channel hydraulics*, McGraw Hill, Kogakusha, Tokyo, 680 pp.

Clapperton, C. M. 1975: The debris content of surging glaciers in Svalbard and Iceland, *J. Glaciol.* 14, 395–406

Clarke, F. W. and Washington, H. S. 1924: The composition of the earth's crust, *U.S. geol. Surv. Prof. Pap.* 127, 117 pp.

Cleaves, E. T., Godfrey, A. E. and Bricker, O. P. 1970: Geochemical balance of a small watershed and its geomorphic implications, *Bull. geol. Soc. Am.* 81, 3015–32

Clement, P. and Vaudour, J. 1968: Observations on the pH of melting snow in the southern French Alps, in *Arctic and alpine environments* (ed. H. E. Wright and W. H. Osburn), Indiana Univ. Press, 205–13

Clifton, H. E. 1976: Wave-formed sedimentary structures—a conceptual model, in *Beach and near-shore sedimentation*, ed. R. A. Davis and R. L. Ethington, *Spec. Publs Soc. econ. Palaeont. Miner.*, Tulsa, 24, 126–48

Cloudsley-Thompson, J. L. and Chadwick, M. J. 1964: *Life in deserts*, Foulis, London, 247 pp.

Cokelet, E. D. 1977: Breaking waves, *Nature, Lond.* 267, 769–74

Colbeck, S. C. and Evans, R. J. 1973: A flow law for temperate glacier ice, *J. Glaciol.*, 12, 71–86

Colby, B. R. 1964a: Practical computation of bed-material discharge, *J. Hydraul. Div. Am. Soc. civ. Engrs* 90, HY2, 217–46

Colby, B. R. 1964b: Discharge of sands and mean velocity relationships in sand-bed streams, *U.S. geol. Surv. Prof. Pap.* 462–A, 47 pp.

Collins, J. I. 1976: Wave modelling and hydrodynamics, in *Nearshore beach processes and sedimentation* ed. R. A. Davis and R. L. Ethington, *Spec. Publs Soc. econ. Palaeont. Miner.*, Tulsa, 24, 54–68

Cook, F. A. and Raiche, V. G. 1962: Freeze-thaw cycles at Resolute, N.W.T., *Geogrl Bull.* 18, 64–78

Cooke, R. U. and Doornkamp, J. C. 1974: *Geomorphology in environmental management: an introduction*, Clarendon Press, Oxford, 413 pp.

Cooke, R. U. and Reeves, R. W. 1977: *Arroyos and environmental change in the American Southwest*, Clarendon Press, Oxford, 213 pp.

Cooke, R. U. and Smalley, I. J. 1968: Salt weathering in deserts, *Nature, Lond.* 220, 1226–7

Cooke, R. U. and Warren, A. 1973: *Geomorphology in deserts*, Batsford, London, 394 pp.

Cooling, L. F. 1945: Development and scope of soil mechanics, in *The principles and application of soil mechanics*, Instn civ. Engrs, London, 1–30

Cooper, A. W. 1960: An example of the role of microclimate in soil genesis, *Soil Sci.* 90, 109–20

Cooper, W. S. 1967: Coastal dunes of California, *Mem. geol. Soc. Am.* 104, 131 pp.

Cooper, H. H. and Rorabaugh, M. I. 1963: Groundwater movement and bank storage due to flood stages in surface streams, *U.S. geol. Surv. Wat.–Supply Irrig. Pap.* 1536–J, 343–66

Corbel, J. 1957: L'érosion climatique des granites et silicates sous climats chauds, *Revue Géomorph. dyn.* 8, 4–8

Corbel, J. 1959: Vitesse de l'érosion, *Z. Geomorph.* 3, 1–28

Corbel, J. 1964: L'érosion terrestre, étude quantitative (methodes-techniques-resultats), *Annls Géogr.* 73, 385–412

Correns, C. W. 1963: Experiments on the decomposition of silicates and discussion of chemical weathering, *Clays Clay Miner.* 5, 443–60

Corte, A. E. 1961: The frost behaviour of soils: laboratory and field data for a new concept: Part 1, Vertical sorting, *U.S. Army Cold Reg. Res. Engng Lab.*, Rep. 85, 22 pp.

Corte, A. E. 1966: Particle sorting by repeated freezing and thawing, *Biul. peryglac.* 15, 175–240

Costin, A. B., Jennings, J. N., Black, H. P. and Thom, B. G. 1964: Snow action on Mount Twynam, Snowy Mountains, Australia, *J. Glaciol.* 5, 219–28

Cotton, C. A. 1945: *Geomorphology. An introduction to the study of landforms*, Whitcombe & Tombs, London & New Zealand, 505 pp.

Crandell, D. R. and Varnes, D. J. 1961: Movement of the Slumgullion earthflow near Lake City, Colorado, *U.S. geol. Surv. Prof. Pap.* 424, 136–9

Crosby, I. B. 1945: Glacial erosion and the buried Wyoming valley of Pennsylvania, *Bull. geol. Soc. Am.* 56, 389–400

Crowther, E. M. 1930: The relationship of climate and geological factors to the composition of clay and the distribution of soil types, *Proc. R. Soc.* 107, 10–30

Crozier, M. J. 1973: Techniques for the morphometric analysis of landslides, *Z. Geomorph.* 17, 78–101

Cruden, D. M. 1976: Major rock slides in the Rockies, *Can. geotech. J.* 13, 8–20

Culling, W. E. H. 1963: Soil creep and the development of hillside slopes, *J. Geol.* 72, 127–61

Culmann, C. 1866: *Graphische Statik*, Zürich

Curray, J. R. 1965: Late Quaternary history, continental shelves of the United States, in *The Quaternary of the United States* (ed. H. E. Wright and D. G. Frey), Princeton Univ. Press, Princeton, N.J., 723–35

Curry, R. R. 1966: Observations on Alpine mudflows in the Ten-Mile Range, central Colorado, *Bull. geol. Soc. Am.* 77, 771–6

Curtis, C. D. 1976a: Chemistry of rock weathering: fundamental reactions and controls, in *Geomorphology and climate* (ed. E. Derbyshire), Wiley, Chichester, 25–57

Curtis, C. D. 1976b: Stability of minerals in surface weathering reactions, *Earth Surface Processes* 1, 63–70

Czeppe, Z. 1960: Annual course and the morphological effect of the vertical frost movements of soil at Hornsund, Vestspitsbergen, *Bull. Acad. pol. Sci., Sér. sci. géol. géogr.* 8, 145–8

Czudek, T. and Demek, J. 1970: Thermokarst in Siberia and its influence on the development of lowland relief, *Quat. Res.* 1, 103–20

Czudek, T. and Demek, J. 1973: Die Reliefentwicklung während der Dauerfrostbodendegradation, *Rospr. čsl. Akad. Věd.* 83, 69 pp.

Dahl, R. 1965: Plastically sculptured detail forms on rock surfaces in northern Nordland, Norway, *Geogr. Annlr* 47, 83–140

Dahl, R. 1967: Post-glacial micro-weathering of bedrock surfaces in the Narvik district of Norway, *Geogr. Annlr*, A, 49, 155–66

Darcy, H. 1856: *Les fontaines publiques de la ville de Dijon*, Dalmont, Paris

Davies, J. L. 1973: *Geographical variation in coastal development*, Hafner, New York, 204 pp.

Davis, S. N. 1964: Silica in streams and ground water, *Am. J. Sci.* 262, 870–91

Davis, W. M. 1899: The geographical cycle, *Geogrl J.* 14, 481–504

Davison, C. 1888: Note on the movement of some material, *Q.J. geol. Soc. Lond.* 4, 232–8

Davison, C. 1889: On the creeping of the soil cap through the action of frost, *Geol. Mag.* 6, 255

Dean, R. G. 1973: Hueristic models of sand transport in the surf zone, *Proc. Conf. Engng Dynamics Coastal Zone*, Sydney, 208–14

Dedkov, A. P. and Duglav, V. A. 1967: Slow movements of soil masses on grassed slopes, *Izv. Akad. Nauk SSSR, Ser. geogr.* 4, 90–3; Boston Spa Translation Programme, National Lending Library, RTS 7548

Demek, J. 1978: Periglacial geomorphology: present problems and future prospects, in *Geomorphology— present problems and future prospects* (ed. C. Embleton *et al.*, Oxford Univ. Press), 139–53

Denny, C. S. 1936: Periglacial phenomena in southern Connecticut, *Am. J. Sci.* 32, 322–42

Derbyshire, E. (ed.) 1976: *Geomorphology and climate*, Wiley, Chichester, 512 pp.

Dickson, R. A. and Crocker, R. C. 1953: A chronosequence of soils and vegetation near Mt Shasta, California. II. The development of the forest floors and the carbon and nitrogen profiles of the soils, *J. Soil Sci.* 4, 142–54

Dingler, J. R. and Inman, D. L. 1976: Wave-formed ripples in nearshore sands, *Proc. 15th Conf. Coastal Engng*, Honolulu, 2109–26

Douglas, I. 1969: Field methods of water hardness determination, *Tech. Bull. Br. Geomorph. Res. Grp* 1, 35 pp.

Douglas, I. 1976: Erosion rates and climate: geomorphological implications, in *Geomorphology and climate* (ed. E. Derbyshire), Wiley, Chichester, 269–87

Draper, L. 1967: Wave activity at the sea bed around northwestern Europe, *Mar. Geol.* 5, 133–40

Dreimanis, A. 1976: Tills, their origin and properties, in *Glacial till* (ed. R F. Legget), *Spec. Publs k. Soc. Can.* 12, 11–49

Dreimanis, A. and Vagners, U. J. 1971: Bimodal distribution of rock and mineral fragments in basal tills, in *Till, a symposium* (ed. R. P. Goldthwait), Ohio State Univ. Press, 237–50

Du Boys, P. 1879: Études du régime du Rhône et l'action exercée par les eaux sur un lit à fond de graviers indéfiniment affouillable, *Annls Ponts Chauss.*, Ser 5, 18, 141–95

Duncan, J. R. 1964: The effects of water table and tide cycles on swash-backwash sediment distribution and beach profile, *Mar. Geol.* 2, 186–97

Dunin, F. X. 1976: Infiltration: its simulation for field conditions, in *Facets of hydrology* (ed. J. C. Rodda), Wiley, London, 199–229

Dunn, J. R. and Hudec, P. P. 1965: The influence of clays on water and ice in rock pores, *Publs Phys. Res. Dep., N.Y. St. Dep.* RR 65–5, 149 pp.

Dunne, T. 1970: Runoff production in a humid area, *Rep. U.S. Dep. Agric., Agric. Res. Serv.* 41–160, 108 pp.

Dunne, T. 1978: Field studies of hillslope flow processes, in *Hillslope hydrology* (ed. M. J. Kirkby), Wiley, Chichester, 227–93

Dunne, T. and Black, R. D. 1970a: An experimental investigation of runoff processes in permeable soils, *Wat. Resour. Res.* 6, 478–90

Dunne, T. and Black, R. D. 1970b: Partial area contributions to storm runoff in small New England watersheds, *Wat. Resour. Res.* 6, 1296–311

Dunne, T. and Black, R. D. 1971: Runoff processes during snowmelt, *Wat. Resour. Res.* 7, 1160–72

Dylik, J. 1951: Some periglacial structures in Pleistocene deposits of Middle Poland, *Bull. Soc. Lettr. Łodz* 3, 2

Dyson, J. L. 1937: Snowslide striations, *J. Geol.* 45, 549–57

Eckel, E. B. (ed.) 1958: Landslides and engineering practice, *Spec. Rep. Highw. Res. Bd*, Washington, D.C., 29, 232 pp.

Edelman, C. H., Florshütz, F. and Jeswiet, J. 1936: Über spät-pleistozäne und frühholzäne Kryoturbate Ablagerungen in den östlichen Niederlanden, *Verh. geol.-mijnb. Genoot. Ned., Geol. ser.* 11, 301–47

Eide, O. and Bjerrum, L. 1954: The slide at Bekkelaget, *Proc. European Conf. Stability Earth Slopes*, Stockholm, 3, 88–100

Einstein, H. A. 1950: The bedload function for sediment transportation in open channel flows, *Tech. Bull. U.S. Dep. Agric.* 1026, 70 pp.

Einstein, H. A. and El-Samni, E. A. 1949: Hydrodynamic forces on a rough wall, *Rev. mod. Phys.* 21, 520–4

Einstein, H. A. and Li, H. 1956: The viscous sub-layer along a smooth boundary, *J. Engng Mech. Div. Am. Soc. civ. Engrs* 82 (EMS), 945-1–945-27

Einstein, H. A. and Shen, H. W. 1964: A study of meandering in straight alluvial channels, *J. geophys. Res.* 69, 5239–47

Ekblaw, W. E. 1918: The importance of nivation as an erosive factor, and of soil flow as a transporting agency, in northern Greenland, *Proc. natn. Acad. Sci. U.S.A.* 4, 288–93

Elias, F. 1963: *Precipitaciones maximas en España: regimen de intensidades y frecuencias*, Direccion General de Agricultura, Servicio de Conservacion de Suelos, Madrid, 123 pp.

Ellison, W. D. 1947: Soil erosion studies—Part I, *Agric. Engng* 28(4), 145–6

Ellwood, J., Evans, P. and Wilson, I. G. 1975: Small-scale aeolian bedforms, *J. sedim. Petrol.* 45, 554–61

Embleton, C. and King, C. A. M. 1975a: *Glacial geomorphology*, Arnold, London, 573 pp.

Embleton, C. and King, C. A. M. 1975b: *Periglacial geomorphology*, Arnold, London, 203 pp.

Emery, K. O. 1941: Rate of surface retreat of sea-cliffs, based on dated inscriptions, *Science, N.Y.* 93, 617–18

Emery, K. O. 1944: Brush fires and rock exfoliation, *Am. J. Sci.* 243, 506–8

Emery, K. O. 1960: Weathering of the Great Pyramid, *J. sedim. Petrol.* 30, 140–3

Emery, K. O. and Foster, J. F. 1948: Water tables in marine beaches, *J. mar. Res.* 7, 644–53

Emery, K. O. and Gale, J. F. 1951: Swash and swash mark, *Trans. Am. geophys. Un.* 32, 31–6

Emery, K. O. and Tschudy, R. H. 1941: Transportation of rock by kelp, *Bull. geol. Soc. Am.* 52, 855–62

Emmett, W. M. 1970: The hydraulics of overland flow, *U.S. geol. Surv. Prof. Pap.* 662-A, 46 pp.

Engelund, F. 1974a: Flow and bed topography in channel bends, *J. Hydraul. Div. Am. Soc. civ. Engrs* 100, HY11, 1631–48

Engelund, F. 1974b: Experiments in curved alluvial channels, *Prog. Rep. Inst. Hydrodyn. Hydraul. Engng*, Technical University Denmark, 34, 31–6

Evans, I. S. 1970: Salt crystallisation and rock weathering: a review, *Revue Géomorph. dyn.* 19, 153–77

Evans, O. F. 1939: Sorting and transportation of material in swash and backwash, *J. sedim. Petrol.* 9, 28–31

Evenson, E. B., Dreimanis, A. and Newsome, J. W. 1977: Subaquatic flow tills: a new interpretation for the genesis of some laminated till deposits, *Boreas* 6, 115–33

Everett, D. H. 1961: The thermodynamics of frost damage to porous solids, *Trans. Faraday Soc.* 57, 541–51

Everett, K. R. 1963: Slope movement, Neotama valley, Southern Ohio, *Rep. Inst. Polar Stud.* 6

Exner, F. M. 1925: Über die Wechselwirkung zwischen Wasser und Gescheibe in Flüssen, *Sber. Akad. Wiss. Wien*, Pt. IIa, Bd 134, 199 pp.

Eyles, R. J. and Ho, R. 1970: Soil creep on a humid tropical slope, *J. trop. Geogr.* 31, 40–2

Fairbairn, H. W. 1943: Packing in ionic minerals, *Bull. geol. Soc. Am.* 54, 1305–74

Fairbridge, R. W. 1961: Eustatic changes in sea level, in *Physics and chemistry of the Earth* (ed. L. H. Ahrens *et al.*), 4, 99–185

Fanshawe, H. 1962: Soils in the Shek Pik Valley, *Symp. Hong Kong soils, Hong Kong Joint Group*, 53–6

Fellenius, W. 1927: *Erdstatische Berechnungen mit Reibung und Kohäsion*, Ernst, Berlin, 1st ed. (4th ed. 1948)

Fenner, C. N. 1950: The chemical kinetics of the Katman eruption, *Am. J. Sci.* 248, 593–627, 697–725

Field, W. G. 1968: Effects of density ratio on sedimentary similitude, *J. Hydraul. Div. Am. Soc. civ. Engrs* 94, HY3, 705–18

Finlayson, B. J. 1976: Measurement of geomorphic processes in a small drainage basin, Unpubl. Ph.D. thesis, Univ. of Bristol

Fisher, J. E. 1955: Internal temperatures of a cold glacier and conclusions therefrom, *J. Glaciol.* 2, 583–91

Fisher, J. E. 1963: Two tunnels in cold ice at 4000 m on the Breithorn, *J. Glaciol.* 4, 513–20

Fisher, O. 1866: On the disintegration of a chalk cliff, *Geol. Mag.* 3, 354–6

Flaxman, E. M. 1963: Channel stability in undisturbed cohesive soils, *J. Hydraul. Div. Am. Soc. civ. Engrs* 89, HY2, 87–96

Flemal, R. C., Odom, E. E., and Vail, R. G. 1972: Stratigraphy and origin of the paha topography of northwestern Illinois, *Quat. Res.* 2, 232–43

Fleming, R. W. and Johnson, A. M. 1971: Soil creep in the vicinity of Stanford University, *U.S. Army Cold Reg. Res. Engng Lab. Tech. Rep.*, 10 pp.

Fleming, R. W. and Johnson, A. M. 1973: Measurement of seasonal soil creep, *Bull. Ass. Engng Geol.* 10, 83–93

Fleming, R. W. and Johnson, A. M. 1975: Rates of seasonal creep of silty clay soil, *Q.J. Engng Geol.* 8, 1–29

Fookes, P. G. and Collis, L. 1976: Cracking and the Middle East, *Concrete* 9, 14–19

Fookes, P. G. and Horswill, P. 1969: Discussion on engineering grade zones, *Proc. Conf. for in situ Testing of Soils & Rocks, Instn civ. Engrs*, London, 53–7

Ford, S. E. H. and Elliott, P. G. 1965: Investigation and design of the Plover Cove Water Scheme, *Proc. Instn civ. Engrs* 32, 255–93

Foster, G. R. and Meyer, L. D. 1972: A closed-form soil erosion equation for upland areas, *Sedimentation: Symposium to honour H. A. Einstein* (ed. H. W. Shen), Colorado State Univ., 12, 1–9

Foster, G. R. and Meyer, L. D. 1975: Mathematical simulation of upland erosion by fundamental erosion mechanics, *Rep. U.S. Dep. Agric. ARS-S-40*, 190–206

Fournier, F. 1960: *Climat et érosion: la relation entre l'érosion du sol par l'eau et les précipitations atmosphériques*, Paris, 201 pp.

Fox, W. T. and Davis, R. A. 1976: Weather patterns and coastal processes, in *Nearshore beach processes and sedimentation* (ed. R. A. Davis and R. L. Ethington), *Spec. Publs Soc. econ. Palaeont. Miner.* 24, 1–23

Francis, E. A. 1975: Glacial sediments: a selective review, in *Ice Ages: ancient and modern* (ed. A. E. Wright and F. Moseley), Liverpool, 43–68

Francis, R. A. 1973: Experiments on the motion of solitary grains along the bed of a water stream, *Proc. R. Soc.*, A, 332, 443–71

Freeze, R. A. 1972: Role of subsurface flow in generating subsurface runoff from upstream source areas, *Wat. Resour. Res.* 8, 1272–83

Friedkin, J. F. 1945: A laboratory study of the meandering of alluvial rivers, in *River morphology* (ed. S. A. Schumm), Dowden, Hutchinson and Ross, 237–81

Friedmann, J. D. *et al.* 1971: Observations on Icelandic polygon surfaces and palsa areas. Photographic interpretation and field studies, *Geogr. Annlr* A, 53, 115–45

Fristrup, B. 1953: Wind erosion within the arctic deserts, *Geogr. Tidsskr.* 53, 51–65

Frye, J. C. 1950: Origin of the Kansas Great Plains depressions, *Bull. Kans. Univ. geol. Surv.* 86, pt. 1, 20 pp.

Fukuoka, M. 1953: Landslides in Japan, *Proc. 3rd int. Conf. Soil Mech. Fdn Engng*, Zürich, 2, 234–8

Gage, M. 1966: Franz Josef Glacier, *Ice* 20, 26–7

Gallagher, B. 1971: Generation of surf beat by non-linear wave interaction, *J. Fluid Mech.* 49, 1–20

Galvin, C. J. 1968: Breaker-type classification on three laboratory beaches, *J. geophys. Res.* 73, 3651–9

Gardner, J. 1969: Snowpatches: their influence on mountain wall temperatures and the geomorphic implications, *Geogr. Annlr* A, 51, 114–20

Gardner, J. 1970a: Rockfall—a geomorphic process in high mountain terrain, *Albertan Geogr.* 6, 15–20

Gardner, J. 1970b: Geomorphic significances of avalanches in the Lake Louise area, Alberta, Canada, *Arct. alp. Res.* 2, 135–44

Gardner, W. R. 1964: Relation of root distribution to water uptake and availability, *J. Agron.* 56, 41–54

Garrels, R. M. and Christ, C. L. 1965: *Solutions, minerals and equilibria*, Harper & Row, New York, 450 pp.

Garrels, R. M. and Mackenzie, F. T. 1971: *Evolution of sedimentary rocks*, Norton, New York, 397 pp.

Geiger, R. 1965: *The climate near the ground*, Cambridge, Mass., 611 pp.

Geikie, A. 1880: Rock weathering, as illustrated in Edinburgh churchyards, *Proc. R. Soc. Edinb.* 10, 518–32

Geikie, J. 1877: The movement of the 'soil cap', *Nature, Lond.* 15, 397–8

Gerwitz, A. and Page, E. R. 1974: An empirical mathematical model to describe plant root systems, *J. appl. Ecol.* 11, 773–81

Gessler, J. 1970: Self-stabilizing tendencies of alluvial channels, *J. WatWays Harb. Div. Am. Soc. civ. Engrs* 96, WW2, 235–49

Gilbert, G. K. 1880: Land sculpture, *U.S. Geographical and Geological Survey of the Rocky Mountain Region*, 93–7, 100–8

Gilbert, G. K. 1904: Domes and domed structures of the High Sierra, *Bull. geol. Soc. Am.* 15, 29–36

Gilbert, G. K. 1906: Crescentic gouges on glaciated surfaces, *Bull. geol. Soc. Am.* 17, 303–16

Gile, L. H. 1966: Coppice dunes and the Rotura soil, *Proc. Soil Sci. Soc. Am.* 30, 657–60

Glass, L. J. and Smerdon, E. T. 1967: The effects of rainfall on the velocity distribution in shallow channel flow, *Trans. Am. Soc. agric. Engrs* 10, 330–6

Glen, J. W. 1955: The creep of polycrystalline ice, *Proc. R. Soc.* A, 228, 519–38

Glen, J. W. 1958: The mechanical properties of ice. I. The plastic properties of ice, *Adv. Phys.* 7, 254–65

Glen, J. W., Donner, J. J. and West, R. G. 1957: On the mechanism by which stones in till become orientated, *Am. J. Sci.* 255, 194–205

Goldich, S. S. 1938: A study of rock weathering, *J. Geol.* 46, 17–58

Goldsmith, V. 1976: Wave climate models for the continental shelf: critical links between shelf hydraulics and shoreline processes, in *Beach and nearshore sedimentation* (ed. R. A. Davis and R. L. Ethington), *Spec. Publs Soc. econ. Palaeont. Miner.* 24, 24–47

Goldthwait, R. P. 1951: Development of end moraines in east-central Baffin Island, *J. Geol.* 59, 567–77

Goldthwait, R. P. 1956: Formation of ice cliffs, *Tech. Rep. Snow Ice Permafrost Res. Establ.* 39, 139–50

Goldthwait, R. P. 1961: Regimen of an ice cliff on land in north-west Greenland, *Folia geogr. dan.* 9, 107–15

Goldthwait, R. P. 1971: Introduction to till, today, in *Till, a symposium* (ed. R. P. Goldthwait), Ohio State Univ. Press, 3–26

Goodchild, J. G. 1890: Notes on some observed rates of weathering of limestones, *Geol. Mag.* 27, 463–6

Gordon, M. and Tracey, J. I. 1952: Origin of the Arkansas bauxite deposits. Problems of clay and laterite genesis, *Trans. Am. Inst. Min. metall. Engrs* 34, 12–21

Gossling, F. 1935: The structure of Bower Hill, Nutfield, *Proc. Geol. Ass.* 46, 360–90

Gossling, F. and Bull, A. J. 1948: The structure of Tilburstow Hill, Surrey, *Proc. Geol. Ass.* 59, 131–40

Goudie, A. S. 1970: Input and output considerations in estimating rates of chemical denudation, *Earth Sci. J.* 4, 59–65

Goudie, A. S. 1974: Furthur experimental investigation of rock weathering by salt and other mechanical processes, *Z. Geomorph., Suppl. Bd* 21, 1–12

Goudie, A. S. 1977: Sodium sulphate weathering and the disintegration of Mohenjo-Daro, Pakistan, *Earth Surface Processes* 2, 75–86

Goudie, A. 1978: *Environmental change*, Oxford Univ. Press, 244 pp.

Goudie, A. S., Cooke, R. U. and Evans, I. S. 1970: Experimental investigation of rock weathering by salts, *Area* 4, 42–8

Grant, U. S. 1948: Influence of the water table on beach aggradation and degradation, *J. mar. Res.* 7, 655–60

Grant, W. H. 1969: Abrasion pH, an index of chemical weathering, *Clays Clay Miner.* 17, 151–5

Gray, W. M. 1965: Surface spalling by thermal stresses in rocks, *Proc. Rock Mech. Symp., Toronto. Dep. Mines Tech. Surv. Ottawa*, 85–106

Grawe, O. R. 1936: Ice as an agent of rock weathering, *J. Geol.* 44, 173–82

Gray, J. T. 1973: Geomorphic effects of avalanches and rock-falls on steep mountain slopes in the central Yukon Territory, in *Research in polar and alpine geomorphology* (ed. B. D. Fahey and R. D. Thompson), 3rd Guelph Symposium on geomorphology, 107–17

Greeley, R., Iverson, J. D., Pollack, J. B., Udorich, N. and White, B. 1974: Wind tunnel simulations of light and dark streaks on Mars, *Science, N.Y.* 183, 847–9

Gregory, K. J. and Park, C. 1974: Adjustment of river channel capacity downstream from a reservoir, *Wat. Resour. Res.* 10, 870–3

Gregory, K. J. and Walling, D. E. 1968: The variation of drainage density within a catchment, *Bull. int. Ass. scient. Hydrol.* 13, 61–8

Gregory, K. J. and Walling, D. E. 1973: *Drainage basin form and process*, Arnold, London, 456 pp.

Griggs, D. T. 1936a: Deformation of rocks under high confining pressures, *J. Geol.* 44, 541–77

Griggs, D. T. 1936b: The factor of fatigue in rock exfoliation, *J. Geol.* 44, 781–96

Gripp, K. 1926: Über Frost und Strukturboden auf Spitzbergen, *Z. Ges. Erdk. Berlin* 61, 351–4

Gripp, K. 1929: Glaciologische und geologische Ergebnisse der Hamburgischen Spitzbergen-Expedition 1927, *Abh. Geb. Naturw., Hamburg*, 22, 146–249

Grissinger, E. H. and Asmussen, L. E. 1963: Discussion of 'Channel stability in undisturbed cohesive soils' by E. M. Flaxman, *J. Hydraul. Div. Am. Soc. civ. Engrs* 89, HY6, 259–64

Groome, G. E. and Williams, H. 1965: The solution of limestone in South Wales, *Geogrl J.* 131, 37–41

Grove, J. M. 1972: The incidence of landslides, avalanches and floods in Western Norway during the Little Ice Age, *Arct. alp. Res.* 4, 131–8

Gruner, J. W. 1950: An attempt to arrange silicates in the order of reaction series at relatively low temperatures, *Am. Miner.* 35, 137–48

Guillien, Y. and Lautridou, J-P. 1970: Recherches de gélifraction expérimentale du Centre de Géomorphologie. 1. Calcaires de Charentes, *Bull. Centre Géomorph. Caen* 5, 45 pp.

Guza, R. T. and Inman, D. L. 1978: Edge waves and surf beat, *J. geophys. Res.* 83 (C4), 1913–20

Haantjens, H. A. and Bleeker, P. 1970: Tropical weathering in the Territory of Papua and New Guinea, *Aust. J. Soil. Res.* 8, 157–77

Haefeli, R. 1948: The stability of slopes acted upon by parallel seepage, *Proc. 2nd int. Conf. Soil Mech. Fdn Engng*, Rotterdam, 1, 134–48

Haefeli, R. 1953: Creep problems in soils, snow and ice, *Proc. 3rd int. Conf. Soil Mech.*, Zürich, 3, 238–51

~~Haefeli, R. 1963: Observations and measurements on the cold ice sheet on Jungfraujoch, *Bull. int.*~~
    Ass. scient. Hydrol. 8, 2, 122

Hagiwara, T. 1938: Mud avalanches on Mt Tokuba, Ibaraki Prefecture, *Bull. Earthq. Res. Inst. Tokyo Univ.* 16, 779–83

Hallsworth, E. G. 1952: An interpretation of the soil formations found on basalt in the Richmond-Tweed region of New South Wales, *Aust. J. agric. Res.* 2, 411

Hallsworth, E. G. 1963: An examination of some of the factors affecting movement of clay in an artificial soil, *J. Soil Sci.* 14, 360–71

Hanna, S. R. 1969: The formulation of longitudinal sand dunes by large helical eddies in the atmosphere, *J. appl. Met.* 8, 874–83

Harr, R. D. 1977: Water flux in soil and subsoil on a steep forested slope, *J. Hydrol.* 33, 37–58

Harradine, F. and Jenny, H. 1958: Influence of parent material and climate on texture and nitrogen and carbon contents of virgin California soils. 1. Texture and nitrogen content of soils, *Soil Sci.* 85, 235–43

Harris, C. 1972: Processes of soil movement in turf-banked solifluction lobes, Okstindan, northern Norway, *Inst. Br. Geogr. Spec. Publ.* 5, 155–74

Harris, C. 1974: Autumn, winter and spring soil temperatures in Okstindan, Norway, *J. Glaciol.* 13, 521–33

Harris, S. E. 1943: Friction cracks and the direction of glacial movement, *J. Geol.* 51, 244–58

Harrison, J. V. and Falcon, N. L. 1938: An ancient landslip at Saidmarreh in southwestern Iran, *J. Geol.* 46, 296–309

Hartshorn, J. H. 1958: Flowtill in southeastern Massachusetts, *Bull. geol. Soc. Am.* 69, 477–82

Hastenrath, S. L. 1967: The barchans of the Arequipa region, southern Peru, *Z. Geomorph.* 11, 300–31

Hay, R. C. 1959: Origin and weathering of late Pleistocene ash deposits on St Vincent, B.W.I., *J. Geol.* 67, 65–87

Hay, R. C. 1960: Rate of clay formation and mineral alteration in a 4000-year-old volcanic ash soil on St Vincent, B.W.I., *Am. J. Sci.* 258, 354–68

Hay, R. C. and Jones, B. F. 1972: Weathering of basaltic tephra on the island of Hawaii, *Bull. geol. Soc. Am.* 83, 317–32

Heede, B. H. 1971: Characteristics and processes of soil piping in gullies, *U.S. Dep. Agric., Forest Serv. Res. Pap.* RM-68, 15–21

Heim, A. 1882: Der Bergsturz von Elm, *Z. dt. geol. Ges.* 34, 74–115

Heim, A. 1932: Bergsturz und Menschenleben, *Vjschr. naturf. Ges. Zürich,* 77 pp.

Henderson, F. M. 1963: Stability of alluvial channels, *Trans. Am. Soc. civ. Engrs* 128, 657–86

Henkel, D. J. and Skempton, A.W. 1955: A landslide at Jacksfield, Shropshire, in an over-consolidated clay, *Proc. Conf. Stability of Earth Slopes* (Stockholm), 1, 90–101

Hewitt, K. 1968: The freeze-thaw environment of the Karakoram Himalaya, *Can. Geogr.* 12, 85–98

Hewitt, K. 1972: The mountain environment and geomorphic processes, in *Mountain geomorphology,* Vancouver, 17–34

Hewlett, J. D. 1961: Watershed management, *Rep. U.S. Forest Serv., Southeast Forest Exp. Stn,* 61–6

Hewlett, J. D. and Hibbert, A. R. 1967: Factors affecting the response of small watersheds to precipitation in humid areas, in *Forest hydrology* (ed. W. E. Sopper and H. W. Lull) Pergamon, 275–90

Hey, R. D. and Thorne, C. R. 1975: Secondary flows in river channels, *Area* 7, 191–5

Hickin, E. J. 1969: A newly identified process of point-bar formation in natural streams, *Am. J. Sci.* 272, 761–99

Hidy, G. M. 1967: *The winds, the origins and behaviour of atmospheric motion,* Van Nostrand, Princeton N.J., 174 pp.

Higashi, A. and Corte, A. E. 1972: Growth and development of perturbations on the soil surface due to the repetition of freezing and thawing, *Hokkaido Univ., Faculty of Engng Mem.,* Suppl. 13, 49–63

High, C. and Hanna, F. K. 1970: A method for the direct measurement of erosion on rock surfaces, *Tech. Bull. Br. geomorph. Res. Grp* 5, 25 pp.

Hill, G. W. and Hunter, R. E. 1976: Interaction of biological and geological processes in the beach and nearshore environments, North Padre Island, Texas, in *Beach and nearshore sedimentation* (ed. R. A. Davis and R. L. Ethington), *Spec. Publs. Soc. econ. Palaeont. Miner., Tulsa*, 24, 169–87

Hill, G. W. and Hunter, R. E. 1973: Burrows of the ghost crab *Ocypode quadrata (Fabricius)* on the barrier islands, south central Texas, *J. sedim. Petrol.* 43, 24–30

Hirschwald, J. 1908: *Die Prüfung der natürlichen Bausteine auf ihre Wetterbeständigkeit*, Berlin, quoted in Ollier, C. D. 1969

Hissink, D. J. 1938: The reclamation of the Dutch saline soils and their further weathering under humid climatic conditions of Holland, *Soil Sci.* 45, 83–94

H.M.S.O. 1969: *A selection of technical reports submitted to the Aberfan Tribunal*, Items 1–7 (H.M.S.O., London), 218 pp.

Hodgkin, E. P. 1964: Rate of erosion of intertidal limestone, *Z. Geomorph.* 8, 385–92

Hoek, E. 1973: Methods for the rapid assessment of the stability of three-dimensional rock slopes, *Q.J. Engng Geol.* 6, 243–56

Holdsworth, G. and Bull, C. 1970: The flow law of cold ice; investigations on Meserve glacier, Antarctica, *Int. Symp. Antarct. Glaciol. Exploration* (Hanover, N.H., 1968), *Publs int. Ass. scient. Hydrol.* 86, 204–16

Holland, T. H. 1894: Report on the Gohna landslip, Garhwal, *Rec. geol. Surv. India* 27, 54–64

Holmsen, P. 1953: Landslides in Norwegian quick clays, *Géotechnique* 3, 187–94

Hopkins, D. M. and Sigafoos, R. S. 1951: Frost action and vegetation patterns on Seward Peninsula, Alaska, *Bull. U.S. geol. Surv.* 974-C, 51–100

Horton, J. H. and Hawkins, R. H. 1965: Flow path of rain from the soil surface to the water table, *Soil Sci.* 100, 377–83

Horton, R. E. 1933: The role of infiltration in the hydrological cycle, *Trans. Am. geophys. Un.* 14, 446–60

Horton, R. E. 1938: The interpretation and application of runoff plot experiments with reference to soil erosion problems, *Proc. Soil Sci. Soc. Am.* 3, 340–9

Horton, R. E. 1945: Erosional development of streams and their drainage basins: hydrophysical approach to quantitative morphology, *Bull. geol. Soc. Am.* 56, 275–330

Horton, R. E., Leach, H. R. and Van Vliet, R. 1934: Laminar sheet flow, *Trans. Am. geophys. Un., Hydrology*, 2, 393–404

Hoyer, M. C. 1971: Puget Peak avalanche, Alaska, *Bull. geol. Soc. Am.* 82, 1267–84

Holmes, C. D. 1941: Till fabric, *Bull. geol. Soc. Am.* 52, 1299–354

Holmes, C. D. 1944: Hypothesis of subglacial erosion, *J. Geol.* 52, 184–90

Hoppe, G. 1952: Hummocky moraine regions, with special reference to the interior of Norrbotten, *Geogr. Annlr* 34, 1–72

Howard, J. D. and Dorjes, J. 1972: Animal-sediment relationships in two beach-related tidal flats: Sapelo Island, Georgia, *J. sedim. Petrol.* 42, 608–23

Hsü, K. J. 1975: Catastrophic debris streams (Sturzstroms) generated by rockfalls, *Bull. geol. Soc. Am.* 86, 129–40

Hsu, S. A. 1973: Computing eolian sand transport from shear velocity measurements, *J. Geol.* 81, 739–43

Huang, W. H. and Kiang, W. C. 1972: Laboratory dissolution of plagioclase felspars in water and organic acids at room temperature, *Am. Miner.* 57, 1849–59

Hubbell, D. W. and Sayre, W. 1964: Sand transport studies with radioactive tracers, *J. Hydraul. Div. Am. Soc. civ. Engrs* 90, HY3, 36–98

Hudson, N. 1971: *Soil conservation*, Batsford, London, 320 pp.

Huggett, R. J. 1976: Lateral translocation of soil plasma through a small valley basin in the Northaw Great Wood, Hertfordshire, *Earth Surface Processes* 1, 99–110

Hume, W. F. 1925: *The geology of Egypt. Vol. I: Surface features. Ch. 2: Temperature variations and their results*, Govt. Printer, Cairo

Hung, C. S. and Shen, H. W. 1971: Research in stochastic models for bed-load transport, *Symposium on statistical hydrology, Int. Ass. for Statistics phys. Sci.*, Tucson, Arizona, 13-1–13-16

Huntley, D. A. and Bowen, A. J. 1975: Comparison of the hydrodynamics of steep and shallow beaches, in *Nearshore sediment dynamics and sedimentation* (ed. J. Hails and A. P. Carr), Wiley, London, 69–109

Huntley, D. A., Guza, R. T. and Bowen, J. A. 1977: A universal form for shoreline run-up spectra? *J. geophys. Res.* 82, 2577–81

Hutchinson, J. N. 1967: The free degradation of London Clay cliffs, *Proc. geotech. Conf. Oslo* 1, 113–18

Hutchinson, J. N. 1968: Field meeting on the coastal landslides of Kent, *Proc. Geol. Ass.* 79, 227–37

Hutchinson, J. N. 1969: A reconsiderations of the coastal landslides of Folkestone Warren, Kent, *Géotechnique* 19, 6–38

Hutchinson, J. N. 1970: A coastal mudflow on the London Clay cliffs at Beltinge, North Kent, *Géotechnique* 20, 412–38

Hutchinson, J. N. 1971a: Mass movement, in *Encyclopaedia of Earth Sciences* (ed. R. W. Fairbridge), Reinhold, New York, 688–95

Hutchinson, J. N. 1971b: Field and laboratory studies of a fall in Upper Chalk cliffs at Jess Bay, Isle of Thanet, *Roscoe Memorial Symp.* (Univ. of Cambridge, 29–31 March 1971), 12 pp.

Hutchinson, J. N. 1974: Periglacial solifluxion: an approximate mechanism for clayey soils, *Géotechnique* 24, 438–43

Hutchinson, J. N. and Bhandari, R. K. 1971: Undrained loading, a fundamental mechanism of mudflows and other mass movements, *Géotechnique* 21, 353–8

Hutchinson, J. N. and Brunsden, D. 1975: Mudflows: a review and classification (Abstract), *Q.J. Engng Geol.* 7, 1–5

Hutchinson, J. N., Prior, D. B. and Stevens, N. 1974: Potentially dangerous surges in an Antrim mudslide, *Q.J. Engng Geol.* 7, 363–76

Hvorslev, M. J. 1960: Physical components of the shear strength of saturated clay, *Proc. Am. Soc. civ. Engrs* (Res. Conf. on the Shear Strength of Cohesive Soils), 1–19

Imeson, A. C. and Kwaad, F. J. 1976: Some effects of burrowing animals and slope processes in the Luxembourg Ardennes, *Geogr. Annlr* A, 58, 317–28

Ingle, J. C. 1966: *The movement of beach sand*, Elsevier, Amsterdam, 221 pp.

Inglis, C. C. 1968: Discussion of 'Systematic evaluation of river régime' by C. R. Neill and V. J. Galey, *J. WatWays Harb. Div. Am. Soc. civ. Engrs*, 94, WW1, 109–14

Inman, D. L. and Bagnold, R. A. 1963: Littoral processes, in *The sea* (ed. M. N. Hill) vol. 3, Interscience, New York, 529–53

Inman, D. L. and Tunstall, E. B. 1972: Phase-dependent roughness control of sand movement, *Proc. 13th Conf. coastal Engng*, Vancouver, 1155–71

Isaacs, J. D., Williams, E. A. and Eckart, C. 1951: Total reflection of surface waves by deep water, *Trans. Am. geophys. Un.* 32, 37–40

Ivanov, D. L. 1899: The Ufa sinks. Caving on the Samara–Zlatoust rail road, *Geol. -rekhn. Issled.* 1894–1896, gg. SPb

Iveronova, M. I. 1964: Stationary studies of the recent denudation processes on the slopes of the River Tchan–Kizclsu basins, Tersky Alatau Ridge, Tien Shan, *Z. Geomorph., Suppl. Bd* 5, 207–12

Jackson, M. L., Tyler, S. A., Bourbeau, G. A. and Pennington, R. P. 1948: Weathering sequence of clay-size minerals in soils and sediments. 1. Fundamental generalizations, *J. phys. Colloid Chem.* 52, 1237–60

Jackson, R. G. 1975: Hierarchical attributes and a unifying model of bedforms composed of cohesionless material and produced by shearing flow, *Bull. geol. Soc. Am.* 86, 1523–33

Jaeger, J. C. 1959: The frictional properties of joints in rock, *Geofis. pura appl.* 43, 148–58

Jahn, A. 1960: Some remarks on the evolution of slopes on Spitsbergen, *Z. Geomorph., Suppl. Bd* 1, 49–58

Jahn, A. 1972: Niveo-eolian processes in the Sudetes Mountains, *Geogr. Polonica* 23, 93–110

Jahn, A. 1975: *Problems of the periglacial zone*, PWN, Warsaw, 223 pp.

Jahns, R. H. 1943: Sheet structure in granites: its origin and use as a measure of glacial erosion in New England, *J. Geol.* 51, 71–98

Jarvis, R. S. 1972: New measures of the topological structure of dendritic drainage networks, *Wat. Resour. Res.* 8, 1265–71

Jenny, H. 1935: The clay content of the soil as related to climatic factors, particularly temperature, *Soil Sci.* 40, 111–28

Jenny, H. 1941: *Factors in soil formation*, McGraw-Hill, New York, 281 pp.

Jenny, H. and Leonard, C. D. 1934: Functional relationships between soil properties and rainfall, *Soil Sci.* 38, 363–81

Joffe, J. S. 1949: *The ABC of soils*, Pedology Publs, New Brunswick, N.J., Somerville, N.J., Somerset Press, 662 pp.

Johnson, A. 1970: *Physical processes in geology*, Freeman, Cooper, San Francisco, 577 pp.

Johnson, A. M. and Hampton, M. A. 1969: Subaerial and subaqueous flow of slurries, *Final Rep. Contract No. 14-08-001-10884, U.S. geol. Surv.*

Johnson, W. D. 1904: The profile of maturity in Alpine glacial erosion, *J. Geol.* 12, 569–78

Jones, A. 1971: Soil piping and stream channel initiation, *Wat. Resour. Res.* 7, 602–10.

Jones, F. O., Embody, D. R. and Peterson, W. L. 1961: Landslides along the Columbia river valley, northeastern Washington, *U.S. geol. Surv. Prof. Pap.* 367, 11 pp.

Jordan, C. F. and Kline, J. R. 1972: Mineral cycling: some basic concepts and their application in a tropical rainforest, *A. Rev. ecol. System* 3, 33–50

Judson, S. and Ritter, D. F. 1964: Rates of regional denudation in the United States, *J. geophys. Res.* 69, 3395–401

Kalin, M. 1971: The active push moraine of the Thompson glacier, *Axel Heiberg Island Res. Rep.* (McGill Univ., Montreal), 'Glaciology' 4, 68 pp.

Kalinske, A. A. 1943: The role of turbulence in river hydraulics, *Bull. Univ. Iowa Stud. Engng* 27, 266–79

Kalinske, A. A. 1947: Movement of sediment as bed load in rivers, *Trans. Am. geophys. Un.* 28, 615–20

Kamb, B. and LaChapelle, E. 1964: Direct observation of the mechanism of glacier sliding over bedrock, *J. Glaciol.* 5, 159–72

Kaplar, C. W. 1970: Phenomenon and mechanism of frost-heaving, in *Frost action: bearing, thrust, stabilization and compaction*, Highway Research Board Record 304, 1–13

Kaye, C. A. 1959: Shoreline features and Quaternary shoreline changes, Puerto Rico, *U.S. geol. Surv. Prof. Pap.* 317-B, 49–140

Keeble, A. B. 1971: Freeze-thaw cycles and rock weathering in Alberta, *Albertan Geogr.* 1971, 1–12

Keeley, J. R. and Bowen, A. J. 1977: Longshore variations in longshore currents, *Can. J. Earth Sci.* 14, 1897–903

Keller, W. D. 1954: Bonding energies of some silicate minerals, *Am. Miner.* 39, 783–93

Keller, W. D. 1957: *The principles of chemical weathering*, Lucas Bros., Columbia, Missouri, 111 pp.

Keller, W. D. and Frederickson, A. F. 1952: The role of plants and colloidal acids in the mechanism of weathering, *Am. J. Sci.* 250, 594–608

Kelvin, W. T., Baron 1899: The age of the Earth as an abode fitted for life, *J. Trans. Vict. Inst.* 31, 11–35 (Presidential Address 1897, esp. para. 17)

Kemp, P. H. 1975: Wave asymmetry in the nearshore zone and breaker area, in *Nearshore sediment dynamics and sedimentation* (ed. J. Hails and A. P. Carr), Wiley, London, 47–67

Kemp, P. H. and Plinston, D. T. 1968: Beaches produced by waves of low phase difference, *J. Hydraul. Div. Am. Soc. civ. Engrs* 94, HY5, 1183–95

Kennedy, J. F. and Brooks, N. H. 1963: Laboratory study of an alluvial stream at constant discharge, *Misc. Publs U.S. Dep. Agric. Res. Serv.* 970, 320–30

Kennedy, J. F. and Locher, F. A. 1972: Sediment suspension by water waves, in *Waves on beaches and resulting sediment transport* (ed. R. E. Meyer), Academic Press, New York, 249–95

Kenney, T. C. 1956: An examination of the methods of calculating the stability of slopes, Unpubl. M.Sc. thesis, Univ. of London

Kent, P. 1966: The transport mechanism of catastrophic rockfalls, *J. Geol.* 74, 79–83

Keppel, R. V. and Renard, K. G. 1962: Transmission losses in ephemeral stream beds, *J. Hydraul. Div. Am. Soc. civ. Engns* 88, HY3, 59–68

Kerr, W. C. 1881: On the action of frost in the arrangement of superficial earthy material, *Am. J. Sci.* 31 (3rd Ser.), 345–58

Kesseli, J. E. 1943: Disintegrating soil slips of the Coast Ranges of Central California, *J. Geol.* 51, 342–52

Kessler, D. W. (ed.) 1940: Physical, mineralogical and durability studies on the building and monumental granites of the United States, *J. Res. nain, Bur. Stand.* 1320, 161–206

Keulegan, G. H. 1938: Laws of turbulent flows in open channels, *J. Res. natn. Bur. Stand.* 21, 707–41

Keulegan, G. H. 1944: Spatially varied discharge over a sloping plane, *Trans. Am. geophys. Un.* 25, 156–8

King, C. A. M. 1951: Depth of disturbance of sand on beaches by waves, *J. sedim. Petrol.* 21, 131–40

King, C. A. M. 1972: *Beaches and coasts*, Arnold, London, 570 pp.

King, L. C. 1953: Canons of landscape evolution, *Bull. geol. Soc. Am.* 64, 721–52

King, L. C. 1962: *The morphology of the Earth*, Oliver and Boyd, Edinburgh, 699 pp.

Kirkby, A. V. and Kirkby, M. J. 1974: Surface wash at the semi-arid break slope, *Z. Geomorph., Suppl. Bd* 21, 151–76

Kirkby, M. J. 1963: A study of rates of erosion and mass movement on slopes, with special reference to Galloway, Unpubl. Ph.D. thesis, Univ. of Cambridge

Kirkby, M. J. 1967: Measurement and theory of soil creep, *J. Geol.* 75, 359–78

Kirkby, M. J. 1969: Infiltration, throughflow and overland flow, in *Water, Earth and Man* (ed. R. J. Chorley) Methuen, London, 215–27

Kirkby, M. J. 1971: Hillslope process-response models based on the continuity equation, *Inst. Br. Geogr. Spec. Publ.* 3, 15–30

Kirkby, M. J. 1976: Tests of the random network model and its application to basin hydrology, *Earth Surface Processes* 1(3), 197–213

Kirkby, M. J. 1977a: Maximum sediment efficiency as a criterion for alluvial channels, in *River channel changes* (ed. K. J. Gregory), Wiley, Chichester, 429–42

Kirkby, M. J. 1977b: Soil development models as a component of slope models, *Earth Surface Processes* 2, 203–30

Kirby, M. J. 1978: The stream head as a significant geomorphic threshold, *Department of Geography, Univ. of Leeds, Working Paper* 216, 17 pp.

Kirkby, M. J. and Weyman, D. R. 1973: Measurement of contributing area in very small drainage basins, *Univ. of Bristol, Geogr. Seminar Papers*, Series B, 3, 12 pp.

Klimaszewski, M. (ed.) 1963: *Problems of geomorphological mapping*, Int. geogr. Un., Sub-Commn Geomorph. Mapping (Poland, 3–12 May 1962), *Pol. Acad. Sci., Geogr. Stud.* 46, 179 pp.

Klinge, H. 1977: Preliminary data on nutrient release from decomposing leaf litter in a neotropical rainforest, *Amazoniana* 6, 193–202

Klinge, H. and Herrera, R. 1977: Composite root mass in tropaquods under Amazon caatinga stands in Southern Venezuela, *Proc. 4th int. Symp. on Tropical Ecology*, Panama, 16 pp.

Knapp, B. J. 1973: Hillslope throughflow observation and the problem of modelling, *Inst. Br. Geogr. Spec. Publ.* 6, 23–31

Kojan, E. 1968: Mechanics and rates of natural soil creep, *Proc. 5th A. Symp. Engng Geol. Soils Engng* (Pocatello, Idaho), 233–53

Komar, P. D. and Inman, D. L. 1970: Longshore sand transport on beaches, *J. geophys. Res.* 75, 5914–27

Komar, P. D. 1971: The mechanics of sand transport on beaches, *J. geophys. Res.* 76, 713–21

Komar, P. D. 1976: *Beach processes and sedimentation*, Prentice-Hall, Englewood Cliffs, 429 pp.

Komar, P. D. 1977a: Selective longshore transport rates of different grain-size fractions within a beach, *J. sedim. Petrol.* 47, 1444–53

Komar, P. D. 1977b: Beach sand transport: distribution and total drift, *J. Watways Port Coast. Ocean Div. Am. Soc. civ. Engrs* 103 (WW2), 225–39

Komar, P. D. and Miller, M. C. 1973: The threshold of sediment movement under oscillatory water waves, *J. sedim. Petrol.* 43, 1101–10

Koppejan, A. W., Wamelen, B. M. and Weinberg, L. J. 1948: Coastal flowslides in the Dutch province of Zeeland, *Proc. 2nd int. Conf. Soil Mech. Fdn Engng*, Rotterdam, 5, 89–96

Kostyaev, A. G. 1969: Wedge- and fold-like diagenetic disturbances in Quaternary sediments and their paleogeographic significance, *Biul. peryglac.* 19, 231–70

Krauskopf, K. B. 1967: *Introduction to geochemistry*, McGraw-Hill, New York, 721 pp.

Kudrass, H. R. 1974: Experimental study of nearshore transportation of pebbles with attached algae, *Mar. Geol.* 16, M9–M12

Kuetner, J. 1971: Cloud bands in the Earth's atmosphere, observations and theory, *Tellus* 23, 404–26

Kwaad, F. J. P. M. 1970: Experiments on the granular disintegration of granite by salt action, *Univ. Amsterdam Fys. Geogr. en Bodernkundig Lab.* 16, 29 pp.

Labasse, H. 1965: Ground stress in longwall and room-and-pillar mining, *Dep. Mines tech. Surv. (Ottawa), Proc. Rock Mech. Symp. Toronto*, 47–64

Lacey, G. 1929: Stable channels in alluvium, *Proc. Instn civ. Engrs* 229, 259–92

Lachenbruch, A. H. 1962: Mechanics of thermal contraction cracks and ice-wedge polygons in permafrost, *Geol. Soc. Am. Spec. Pap.* 70, 69 pp.

Ladd, G. E. 1935: Landslides, subsidence and rockfalls, *Proc. Am. Rly Engng Ass.* 36, 1091–162

Lamb, D. W. 1962: Decomposed granite as fill material with particular reference to earth dam construction, *Symp. Hong Kong Soils, Hong Kong Joint Group*, 57 pp.

Lamplugh, G. W. 1911: On the shelly moraine of the Sefström glacier and other Spitsbergen phenomena illustrative of British glacial conditions, *Proc. Yorks. geol. Soc.* 17, 216–41

Lane, E. W. 1953: Design of stable channels, *Proc. Am. Soc. civ. Engrs* 79, 280-1–280-31

Lane, E. W. and Borland, W. M. 1954: River-bed scour during floods, *Trans. Am. Soc. civ. Engrs* 79, 303–13

Lane, E. W., Carlson, E. J. and Hanson, O. S. 1949: Low temperature increases sediment transportation in Colorado River, in *Civil engineering*, American Society Civil Engineers 19, 9, 45–6

Lane, L. J., Woolhiser, D. A. and Yevjevich, V. 1976: Influence of simplifications in watershed geometry on simulation of surface runoff, *Colorado St Univ. Hydrol. Pap.* 81, 52 pp.

Langbein, W. B. and Leopold, L. B. 1968: River channel bars and dunes—theory of kinematic waves, *U.S. geol. Surv. Prof. Pap.* 122L, L1–20

Lapparent, V. de 1941: Logique des minéraux du granite, *Revue scient.*, Paris, 284–92

Laursen, E. M. 1952: Observations on the nature of scour, *Proc. 5th Hydraul. Conf., Bull.* 34, Univ. of Iowa, 179–97

Lee, W. H. K. 1963: Heat flow data analysis, *Rev. Geophys.* 1, 449–79

Lee, W. H. K. and Uyeda, S. 1965: Review of heat flow data, in *Terrestrial heat flow* (ed. W. H. K. Lee), *Am. Geophys. Un., Geophys. Monogr.* 8, 87–190

Leffingwell, E. K. 1915: Ground ice wedges, *J. Geol.* 23, 635–54

Lehmann, D. S. 1963: Some principles of chelation chemistry, *Proc. Soil Sci. Soc. Am.* 27, 167–70

Lehmann, O. 1933: Morphologische Theorie der Verwitterung von Steinschlag Wänden, *Vjschr. naturf. Ges. Zürich* 87, 83–126

Lelong, F. 1966: Régime des nappes phréatiques contenues dans les formations d'altération tropicale, *Conséquences pour la pédogenèse, Sci. Terre* 11, 203–44

Lelong, F. and Millet, G. 1966: Sur l'origine des minéraux micacés des altérations latéritiques. Diagenèse régressive—minéraux en transit, *Bull. Serv. Carte géol. Als. Lorr.* 19, 271–87

Leneuf, N and Aubert, G. 1960: Essai d'évaluation de la vitesse de ferrallitisation, *Proc. 7th int. Conf. Soil Sci.*, 225–8

Leopold, L. B., 1978: El Asunto del Arroyo, in *Geomorphology: present problems and future prospects* (ed. C. Embleton *et al.*), Oxford University Press, 25–39

Leopold, L. B. and Emmett, W. V. 1972: Some rates of geomorphological processes, *Geogr. pol.* 23, 5–9

Leopold, L. B. and Langbein, W. B. 1963: Association and indeterminacy in geomorphology, in *The fabric of geology* (ed. C. C. Albritton), Freeman, Cooper, Stanford, California, 184–92

Leopold, L. B. and Maddock, T. 1953: The hydraulic geometry of stream channels and some physiographic implications, *U.S. geol. Surv. Prof. Pap.* 252, 56 pp.

Leopold, L. B. and Wolman, M. G. 1957: River channel patterns—braided, meandering and straight, *U.S. geol. Surv. Prof. Pap.* 282B, B39–B85

Leopold, L. B., Wolman, M. G. and Miller, J. 1964: *Fluvial processes in geomorphology*, Freeman, San Francisco, 522 pp.

Lepp, H. 1973: *Dynamic earth: an introduction to earth science*, McGraw-Hill, New York, 485 pp.

Lettau, K. and Lettau, H. 1969: Bulk transport of sand by the barchans of the Pampa La Joya in southern Peru, *Z. Geomorph.* 13, 182–95

Lewin, J. 1972: Late-stage meander growth, *Nature, Lond., Phys. Sci.* 240, 116

Lewin, J. 1976: Initiation of bed forms and meanders in coarse-grained sediment, *Bull. geol. Soc. Am.* 75, 767–74

Lewis, A. 1975: Slow movements of earth under tropical rainforest conditions, *Geology* 2, 9–10

Lewis, P. F. 1960: Linear topography in southwestern Palouse, Washington–Oregon, *Ann. Ass. Am. Geogr.* 50, 98–111

Lewis, W. V. 1936: Nivation, river grading and shoreline development in south-east Iceland, *Geogrl J.* 88, 431–47

Lewis, W. V. 1938: A meltwater hypothesis of cirque formation, *Geol. Mag.* 75, 249–65

Lewis, W. V. 1939: Snow-patch erosion in Iceland, *Geogrl J.* 94, 153–61

Lewis, W. V. 1940: The function of meltwater in cirque formation, *Geogrl Rev.* 30, 64–83

Lewis, W. V. 1954: Pressure release and glacial erosion, *J. Glaciol.* 2, 417–22

Lewis, W. V. and Miller, M. M. 1955: Kaolin model glaciers, *J. Glaciol.* 2, 533–8

Li, R. M., Simons, D. B. and Carder, D. R. 1976: Mathematical modelling of overland flow for soil erosion, *National Soil Erosion Conf., Purdue University, Lafayette*, 25–26 May 1976

Li, R. M., Simons, D. B. and Stevens, M. A. 1976: Morphology of cobble streams in small watersheds, *J. Hydraul. Div. Am. Soc. civ. Engrs* 102, HY8, 1101–17

Liggett, J. A. 1975: Stability, in *Unsteady flow in open channels* (ed. K. Mahmood and V. Yevjevich), Water Resources Publications, Fort Collins, Colorado, 1, 259–82

Lindsay, J. F. 1973: Ventifact evolution in Wright Valley, Antarctica, *Bull. geol. Soc. Am.* 84, 1791–8

Linell, K. A. and Kaplar, C. W. 1959: The factor of soil and material type in frost action, *Natl Acad. Sci.—Natl Res. Coun. Highw. Res. Bd Bull.* 225, 81–128

Linton, D. L. 1955: The problem of tors, *Geogrl J.* 121, 470–87

Linton, D. L. 1963: The forms of glacial erosion, *Trans. Inst. Br. Geogr.* 33, 1–28

Little, A. L. 1969: The engineering classification of residual tropical soils, *Proc. 7th int. Conf. Soil Mech. Fdn Engng*, Mexico 1, 1–10

Livingstone, D. A. 1963: Chemical composition of rivers and lakes, *U.S. geol. Surv. Prof. Pap.* 440–G, 64 pp.

Lliboutry, L. A. 1953: Internal moraines and rock glaciers, *J. Glaciol*, 2, 296

Lliboutry, L. A. 1959: Une théorie du frottement du glacier sur son lit, *Annls Géophys.* 15, 250–65

Lliboutry, L. A. 1964: Subglacial 'super-cavitation' as a cause of the rapid advances of glaciers, *Nature, Lond.* 202, 77

Lliboutry, L. A. 1964–65: *Traité de glaciologie*, Masson, Paris, 2 vols., 1040 pp.

Lliboutry, L. A. 1968: General theory of subglacial cavitation and sliding of temperate glaciers, *J. Glaciol.* 7, 21–58

Lliboutry, L. A. 1969: The dynamics of temperate glaciers from the detailed viewpoint, *J. Glaciol.* 8, 185–205

Lohnes, R. A. and Handy, R. L. 1968: Slope angles in friable loess, *J. Geol.* 76, 247–58

Longuet-Higgins, M. S. 1953: Mass transport in water waves, *Phil. Trans. R. Soc.* A, 245, 535–81

Longuet-Higgins, M. S. 1972: Recent progress in the study of longshore currents, in *Waves on beaches* (ed. R. E. Meyer), 203–48

Longuet-Higgins, M. S. 1976: Recent developments in the study of breaking waves, *Proc. 15th Conf. Coastal Engng*, Honolulu, 441–60

Longuet-Higgins, M. S. and Stewart, R. W. 1962: Radiation stress and mass transport in gravity waves, with application to surf beats, *J. Fluid Mech.* 13, 481–504

Loughnan, F. C. 1962: Some considerations in the weathering of the silicate minerals, *J. sedim. Petrol.* 32, 289–90

Loughnan, F. C. 1969: *Chemical weathering of the silicate minerals*, Elsevier, New York, 154 pp.

Lovering, T. S. 1959: Geologic significance of accumulator plants in rock weathering, *Bull. geol. Soc. Am.* 70, 781–800

Luckman, B. H. 1971: The role of snow avalanches in the evolution of alpine talus slopes, *Inst. Br. Geogr. Spec. Publ.* 3, 93–110

Lumb, P. 1962: The properties of decomposed granite, *Géotechnique* 12, 226–43

Lumb, P. 1965: The residual soils of Hong Kong, *Géotechnique* 15, 180–94

Lyles, L. and Krauss, R. K. 1971: Threshold velocities and initial particle motion as influenced by air turbulence, *Trans. Am. Soc. agric. Engrs* 14, 563–6

Maag, H. 1969: Ice-dammed lakes and marginal glacial drainage on Axel Heiberg Island, *Axel Heiberg Island Res. Rep.* (McGill Univ. Montreal), 147 pp.

McCabe, L. H. 1939: Nivation and corrie erosion in West Spitsbergen, *Geogrl J.* 94, 447–65

McCall, J. G. 1952: The internal structure of a cirque glacier, *J. Glaciol.* 2, 122–30

McCall, J. G. 1960: The flow characteristics of a cirque glacier and their effect on glacial structure and cirque formation, in *Norwegian cirque glaciers* (ed. W. V. Lewis), *R. geogr. Soc. Res. Ser.* 4, 39–62

McClean, R. 1967: Origin and development of ridge-furrow systems in beachrock in Barbados, West Indies, *Mar. Geol.* 5, 181–93

MacClintock, P. 1953: Crescentic crack, crescentic gouge, friction crack, and glacier movement, *J. Geol.* 61, 186

McConnell, R. G. and Brock, R. W. 1904: The great landslide at Frank, Alberta, *A. Rep. Dep. Interior 1902–3, Canada, Sess. Pap.* 25, 4–17

MacDonald, G. A. and Alcaraz, A. 1956: Nuées ardentes of the 1948–1953 eruption of Hibok Hibok, *Bull. volcan.* 2, 169–78

MacGregor, A. G. 1952: Eruptive mechanisms, Mt Pelée, the Soufrière of St Vincent and the Valley of Ten Thousand Smokes, *Bull. volcan.* 2, 49–74

Mackay, J. R. 1962: Pingos of the Pleistocene Mackenzie River delta, *Geogrl Bull.* 18, 21–63

Mackay, J. R. 1971: The origin of massive icy beds in permafrost, western Arctic coast, Canada, *Can. J. Earth Sci.* 8, 397–421

Mackay, J. R. 1972: The world of underground ice, *Ann. Ass. Am. Geogr.* 62, 1–22

MacKechnie, W. R. 1967: Some consolidation characteristics of a residual mica schist, *Proc. 4th Reg. Conf. Afr. Soil Mech. Fdn Engng*, Cape Town, 135–9

McKee, E. D. 1966: Structures of dunes at White Sands National Monument, New Mexico (and a comparison with structures of dunes from other selected areas), *Sedimentology* 7, 1–69

McKee, E. D., Douglass, J. R. and Rittenhouse, S. 1971: Deformation of lee-side laminae in eolian dunes, *Bull. geol. Soc. Am.* 82, 359–78

McKee, E. D. and Tibbitts, G. C. 1964: Primary structures of a seif dune and associated deposits in Libya, *J. sedim. Petrol.* 34, 5–17

Mackey, S. and Yamashita, T. 1967: Soil conditions and their influence on foundations of waterfront structures in Hong Kong, *Proc. 3rd Reg. Conf. Soil Mech. Fdn Engng*, Haifa, 1, 220–4

McTaggart, K. C. 1960: The mobility of *nuées ardentes*, *Am. J. Sci.* 258, 369–82

Maddock, T. 1970: Indeterminate hydraulics of alluvial channels, *J. Hydraul. Div. Am. Soc. civ. Engrs* 96, HY11, 2309–23

Madsen, O. S. 1974: Stability of sand bed under breaking waves, *Proc. 14th Conf. Coastal Engng*, Copenhagen, 776–94

Marshall, C. E. 1964: *The physical chemistry and mineralogy of soils. Vol. 1: soil materials*, Wiley, New York, 358 pp.

Marshall, J. 1971: Drag measurements in roughness arrays of varying density and distribution, *Agric. Met.* 8, 269–92

Martinec, J. 1975: Subsurface flow from snowmelt traced by tritium, *Wat. Resour. Res.* 11, 496–8

Martinec, J. 1976: Snow and ice, in *Facets of hydrology* (ed. J. C. Rodda), Wiley, New York, 85–118

Martini, A. 1967: Preliminary experimental studies on frost weathering of certain rock types from the West Sudetes, *Biul. peryglac.* 16, 147–94

Mason, C. C. and Folk, R. L. 1958: Differentiation of beach, dune, and aeolian flat environments by size analysis, Mustang Island, Texas, *J. sedim. Petrol.* 28, 211–26

Mason, B. 1968: *Principles of geochemistry*, Wiley, New York, 310 pp.

Mathews, W. H. 1959: Vertical distribution of velocity in Salmon glacier, British Columbia, *J. Glaciol.* 3, 448–54

Matthes, F. E. 1900: Glacial sculpture of the Bighorn Mountains, Wyoming, *U.S. geol. Surv. 21st A. Rep.* (1899–1900), 167–90

Matthes, F. E. 1930: Geologic history of the Yosemite Valley, *U.S. geol. Surv. Prof. Pap.* 160, 137 pp.

Matthes, F. E. 1938: Avalanche sculpture in the Sierra Nevada of California, *Bull. int. Ass. scient. Hydrol.* (Riga), 23, 631–7

Matthias, G. F. 1967: Weathering rates of the Portland Arkose tombstones, *J. geol. Educ.* 15, 140–4

Matyas, E. L. 1967: Air and water permeability of compacted soils, *Symp. Perm. Capill. Soils, Am. Soc. Test. Mat. Spec. Tech. Publ.* 417, 160–75

Matyas, E. L. 1969: Some engineering properties of Sasumna clay, *Proc. 7th int. Conf. Soil. Mech. Fdn Engng*, Mexico, 1, 143–51

Meade, R. H. 1969: Errors in using modern streamload data to estimate natural rates of denudation, *Bull. geol. Soc. Am.* 80, 1265–74

Meadows, G. A. 1976: Time-dependent fluctuations in nearshore currents, *Proc. 15th Conf. Coastal Engng*, Honolulu, 660–80

Meier, M. F. 1960: Mode of flow of Saskatchewan Glacier, Alberta, Canada, *U.S. geol. Surv. Prof. Pap.* 351, 70 pp.

Mellor, M. 1970: Phase composition of pore water in cold rocks, *Res. Rep. Cold Reg. Res. Engng Lab.* (*U.S. Army Corps Engrs*) 292, 61 pp.

Mellor, M. and Smith, J. H. 1967: Creep of ice and snow, in *Physics of snow and ice: international conference on low temperature science* (ed. H. Oura), Sapporo, 1966, 1, 843–55

Mellor, M. and Testa, R. 1969: Creep of ice under low stress; Effect of temperature on the creep of ice, *J. Glaciol.* 8, 131–52

Melton, M. A. 1957: An analysis of the relations among elements of climate, surface properties and geomorphology, *Office of Naval Research Tech. Rep.* II (Project NR. 389–042), 102 pp.

Melton, M. A. 1958: Correlation structure of morphometric properties of drainage systems and their controlling agents, *J. Geol.* 66, 442–60

Melton, M. A. 1965: Debris-covered hillslopes of the Southern Arizona desert—consideration of their stability and sediment contribution, *J. Geol.* 73, 715–29

Menard, H. W. 1961: Some rates of regional erosion, *J. Geol.* 69, 154–61

Merrill, G. P. 1897: *A treatise on rocks, rock-weathering and soils*, Macmillan, London. Reprinted 1904, 1921. 400 pp.

Merrill, G. P. 1900: Sandstone disintegration through the formation of interstitial gypsum, *Science N.Y.* 11, 850–1

Meyer, L. D. 1971: Soil erosion by water on upland areas, in *River mechanics* (ed. H. W. Shen), Ft. Collins, Colo., 1, 27·1–27·24

Meyer, L. D., Foster, G. R. and Romkens, J. M. 1975: Source of soil eroded by water from upland slopes, *Proc. Sediment Yield Workshop, U.S. Dep. Agric.* ARS-S-40, 177–89

Meyer, L. D. and Wischmeir, W. H. 1969: Mathematical simulation of the process of soil erosion by water, *Trans. Am. Soc. agric. Engrs* 12, 754–8

Meyer-Peter, E. and Muller, R. 1948: Formulas for bed-load transport, *Int. Ass. Hydraul. Res.*, 2nd Meeting, Stockholm, 39–64

Mickelson, D. M. 1971: Glacial geology of the Burroughs glacier area, south-east Alaska, *Ohio St Univ., Inst. Polar Stud. Rep.* 40, 149 pp.

Milhara, H. 1959: Raindrops and soil erosion, *Bull. natn. Inst. agric. Sci.* A, 1, 29–52

Miller, H. 1884: On boulder-glaciation, *Proc. R. phys. Soc. Edinb.* 8, 156–89

Miller, C. D. and Barcilon, A. 1976: The dynamics of the littoral zone, *Rev. Geophys. Space Phys.* 14, 81–91

Miller, R. L. 1976: Role of vortices in surf zone prediction: sedimentation and wave forces, in *Nearshore beach processes and sedimentation* (ed. R. A. Davis and R. L. Ethington), *Spec. Publs Soc. econ. Palaeont. Miner.*, Tulsa, 24, 92–114

Miller, R. L. and Ziegler, J. M. 1958: A model relating dynamics and sediment pattern in equilibrium in the region of shoaling waves, breaker zone, and foreshore, *J. Geol.* 66, 417–41

Miotke, F. D. 1968: Karstmorphologie Studien in der glazial-überformten Höhenstufe der Picos de Europa, Nordspanien, *Jb. geogr. Ges. Hannover* 4, 161 pp.

Mohr, E. C. J. and Van Baren, F. A. 1954: *Tropical soils*, Interscience, London, 496 pp.

Morales, B. 1966: The Huascarán avalanche in the Santa valley, Peru, *Proc. int. Symp. on scientific aspects of snow and ice avalanches* (Davos, 1965), *Int. Ass. scient. Hydrol. Publ.* 69, 304–15

Morales, B. 1979: *Saharan dust*. Scope 14. Wiley, Chichester, 297 pp.

Moran, S. R. 1971: Glaciotectonic structures in drift, in *Till, a symposium* (ed. R. P. Goldthwait), Ohio State Univ. Press, 127–48

Morton, D. M. and Campbell, R. H. 1974: Spring mudflows at Wrightwood, Southern California, *Q.J. Engng Geol.* 7, 377–84

Moseley, H. 1869: On the descent of a solid body on an inclined plane when subjected to alteration of temperature, *Lond. Edinb. Dubl. Phil. Mag.* 38, 99–118

Mosley, M. P. 1973: Rainsplash and the convexity of badland divides, *Z. Geomorph., Suppl. Bd* 18, 10–25

Mosley, M. P. 1976: An experimental study of channel confluences, *J. Geol.* 84, 535–62

Mudge, M. R. 1965: Rockfall-avalanche and rockslide–avalanche deposits of Sawtooth Ridge, Montana, *Bull. geol. Soc. Am.* 76, 1003–14

Müller, F. 1959: Beobachtungen über Pingos. Detailuntersuchungen in Ostgrönland und in der Kanadischen Arktis, *Meddr Grønland* 153(3), 127 pp.

Müller, L. 1964: The rock slide in the Vaiont valley, *Rock Mech. Engng Geol.* 2, 148–212

Munk, W. H. 1949: Surf beats, *Trans. Am. geophys. Un.* 30, 849–54

Munn, R. E. 1966: *Descriptive micrometeorology*, Academic Press, New York, 245 pp.

Murphy, J. B., Diskin, M. H. and Lane, L. J. 1974: Bed material characteristics and transmission losses in an ephemeral stream, *Arizona Acad. Sci., Hydrol. Sect., Proc. May 5–6 Mtg* (Prescott, Arizona), II, 455–72

Murray, S. P. 1967: Control of grain dispersion by particle size and wave state, *J. Geol.* 75, 612–34

Mutchler, C. K. and Young, R. A. 1975: Soil detachment by raindrops, *Rep. U.S. Dep. Agric.*, ARS-S-40, 113–7

Muzik, I. 1974: Laboratory experiments with surface runoff, *J. Hydraul. Div. Am. Soc. civ. Engrs* 100, HY4, 501–13

Neill, C. 1976: Scour holes in a wandering gravel river, in *Rivers 76*, American Society of Civil Engineers, II, 1301–17

Neumann, A. C. 1966: Observations on coastal erosion in Bermuda and measurements of the boring rate of the sponge *Cliona lampa*, *Limnol. Oceanogr.* 11, 92–108

Nevins, T. H. F. 1969: River training—the single thread channel, *N.Z. Engng*, December, 367–73

Nicholson, F. H. and Granberg, H. B. 1973: Permafrost and snowcover relationships near Schefferville, *North American contribution*, 2nd Int. Permafrost Conf., Yakutsk 1973 (Washington, D.C.), 151–8

Nieuwenhius, P. and Kleindorst, S. 1971: The measurement of small soil displacements on a Dutch hillslope, *Engng Geol.* 5, 1–4

Niki-foroff, C. C. 1949: Weathering and soil evolution, *Soil Sci.* 67, 219–30

Nixon, M. 1959: A study of the bankfull discharges of rivers in England and Wales, *Proc. Instn civ. Engrs* 6322, 157–74

Nobles, L. H. 1966: Slush avalanches in northern Greenland and the classification of rapid mass movements, *Proc. int. Symp. on scientific aspects of snow and ice avalanches* (Davos, 1965), *Int. Ass. scient. Hydrol. Publ.* 69, 267–72

Nortcliff, S. 1976: Spatial variability in semi-arid soils: a case study from Spain. Unpublished paper presented to meeting of *British Geomorphological Research Group*, Sheffield, October 1976

Nortcliff, S. and Thornes, J. B. 1977: Water and cation movement in a tropical rainforest environment. I—objectives, experimental design and preliminary results. *London School of Economics, Graduate School of Geography, Discussion Paper* 62, 18 pp.

Nortcliff, S., Waylen, M. J. and Thornes, J. B. 1979: Tropical weathering: a hydrological approach, *Amazoniana* 8, 82–97

North, W. J. 1954: Size distribution, erosive activities, and gross metabolic efficiency of the marine intertidal snails, *Littorina planaxis* and *L. scutula, Biol. Bull. mar. biol. Lab., Woods Hole* 106, 185–97

Nye, J. F. 1952: The mechanics of glacier flow, *J. Glaciol.* 2, 82–93

Nye, J. F. 1959: A method of determining the strain-rate tensor at the surface of a glacier, *J. Glaciol.* 3, 409–19

Nye, J. F. 1965: The flow of a glacier in a channel of rectangular, elliptic, or parabolic cross-section, *J. Glaciol.* 5, 661–90

Nye, J. F. and Martin, P. C. S. 1968: Glacial erosion, *Assemblée Générale de Berne, Commn Snow Ice* (1967), *Int. Ass. scient. Hydrol., Publ.* 79, 78–86

Nye, P. H. 1955: Some soil-forming processes in the humid tropics. Pt. II: The development of the upper slope member of the catena, *J. Soil Sci.* 6, 51–62

Nye, P. H. and Tinker, P. B. 1977: *Solute movement in the soil-root system*, Blackwell, Oxford, 342 pp.

Okko, V. 1955: Glacial drift in Iceland, its origin and morphology, *Bull. Commn géol. Finl.* 170, 1–133

Ollier, C. D. 1965: Some features of granite weathering in Uganda, *Z. Geomorph.* 4, 43–52

Ollier, C. D. 1967: Spheroidal weathering, exfoliation and constant volume alteration, *Z. Geomorph.* 9, 285–304

Ollier, C. D. 1969: *Weathering*, Longman, London. Reprinted 1975. 304 pp.

Ollier, C. D. 1971: Causes of spheroidal weathering, *Earth Sci. Rev.* 7, 127–41

Ollier, C. D. and Tuddenham, W. G. 1961: Inselbergs of central Australia, *Z. Geomorph.* 5, 257–76

Olson, J. S. 1958: Lake Michigan dune development, *J. Geol.* 66, 254–63, 345–51, 437–83

Ong, H. Ling., Swanson, V. F. and Bisque, R. E. 1970: Natural organic acids as agents of chemical weathering, *U.S. geol. Surv. Prof. Pap.* 700–C, 130–7

Orford, J. D. 1977: A proposed mechanism for storm beach sedimentation, *Earth Surface Processes* 2, 381–400

Orvig, S. 1953: On the variation of the shear stress on the bed of an ice cap, *J. Glaciol.* 2, 242–7

Østrem, G. 1975: Sediment transport in glacial meltwater streams, in *Glaciofluvial and glaciolacustrine sedimentation, Spec. Publs Soc. econ. Palaeont. Miner.*, Tulsa, 23, 101–22

Ovington, J. D. 1965: Nutrient cycling in woodlands, in *Experimental pedology* (ed. E. G. Hallsworth and D. W. Crawford), Butterworth, London, 208–15

Owens, I. F. 1969: Causes and rates of soil creep in the Chilton valley, Cass, New Zealand, *Arct. alp. Res.* 1, 213–20

Palmer, V. J. 1946: Retardance coefficients for low flow in channels lined with vegetation, *Trans. Am. geophys. Un.* 2, II, 35–73

Parfenova, E. I. and Yarilova, E. A. 1965: Mineralogical investigations in soil science, *Israel Programs for Scientific Translations*, Jerusalem

Parker, G. 1976: On the cause and characteristic scales of meandering and braiding in rivers, *J. Fluid Mech.* 76, 475–80

Parsons, D. A. 1949: Depths of overland flow, *Tech. Pap. Soil Conserve. Serv. U.S.* 82, 33 pp.

Parsons, R. B. Balster, C. A. and Ness, A. O. 1970: Soil development and geomorphic surfaces, Willamette Valley, Oregon, *Proc. Soil Sci. Soc. Am.* 34, 485–91

Parsons, R. B., Scholtes, W. H. and Riecken, F. F. 1962: Soils on Indian mounds in northeastern Iowa as benchmarks for studies of soil genesis, *Proc. Soil Sci. Soc. Am.* 26, 491–6

Parthenaides, E. 1971: Erosion and deposition of cohesive materials, in *River mechanics* (ed. D. Shen), Ft Collins, Colorado, Water Resources Publ., 25·1–25·91

Pearce, A. J. 1976: Magnitude and frequency of erosion by Hortonian overland flow, *J. Geol.* 84, 65–80

Pedro, G. 1968: Distribution des principaux types d'altération chimique à la surface du globe, *Rev. Géogr. phys. Géol. dyn.* 10, 457–70

Peel, R. F. 1974: Insolation weathering: some measurements of diurnal temperature changes in exposed rocks in the Tibesti region, central Sahara, *Z. Geomorph., Suppl. Bd* 21, 19–28

Peev, C. D. 1966: Geomorphic activity of snow avalanches, *Proc. int. Symp. on scientific aspects of snow and ice avalanches* (Davos, 1965), *Int. Ass. scient. Hydrol. Publ.* 69, 357–68

Peltier, L. C. 1950: The geographical cycle in periglacial regions as it is related to climatic geomorphology, *Ann. Ass. Am. Geogr.* 40, 214–36

Penck, A. 1879: Die Geschiebeformation Norddeutschlands, *Z. dt. geol. Ges.* 31, 117–203

Penner, E. 1967: Heaving pressure in soils during unidirectional freezing, *Can. geotech. J.* 4, 398–408

Perret, F. A. 1935: The eruption of Mount Pelée 1929–1932, *Publs Carnegie Instn* 458, 126 pp.

Perrin, R. M. S. 1965: The use of drainage water analyses in soil studies, in *Experimental pedology* (ed. E. G. Hallsworth and D. V. Crawford), Butterworth, London, 73–92

Perutz, M. F. 1950: Direct measurement of the velocity distribution in a vertical profile through a glacier, *J. Glaciol.* 1, 382–3

Petryk, A. and Bosmajian, G. 1975: Analysis of flow through vegetation, *J. Hydraul. Div. Am. Soc. civ. Engrs* 101 (HY7), 1105–20

Pettijohn, F. J. 1941: Persistence of heavy minerals and geological age, *J. Geol.* 49, 610–25

Péwé, T. L. 1959: Sand-wedge polygons (tesselations) in the McMurdo Sound region, Antarctica, *Am. J. Sci.* 257, 542–52

Péwé, T. L. 1966: Paleoclimatic significance of fossil ice wedges, *Biul. peryglac.* 15, 65–73

Philip, J. R. 1957: The theory of infiltration, *Soil Sci.* 83, 345–57, 435–48; 84, 163–77, 257–64, 329–39; 85, 278–86, 333–6

Picknett, R. G. 1964: A study of calcite solutions at 10°C, *Trans. Cave Res. Grp Gt Br.* 7, 39–62

Picknett, R. G. 1972: The pH of calcite solutions with and without magnesium carbonate present and the implications concerning rejuvenated aggressiveness, *Trans. Cave Res. Grp Gt Br.* 14, 141–50

Pissart, A. 1964: Contribution expérimentale à la connaissance de la genèse des sols polygonaux, *Annls Soc. géol. Belg.* 87, 213–23

Pissart, A. 1969: Le mécanisme périglaciaire dressant les pierres dans le sol. Résultats d'expériences, *C. r. Acad. Sci. Paris* 268, 3015–17

Pissart, A. 1970: Les phénomènes physiques essentiels liés au gel, les structures périglaciaires qui en résultent et leur signification climatique, *Annls Soc. géol. Belg.* 93, 7–49

Ploey, J. de and Moeyersons, J. 1975: Runoff creep of coarse debris: experimental data in some field observations, *Catena* 2, 275–88

Ploey, J. de, Savat, J. and Moeyersons, J. 1976: The differential impact of some soil loss factors on flow, runoff, creep and rainwash, *Earth Surface Processes*, 1, 151–62

Polynov, B. B. 1937: *The cycle of weathering*, Murby, London. Trans. Alexander Muir, 220 pp.

Ponce, V. M. and Mahmood, K. 1976: Meandering thalwegs in straight alluvial channels, in *Rivers 76*, American Society of Civil Engineers, II, 1418–42

Poser, H. 1948: Boden- und Klimaverhältnisse in Mittel- und Westeuropa wahrend der Würmeiszeit, *Erdkunde* 2, 53–68

Potts, A. S. 1970: Frost action in rocks: some experimental data, *Trans. Inst. Br. Geogr.* 49, 109–24

Prandtl, L. 1925: Bericht über Untersuchungen zur ausgebilten Turbulenz, *Trans. 2nd int. Congr. Applied Mechanics*, Zürich, 62–93

Prescott, J. A. 1949: A climatic index for the leaching factor in soil formation, *J. Soil Sci.* 1, 1–9

Price, R. J. 1969: Moraines, sandar, kames and eskers near Breiðamerkurjökull, Iceland, *Trans. Inst. Br. Geogr.* 46, 17–43

Price, R. J. 1970: Moraines at Fjällsjökull, Iceland, *Arct. alp. Res.* 2, 27–42

Price, R. J. 1973: *Glacial and fluvioglacial landforms*, Oliver & Boyd, Edinburgh, 242 pp.

Prior, D. B. and Coleman, J. M. 1978: Disintegrating, retrogressive landslides on very low-angle subaqueous slopes, Mississippi delta, *Mar. Geotech.* (in press)

Prior, D. B., Stephens, N. and Douglas, G. R. 1970: Some examples of modern debris flows from Northern Ireland, *Z. Geomorph.* 14, 275–88

Pugh, J. C. 1955: Isostatic readjustment in the theory of pediplanation, *Q.J. geol. Soc. Lond.* 111, 361–74

de Quervain, M. R. 1966: On avalanche classification, a further contribution, *Proc. int. Symp. on scientific aspects of snow and ice avalanches* (Davos, 1965), *Int. Ass. scient. Hydrol. Publ.* 69, 410–17

Ragan, R. M. 1968: An experimental investigation of partial area contributions, *Publs Int. Ass. scient. Hydrol., Gen. Ass. Berne*, 76, 241–9

Rahn, P. H. 1971: The weathering of tombstones and its relationship to the topography of New England, *J. geol. Educ.* 19, 112–18

Raistrick, A. and Gilbert, O. L. 1963: Malham Tarn House: its building materials, their weathering and colonization by plants, *Fld Stud.* 1, 89–115

Ramberg, H. 1964: Note on model studies of folding of moraines in piedmont glaciers, *J. Glaciol.* 5, 207–18

Rapp, A. 1959: Avalanche boulder tongues in Lapland, a description of little known forms of periglacial accumulations, *Geogr. Annlr* 41, 34–48

Rapp, A. 1960: Recent development of mountain slopes in Kärkevagge and surroundings, northern Scandinavia, *Geogr. Annlr* 42, 65–200

Rapp, A. 1962: Kärkevagge: some recordings of mass movements in the northern Scandinavian mountains, *Biul. peryglac.* 11, 287–309

Rapp, A. 1974: A review of desertization in Africa: water, vegetation and man, *Secretariat for International Ecology, Sweden*, Report No. 1, 77 pp.

Raudkivi, A. J. 1963: Study of sediment ripple formation, *Proc. Am. Soc. civ. Engrs.* 89, HY6, 15–33

Raudkivi, A. J. 1967: *Loose boundary hydraulics*, Pergamon, Oxford, 331 pp.

Ravina, I. and Zaslavsky, D. 1974: The electrical double layer as a possible factor in desert weathering, *Z. Geomorph., Suppl. Bd* 21, 15–18

Raymond, C. F. 1971: Flow in a transverse section of Athabasca glacier, Alberta, Canada, *J. Glaciol.* 10, 55–84

Reiche, P. 1943: Graphic representation of chemical weathering, *J. sedim. Petrol.* 13, 58–68

Reiche, P. 1950: *A survey of weathering processes and products*, Univ. of New Mexico, Albuquerque, Publs in Geology, 3, 95 pp.

Reiner, M. 1958: *Handbuch der Physik*, Springer, Berlin, 6, 434–50

Reynaud, L. 1973: Flow of a valley glacier with a solid friction law, *J. Glaciol.* 12, 251–8

Reynolds, H. R. 1961: *Rock mechanics*, Lockwood, London, 136 pp.

Rich, L. G. 1973: *Environmental systems engineering*, McGraw Hill, Kogakusha, Tokyo, 448 pp.

Rigsby, G. P. 1960: Crystal orientation in glacier and in experimentally deformed ice, *J. Glaciol.* 3, 589–606

Roberts, H. H., Suhayda, J. N. and Coleman, J. M. 1978: Sediment deformation and transport on low-angle slopes: Mississippi River delta', *9th Geomorph. Symp. Binghampton, N.Y. State* (in press)

Robin, G. de Q., Evans, S. and Bailey, J. T. 1969: Interpretation of radio-echo sounding in polar ice-sheets, *Phil. Trans. R. Soc.* 265, 437–505

Robinson, D. N. 1969: Soil erosion by wind in Lincolnshire, March 1968, *Geography* 54, 351–62

Roch, A. 1966: Les variations de la résistance de la neige, *Proc. int. Symp. on scientific aspects of snow and ice avalanches* (Davos, 1965), *Int. Ass. scient. Hydrol. Publ.* 69, 86–99

Rodda, J. C. 1967: A countrywide study of intense rainfall for the United Kingdom, *J. Hydrol.* 5, 58–69

Rodda, J. C., Downing, R. A. and Law, F. M. 1976: *Systematic hydrology*, Newnes–Butterworth, 399 pp.

Rodier, J. 1960: *L'analyse chimique et physico-chimique de l'eau*, Editions Techniques, Paris, 27 pp.

Rodin, L. E. and Bazilevich, N. I. 1965: *Production and mineral cycling in terrestrial vegetation*, Oliver & Boyd, London

Rogowski, A. S. 1972: Watershed physics—soil variability criteria, *Wat. Resour. Res.* 8, 1015–23

Rosenqvist, T. 1953: Considerations on the sensitivity of Norwegian quick clays, *Géotechnique* 3, 195–200

Roth, E. S. 1965: Temperature and water content as factors in desert weathering, *J. Geol.* 73, 454–68

Rotnicki, K. 1976: The theoretical basis for and a model of the origin of glaciotectonic deformations, *Quest. geogr.* 3, 103–39

Rouse, W. C. 1969: An investigation of the stability and frequency distribution of slopes in selected areas of West Glamorgan, Unpubl. Ph.D. thesis, Univ. of Wales

Rozovskii, I. L. 1961: *Flow of water in bends of open channels*, Israel Program for Scientific Translation, 233 pp.

Rubin, J. 1966: Theory of rainfall uptake by soils initially drier than their field capacity and its application, *Wat. Resour. Res.* 2, 739–94

Rudberg, S. 1962: A report on some field observations concerning periglacial geomorphology and mass movement on slopes in Sweden, *Biul. peryglac.* 11, 311–23

Rudberg, S. 1964: Slow mass movements, processes and slope development in the Norva Storfjall area, Southern Swedish Lapland, *Z. Geomorph., Suppl. Bd* 5, 192–203

Russell, J. C. and Engle, E. G. 1925: Soil horizons in the central prairies, *Rep. 5th Mtg. Am. Soil Surv. Ass.* 6, 1–18

Russell, R. J. 1963: Recent recession of tropical cliff coasts, *Science, N.Y.* 139, 9–15

Russell, W. L. 1929: Drainage alignment in the western Great Plains, *J. Geol.* 37, 249–55

Ruxton, B. P. 1966: The measurement of denudation rates, *Inst. Aust. Geogr., Rep. 5th Mtg*, Sydney, 1–5

Ruxton, B. P. 1968: Measures of the degree of chemical weathering of rocks, *J. Geol.* 76, 518–27

Ruxton, B. P. and Berry, L. 1957: The weathering of granite and associated erosional features in Hong Kong, *Bull. geol. Soc. Am.* 68, 1263–92

Ruxton, B. P. and Berry, L. 1959: The basal rock surface of weathered granitic rocks, *Proc. Geol. Ass.* 70, 285–90

Saito, M. 1965: Forecasting the time of occurrence of a slope failure, *Proc. 6th int. Conf. Soil Mech. Fdn Engng*, Montreal, 2, 537–41

Salisbury, E. J. 1925: Note on the edaphic succession in some dune soils with special reference to the time factor, *J. Ecol.* 13, 322–8

Sanches Furtado, A. F. A. 1968: Altération des granites dans les régions intertropicales sous différents climats, *9th int. Congr. Soil Sci.*, Adelaide, 4, 403–9

Sanger, F. J. 1966: Degree-days and heat conduction in soils, *Proc. Permafrost int. Conf., Lafayette, Indiana 1963. Proc. natn. Acad. Sci. natn. Res. Coun. Publ.* 1287, 563 pp.

Sarnthein, M. and Walger, E. 1974: Der äolische Sandstrom aus der W-Sahara zur Atlantikküste, *Geol. Rundsch.* 63, 1065–87

Saunders, M. K. and Fookes, P. G. 1970: A review of the relationship of rock weathering and climate and its significance to foundation engineering, *Engng Geol.* 4, 289–325

Savage, J. C. and Paterson, W. S. B. 1963: Borehole measurements in the Athabasca glacier, *J. geophys. Res.* 68, 4521–36

Savat, J. 1977: The hydraulics of sheet flow on a smooth surface and the effect of simulated rainfall, *Earth Surface Processes* 2, 125–40

Savigear, R. A. G. 1965: A technique of morphological mapping, *Ann. Ass. Am. Geogr.* 55, 514–38

Schaffer, R. T. 1932: *The weathering of natural building stones*, HMSO, London, 149 pp.

Schalscha, E. B., Appelt, H. and Schatz, A. 1967: Chelation as a weathering mechanism. 1. Effect of complexing agents on the solubilization of iron from minerals and granodiorite, *Geochim. cosmochim. Acta* 31, 587–96

Schatz, A., Schatz, V. and Martin, J. J. 1957: Chelation as a biochemical weathering factor, *Bull. geol. Soc. Am.* 68, 1792–819

Scheffer, F., Mayer, B. and Kalk, E. 1963: Biologische Ursachen der Wüstenlackbildung, *Z. Geomorph.* 7, 112–19

Scheidegger, A. E. 1961: Mathematical models of slope development, *Bull. geol. Soc. Am.* 72, 37–50

Scheidegger, A. E. 1970: *Theoretical geomorphology*, Prentice-Hall, Englewood Cliffs, N. J., 2nd ed., 435 pp.

Schiffman, A. 1965: Energy measurements in the swash-surf zone, *Limnol. Oceanogr.* 10, 255–60

Schmidt, W. 1925: Der Massenaustausch in freier Luft und verwandte Erscheinungen, *Probleme kosm. Phys.* 7, 1–33

Schreckenthal-Schimitschek, G. 1935: Der Einfluss des Bodens auf die Vegetation im Moränengelände des Mittelbergferners (Pitztal, Tirol), *Z. Gletscherk.* 23, 57–66

Schumm, S. A. 1956: The role of creep and rainwash in the retreat of badland slopes, *Am. J. Sci.* 254, 693–706

Schumm, S. A. 1962: Erosion on miniature pediments in Badlands National Monument, South Dakota, *Bull. geol. Soc. Am.* 73, 719–24

Schumm, S. A. 1963: The disparity between present rates of denudation and orogeny, *U.S. geol. Surv. Prof. Pap.* 454–H, 13 pp.

Schumm, S. A. 1964: Seasonal variations of erosion rates and processes on hillslopes in western Colorado, *Z. Geomorph., Suppl. Bd* 5, 215–38

Schumm, S. A. 1967: Rates of surficial rock creep on hillslopes in western Colorado, *Science, N.Y.* 115, 560–1

Schumm, S. A. 1972: *River morphology*, Dowden, Hutchinson and Ross, Stroudsburg, Pennsylvania, 41–2

Schumm, S. A. 1973: Geomorphic thresholds and complex response of drainage systems, in *Fluvial geomorphology* (ed. M. Morisawa), Publications in Geomorphology, State University of New York, Binghampton, New York, 299–309

Schumm, S. A. and Beathard, R. M. 1976: Geomorphic thresholds: an approach to river management, *Rivers 76*, Symposium on Inland Waterways for Navigation, Flood Control and Water Diversions. American Society of Civil Engineers, 1, 707–24

Schumm, S. A. and Chorley, R. J. 1964: The fall of Threatening Rock, *Am. J. Sci.* 262, 1041–54

Schumm, S. A. and Lichty, R. W. 1963: Channel widening and flood-plain construction along the Cimarron River in southwestern Kansas, *U.S. geol. Surv. Prof. Pap.* 352–D, 71–88

Schumm, S. A. and Lichty, R. W. 1965: Time, space and causality in geomorphology, *Am. J. Sci.* 263, 110–19

Scorer, R. S. 1958: *Natural aerodynamics*, Pergamon, London, 312 pp.

Scorer, R. S. 1977: *Environmental aerodynamics*, Ellis Horwood, Chichester, 488 pp.

Selby, M. J. 1972: Antarctic tors, *Z. Geomorph., Suppl. Bd* 13, 73–86

Sevaldson, R. A. 1956: The slide in Lodalen, October 6th 1954, *Géotechnique* 6, 167–82

Sharp, R. P. 1942: Mudflow levees, *J. Geomorph.* 5, 222–7

Sharp, R. P. 1953: Deformation of a vertical bore-hole in a piedmont glacier, *J. Glaciol.* 2, 182–4

Sharp, R. P. 1963: Wind ripples, *J. Geol.* 71, 617–36

Sharp, R. P. 1964: Wind-driven sand in the Coachella Valley, California, *Bull. geol. Soc. Am.* 75, 785–804

Sharp, R. P. and Nobles, L. H. 1953: Mudflow of 1941 at Wrightwood, Southern California, *Bull. geol. Soc. Am.* 64, 547–60

Sharpe, C. F. S. 1938: *Landslides and related phenomena*, Columbia Univ. Press, New York, 137 pp.

Shaw, C. F. 1928: Profile development and the relationships of soils in California, *Proc. 1st int. Conf. Soil Sci.* 4, 291–317

Shen, H. W. and Cheong, F. H. 1971: Dispersion of contaminated bed-load particles, Symposium on statistical hydrology, *Int. Ass. for Statistics phys. Sci.*, Tucson, Arizona

Shen, H. W. and Hung, C. S. 1971: An engineering approach to total bed material load by regression analysis, *Sedimentation Symposium to honour H. A. Einstein* (ed. Shen, H. W.), Fort Collins, Colorado, 14·1–14·17

Shen, H. W. and Komura, K. 1968: Meandering tendencies in straight alluvial channels, *J. Hydraul. Div. Am. Soc. civ. Engrs* 94 (HY6), 997–1016

Shepard, F. P. and Grant, U.S. 1947: Wave erosion along the southern California coast, *Bull. geol. Soc. Am.* 58, 919–26

Sherman, G. D. 1952: The genesis and morphology of the alumina-rich laterite clays, in *Problems of clay and laterite genesis*, American Institute of Mining and Metallurgical Engineers, New York, 154–61

Shields, A. 1936: *Anwendung der Ánlichkeitsmechanik und der Turbulenzforschung auf die Geschiebebewegung*. English Trans. W. P. Ott and J. C. van Uchelen, California Institute of Technology, Pasadena, 26 pp.

Shoda, M. 1966: An experimental study on dynamics of avalanche snow, *Proc. int. Symp. on scientific aspects of snow and ice avalanches* (Davos, 1965), *Int. Ass. scient. Hydrol. Publ.* 69, 215–29

Shreve, R. L. 1966: Sherman landslide, Alaska, *Science, N.Y.* 154, 1639–43

Shreve, R. L. 1966: Statistical law of stream numbers, *J. Geol.* 74, 17–37

Shreve, R. L. 1968a: Leakage and fluidization in air-layer lubrication avalanches, *Bull. geol. Soc. Am.* 79, 653–8

Shreve, R. L. 1968b: The Blackhawk landslide, *Geol. Soc. Am. Spec. Pap.* 108, 47 pp.

Shreve, R. L. and Sharp, R. P. 1970: Internal deformation and thermal anomalies in Lower Blue Glacier, Mount Olympus, Washington, U.S.A., *J. Glaciol.* 9, 65–86

Shulits, S. 1935: The Schoklitsch bed-load formula, *Engineering*, June 21st 1935, 644–6; June 28th 1935, 687

Simons, D. B. and Richardson, E. V. 1966: Resistance to flow in alluvial channels, *U.S. geol. Surv. Prof. Pap.* 422–J, 61 pp.

Simony, F. 1871, Die Gletscher des Dachsteingebirges, *Sber. Akad. wien. math. naturw. Kl.* 63, 501–36

Singh, A. and Mitchell, J. K. 1968: General stress-strain-time functions for soils, *J. Soil Mech. Fdns Div. Am. Soc. civ. Engrs* 94(1), 21–46

Skempton, A. W. 1948: The rate of softening of stiff, fissured clays, *Proc. 2nd int. Conf. Soil Mech. Fdn Engng*, Rotterdam, 2, 50–3

Skempton, A. W. 1953: Soil mechanics in relation to geology, *Proc. Yorks. geol. Soc.* 29, 33–62

Skempton, A. W. 1954: The pore pressure coefficients of A and B, *Géotechnique* 4, 143–7

Skempton, A. W. 1964: Long-term stability of clay slopes, *Géotechnique* 14, 71–200

Skempton, A. W. 1970: First-time slides in over-consolidated clays, *Géotechnique* 20, 320–4

Skempton, A. W. and Brown, J. D. 1961, A landslide in boulder clay at Selset, Yorkshire, *Géotechnique* 11, 280–93

Skempton, A. W. and DeLory, F. A. 1957: Stability of natural slopes in London Clay, *Proc. 4th int. Conf. Soil Mech. Fdn Engng*, London, 2, 378–81

Skempton, A. W. and Hutchinson, J. N. 1969: Stability of natural slopes and embankment sections, *Proc. 7th int. Conf. Soil Mech. Fdn Engng*, Mexico, *State of the Art Volume*, 291–340

Skempton, A. W. and La Rochelle, P. 1965: The Bradwell slip, a short-term failure in London Clay, *Géotechnique* 15, 221–42

Slaymaker, H. O. 1972: Patterns of present sub-aerial erosion and landforms in mid-Wales, *Trans. Inst. Br. Geogr.* 55, 47–67

Slaymaker, H. O. 1978: An overview of geomorphic processes in the Canadian Cordillera, *Z. Geomorph.* 21, 169–86

Smalley, I. J. 1970: Cohesion of soil particles and the intrinsic resistance of simple soil systems to wind erosion, *J. Soil Sci.* 21, 154–61

Smalley, I.J. (ed.) 1977: *Loess, lithology and genesis*, Benchmark Papers in Geology 26, Dowden, Hutchinson & Ross, Halsted Press, New York, 429 pp.

Smalley, I. J. and Unwin, D. J. 1968: The formation and shape of drumlins and their distribution and orientation in drumlin fields, *J. Glaciol.* 7, 377–90

Smerdon, E. T. and Beasley, R. P. 1961: Critical tractive forces in cohesive soils, *Agric. Engng*, Jan. 1961, 26–9

Smith, D. D. and Wischmeir, W. H. 1962: Rainfall erosion, *Adv. Agron.* 14, 109–48

Smith, D. G. 1976: Effects of vegetation on lateral migration of a glacial meltwater river, *Bull. geol. Soc. Am.* 87, 857–60

Smith, D. I. 1972: The solution of limestone in an Arctic environment, *Inst. Br. Geogr. Spec. Publ.* 5, 187–200

Smith, D. I. and Atkinson, T. C. 1976: Process, landforms and climate in limestone regions, in *Geomorphology and climate* (ed. E. Derbyshire), Wiley, London, 367–409

Smith, H. T. U. 1948: Giant glacial grooves in Northwest Canada, *Am. J. Sci.* 246, 503–14

Smith, J. 1960: Cryoturbation data from South Georgia, *Biul. peryglac.* 8, 73–9

Smith, R. E. 1972: The infiltration envelope: results from a theoretical infiltrometer, *J. Hydrol.* 17, 1–21

Smith, T. R. and Bretherton, F. P. 1972: Stability and the conservation of mass in drainage basin evolution, *Wat. Resour. Res.* 8, 1506–24

Smith, T. R. and Dunne, T. 1977: Watershed geochemistry—the control of aqueous solutions by soil materials in a small watershed, *Earth Surface Processes* 2, 421–5

Sonu, C. J., Murray, S. P., Hsu, S. A., Suhayda, J. N. and Waddell, E. 1973: Sea breeze and coastal processes, *Trans. Am. geophys. Un.* 54, 820–33

Souchez, R. A., Lorrain, R. D. and Lemmens, M. M. 1973: Refreezing of interstitial water in a sub-glacial cavity of an alpine glacier as indicated by the chemical composition of the ice, *J. Glaciol.* 12, 453–9

Sparks, B. W. 1960: *Geomorphology*, Longmans,London, 371 pp.

Stalker, A. M. S. 1960: Ice-pressed drift forms and assocated deposits in Alberta, *Bull. geol. Surv. Can.* 57, 38 pp.

Statham, I. 1973: Scree slope development under conditions of surface particle movement, *Trans. Inst. Br. Geogr.* 59, 41–53

Statham, I. 1976: A scree slope rockfall model, *Earth Surface Processes* 1, 43–62

Statham, I. 1977: *Earth surface sediment transport*, Clarendon Press, Oxford, 184 pp.

Steers, J. A. 1953: The East Coast floods January 31–February 1 1953, *Geogrl J.* 119, 280–95

Steinemann, S. 1958: Experimentelle Untersuchungen zur Plastizität von Eis, *Beitr. Geol. Schweiz (Hydrologie)*, 10, 72 pp.

Stevenson, C. M. 1969: The dust fall and severe storms of July 1, 1968, *Weather* 24, 126–32

Stocking, M. A. 1976: Erosion of soils in central Rhodesia, Unpubl. Ph.D. thesis, Univ. of London, 355 pp.

Stocking, M. A. and Elwell, H. A. 1976: Rainfall erosivity over Rhodesia, *Trans. Inst. Br. Geogr. N.S.* 1, 231–45

Stoddart, D. R. 1969: Climatic geomorphology: review and assessment, *Progr. Geogr.* 1, 160–222

Strahler, A. N. 1954: Statistical analysis in geomorphic research, *J. Geol.* 62, 1–25

Strakhov, N. M. 1967: *Principles of lithogenesis*, Vol. 1, Consultants Bureau, N.Y., Oliver & Boyd, London. Trans. J. P. Fitzsimmons, ed. S. I. Tomkeieff and J. E. Hemingway, 245 pp.

Streeruwitz, H. von 1892: Geological report quoted in Merrill (1897) from *4th A. Rep. geol. Surv. Texas*, 144

Sugden, D. E. and John, B. S. 1976: *Glaciers and landscape*, Arnold, London, 376 pp.

Suhayda, J. N. and Pettigrew, N. R. 1977: Observations of wave height and wave celerity in the surf zone, *J. geophys. Res.* 82, 1419–24

Suhayda, J. N. and Prior, D. B. 1978: Explanation of submarine landslide morphology by stability analysis and rheological models, *10th A. Conf. Offshore Technology* (Houston, Texas), 14 pp.

Suklje, L. and Vidmar, S. 1961: A landslide due to long-term creep, *Proc. 5th int. Conf. Soil Mech. Fdn Engng*, Paris, 2, 727–35

Sulebak, J. R. 1969: Mudflow in the low Alpine region, *Norsk geogr. Tidsskr.* 23, 15–23

Sundborg, A. 1956: The river Klarälven, a study of fluvial processes, *Geogr. Annlr* 38, 127–316

Swartzendruber, D. and Hillel, D. 1975: Infiltration and runoff for small field plots under constant intensity rainfall, *Wat. Resour. Res.* 11, 445–51

Sweeting, M. M. 1950: Erosion cycles and limestone caverns in the Ingleborough District, *Geogrl J.* 115, 63–78

Sweeting, M. M. 1958: The karstlands of Jamaica, *Geogrl J.* 124, 184–99

Sweeting, M. M. 1960: The caves of the Buchan area, Victoria, *Z. Geomorph.* 2, 81–91

Sweeting, M. M. 1966: The weathering of limestones, in *Essays in geomorphology* (ed. G. H. Dury), Heinemann, London, 177–210

Swineford, A., and Frye, J. C. 1951: Petrography of the Peoria loess in Kansas, *J. Geol.* 59, 306–22

Taber, S. 1929: Frost heaving, *J. Geol.* 37, 428–61

Taber, S. 1930: The mechanics of frost heaving, *J. Geol.* 38, 303–17

Taber, S. 1943: Perennially frozen ground in Alaska: its origin and history, *Bull. geol. Soc. Am.* 54, 1433–548

Taber, S. 1950: Intensive frost action along lake shores, *Am. J. Sci.* 248, 784–93

Tamburi, A. J. 1974: Creep of single rocks on bedrock, *Bull. geol. Soc. Am.* 85, 351–6

Tamm, O. 1920: Bodenstudien in der Nordschwedischen Nadelwaldregion, *Meddn St. Skogsförs Anst.* 17, 49–300

Taylor, D. W. 1937: Stability of earth slopes, *J. Boston Soc. civ. Engrs* 24, 197–246

Teichert, C. 1939: Corrasion by wind-blown snow in polar regions, *Am. J. Sci.* 237, 146–8

Ter-Stephanian, G. 1965: In-situ determination of the rheological characteristics of soil on slopes, *Proc. 6th int. Conf. Soil Mech. Fdn Engng*, 2, 375–7

Terzaghi, K. 1925: *Erdbaumechanik*, Franz Deuticke, Wien, 399 pp.

Terzaghi, K. 1936: The shearing resistance of saturated soils, *Proc. 1st int. Conf. Soil Mech. Fdn Engng.* 1, 54–66

Terzaghi, K. 1943: *Theoretical soil mechanics*, Wiley, New York, 510 pp.

Terzaghi, K. 1952: Permafrost, *J. Boston Soc. civ. Engrs* 39, 1–50

Terzaghi, K. 1957: Varieties of submarine slope failure, *Publs Norw. geotech. Inst.* (Oslo), 25, 20 pp.

Terzaghi, K. 1958: Design and performance of the Sasamua Dam, *Proc. Instn civ. Engrs* 9, 369–94

Terzaghi, K. 1960: Mechanism of landslides, *Bull. geol. Soc. Am.*, Berkey Volume, 83–122

Terzaghi, K. 1962a: Stability of steep slopes on hard unweathered rock, *Géotechnique* 12, 251–70

Terzaghi, K. 1962b: Dam foundations on sheeted granite, *Géotechnique* 12, 199–208

Terzaghi, K. and Peck, R. B. 1948, reprinted 1967: *Soil mechanics in engineering practice*, Wiley, New York, 566 pp.

Theakstone, W. H. 1967: Basal sliding and movement near the margin of the glacier Østerdalsisen, Norway, *J. Glaciol.* 6, 805–16

Thomas, M. F. 1974: *Tropical geomorphology. A study of weathering and landform development in warm climates*, Macmillan, London, 332 pp.

Thomas, W. N. 1938: Experiments on the freezing of certain building materials, *Tech. Pap. Building Res. D.S.I.R.* 17, 146 pp.

Thorarinsson, S. 1969: Glacier surges in Iceland with special reference to the surges of Brúarjökull, *Can. J. Earth Sci.* 6, 875–82

Thornbury, W. D. 1954: *Principles of geomorphology*, Wiley, New York, 618 pp.

Thornes, J. B. 1971: State, attribute and environment in scree slope studies, *Inst. Br. Geogr. Spec. Publ.* 3, 49–64

Thornes, J. B. 1974: Speculations on the behaviour of stream channel width, *London School of Economics, Department of Geography, Working Paper* No. 49

Thornes, J. B. 1975: Lithological control of hillslope erosion in the Soria area, Duero alto, Spain, *Boln Geol. Min.* 85, 11–19

Thornes, J. B. 1976: Semi-arid erosional systems, *London School of Economics, Department of Geography, Occas. Pap.* No. 7, 79 pp.

Thornes, J. B. 1977a: Hydraulic geometry and river channel change, in *River channel changes* (ed. K. J. Gregory), Wiley, Chichester, 215–26

Thornes, J. B. 1977b: Channel changes in ephemeral streams—observations, problems and models, in *River channel changes* (ed. K. J. Gregory), Wiley, Chichester, 317–35

Thornes, J. B. 1978: The character and problems of contemporary theory in geomorphology, in *Geomorphology: present problems and future prospects* (ed. C. Embleton *et al.*, Oxford University Press), 14–24

Thornes, J. B. and Brunsden, D. 1977: *Geomorphology and time*, Methuen, London, 208 pp.

Thornton, E. B. 1968: A field investigation of sand transport in the surf zone, *Proc. 11th Conf. coastal Engng*, London, 335–51

Thornton, E. B. 1972: Distribution of sediment transport across the surf zone, *Proc. 13th Conf. coastal Engng*, Vancouver, 1049–68

Thorud, D. B. and Duncan, D. P. 1972: Effects of snow removal, litter removal and soil compaction on soil freezing and thawing in a Minnesota Oak Stand, *Proc. Soil Sci. Soc. Am.* 36, 153–7

Thrush, P. W. 1968: A dictionary of mining, mineral and related terms, *Spec. Publs U.S. Bur. Mines*, 1269 pp.

Toebes, G. H. and Sooky, A. A. 1967: Hydraulics of meandering rivers with flood plains, *J. WatWays Harb. Div. Am. Soc. civ. Engrs* 93, 213–36

Tomlinson, M. J. and Holt, J. B. 1953: The foundation of the Bank of China Building, Hong Kong, *Proc. 3rd int. Conf. Soil Mech. Fdn Engng*, Zürich, 1, 466–72

Toms, A. H. 1953: Recent research into the coastal landslides at Folkestone Warren, Kent, England, *Proc. 3rd int. Conf. Soil Mech. Fdn Engng*, Zürich, 2, 288–93

Trainer, F. W. 1973: Formation of joints in bedrock by moving glacial ice, *J. Res. U.S. geol. Surv.* 1, 229–36

Trendall, A. F. 1962: The formation of apparent peneplains by a process of combined laterization and surface work, *Z. Geomorph.* 6, 183–97

Trenhaile, A. S. 1972: The shore platforms of the Vale of Glamorgan, Wales, *Trans. Inst. Br. Geogr.* 56, 127–44

Tricart, J. 1956: Étude expérimentale du problème de la gélivation, *Biul. peryglac.* 4, 285–318

Tricart, J. and Cailleux, A. 1955: *Introduction à la géomorphologie climatique*, Centre Docum. Univ., Paris, 222 pp.

Tricart, J. and Cailleux, A. 1962: *Le modelé glaciaire et nival*, SEDS, Paris, 508 pp.

Troll, C. 1943: Die Frost Wechselhäfigkeit in den Luft- und Bodenklimaten der Erde, *Met. Z.* 60, 161–71

Trombe, F. 1952: *Traité de spéléologie*, Payot, Paris, 367 pp.

Trow, W. A. and Morton, J. D. 1967: Laterite soils at Guardarraya la Republica Dominicana—their development, composition and engineering properties, *Proc. 7th int. Conf. Soil Mech. Fdn Engng*, Mexico, 1, 73–84

Trudgill, S. T. 1976: Rock weathering and climate: quantitative and experimental aspects, in *Geomorphology and climate* (ed. E. Derbyshire) Wiley, Chichester, 59–99

Trudgill, S. T. 1977: *Soils and vegetation systems*, Clarendon Press, Oxford, 180 pp.

Tufnell, L. 1971: Erosion by snow patches in the north Pennines, *Weather* 26, 492–8

Turner, S. F. *et al.* 1943: Groundwater resources of the Santa Cruz Basin, Arizona, *Rep. U.S. geol. Surv.* Tucson, Arizona, 35–53

Twidale, C. R. 1964: A contribution to the general theory of domed inselbergs, *Trans. Inst. Br. Geogr.* 34, 91–113

Twidale, C. R. 1976: *Analysis of landforms*, Wiley, Australasia Pty., Sydney, 572 pp.

U.S. Army Corps of Engineers 1952: Soil mechanics design, stability of slopes and foundations, *U.S. Army Corps Engrs, Engng Man.*, Part 119, Ch. 2

U.S. Dep. of Agriculture, Forest Service 1968: *Snow avalanches: a handbook of forecasting and control measures*, Agriculture Handbook No. 194, Washington D.C., 84 pp.

Van Breeman, N. and Brinkman, R. 1976: Chemical equilibrium and soil formation, in *Soil chemistry, A. Basic elements* (ed. G. H. Bolt and M. G. M. Bruggenwelt), Elsevier, 141–70

Van Burkalow, A. 1945: Angle of repose and angle of sliding friction: an experimental study, *Bull. geol. Soc. Am.* 56, 669–707

Vanoni, V. A. (ed.) 1975: *Sedimentation engineering*, American Society of Civil Engineers, New York, 745 pp.

Vargas, M. 1953a: Correlation between angle of internal friction and angle of shearing resistance in consolidated quick triaxial compression tests on residual clays, *Proc. 3rd int. Conf. Soil Mech. Fdn Engng*, Zürich, 1, 67

Vargas, M. 1953b: Some engineering properties of residual clay soils occurring in Southern Brazil, *Proc. 3rd int. Conf. Soil Mech. Fdn Engng*, Zürich, 1

Vargas, M., Silva, F. G. and Tubio, M. 1965: Residual clay dams in the State of São Paulo, Brazil, *Proc. 6th int. Conf. Soil Mech. Fdn Engng*, 2, 578

Varnes, D. J. 1958: Landslide types and processes, in *Landslides and engineering practice* (ed. E. B. Eckel), *Spec. Rep. Highw. Res. Bd* 29, 20–47

Verstappen, H. Th. 1968: On the origin of longitudinal (*seif*) dunes, *Z. Geomorph.* 12, 200–20

Visher, S. S. 1945: Climatic maps of geologic interest, *Bull. geol. Soc. Am.* 56, 713–76

Vita-Finzi, C. 1973: *Recent Earth history*, Macmillan, London, 138 pp.

Vita-Finzi, C. and Cornelius, P. F. S. 1973: Cliff sapping by molluscs in Oman, *J. sedim. Petrol.* 43, 31–2

Vivian, R. 1966: La catastrophe du glacier Allalin, *Revue Géogr. alp.* 54, 97–112

Vivian, R. 1970: Hydrologie et érosion sous-glaciaires, *Revue Géogr. alp.* 58, 241–64

Vivian, R. and Bocquet, G. 1973: Subglacial cavitation phenomena under the Glacier d'Argentière, Mont Blanc, France, *J. Glaciol.* 12, 439–51

Vivian, R. and Zumstein, J. 1973: Hydrologie sous-glaciaire au glacier d'Argentière (Mont-Blanc, France), *Symposium on the hydrology of glaciers* (Cambridge, 1969), *Int. Ass. scient. Hydrol. Publ.* 95, 53–64

Von Moos, A. 1953: The subsoil of Switzerland, *Proc. 3rd int. Conf. Soil Mech. Fdn Engng*, Zürich, 3, 252–64

Waddell, E. 1976: Swash-groundwater-beach profile interactions, in *Beach and nearshore sedimentation* (ed. R. A. Davis and R. L. Ethington), *Spec. Publs. Soc. econ. Palaeont. Miner.*, Tulsa, 24, 115–25 ·

Wahrhaftig, C. 1965: Stepped topography of the southern Sierra Nevada, California, *Bull. geol. Soc. Am.* 76, 1165–90

Walling, D. E. 1971: Sediment dynamics of small instrumented catchments in south-east Devon, *Rep. Trans. Devon. Ass. Advmt. Sci.* 103, 147–65

Ward, R. 1977: *Principles of hydrology*, McGraw Hill, London (2nd edn), 367 pp.

Ward, W. H. 1945: The stability of natural slopes, *Geogrl J.* 111, 170–91

Warnke, D. A. 1970: Glacial erosion, ice-rafting and glacial-marine sediments: Antarctica and the Southern Ocean, *Am. J. Sci.* 269, 276–94

Warren, A. 1972: Observations on dunes and bimodal sands in the Ténéré desert, *Sedimentology* 19, 37–44

Warren, A. 1976a: Dune trend and the Ekman Spiral, *Nature, Lond.* 259, 653–4

Warren, A. 1976b: Morphology and sediments of the Nebraska Sand Hills in relation to Pleistocene winds and the development of aeolian bedforms, *J. Geol.* 84, 685–700

Washburn, A. L. 1947: Reconnaissance geology of portions of Victoria Island and adjacent regions, Arctic Canada, *Am. geol. Soc. Mem.* 22, 142 pp.

Washburn, A. L. 1967: Instrumental observations of mass wasting in the Mesters Vig district, north-east Greenland, *Meddr Grønland* 166(4), 296 pp.

Washburn, A. L. 1969: Weathering, frost action and patterned ground in the Mesters Vig district, north-east Greenland, *Meddr Grønland* 176(4), 303 pp.

Washburn, A. L. 1973: *Periglacial processes and environments*, Arnold, London, 320 pp.

Washburn, A. L. and Goldthwait, R. P. 1958: Slushflows, *Bull. geol. Soc. Am.* 69, 1657–8

Washburn, A. L., Smith, D. D. and Goddard, R. H. 1963: Frost cracking in a middle-latitude climate, *Biul. peryglac.* 12, 175–89

Watson, E. 1966: Two nivation cirques near Aberystwyth, Wales, *Biul. peryglac.* 15, 79–101

Wayland, E. J. 1947: The study of past climates in tropical Africa, *Proc. pan-Afr. Congr. Prehist.*, 59–66

Weertman, J. 1957: On the sliding of glaciers, *J. Glaciol.* 3, 33–8

Weertman, J. 1962: Catastrophic glacier advances, *Ass. int. Hydrol. scient.*, *Commn Neiges Glaces* (Obergurgl, 1962), 31–9

Weertman, J. 1964: The theory of glacier sliding, *J. Glaciol.* 5, 287–303

Weertman, J. 1966: Effect of a basal water layer on the dimensions of ice sheets, *J. Glaciol.* 6, 191–207

Weertman, J. and J. R. 1964: *Elementary dislocation theory*, Macmillan, New York, 213 pp.

Weinert, H. H. 1961: Climate and weathered Karoo dolerites, *Nature, Lond.* 191, 325–9

Weinert, H. H. 1964: Basic igneous rocks in road foundations, *C.S.I.R. Res. Rep.* 218, *Natl Inst. Road Res. Bull.* 5, 47

Weinert, H. H. 1965: Climate factors affecting the weathering of igneous rocks, *Agric. Met.* 2, 27–42

Weise, O. P. 1974: Zur Hangentwicklung und Flächenbildung im Trockengebiet des iranischen Hochlandes, *Würtzb. geogr. Arb.* 42, 325 pp.

Weller, G. E. and Schwerdtfeger, P. 1971: New data on the thermal conductivity of natural snow, *J. Glaciol.* 10, 309–11

Wellman. H. W. and Wilson, A. T. 1965: Salt weathering, a neglected geological erosion agent in coastal and arid environments, *Nature, Lond.* 205, 1097–8

Wentworth, C. K. 1943: Soil avalanches in Oahu, Hawaii, *Bull. geol. Soc. Am.* 54, 53–64

West, G. and Dumbleton, J. J. 1970: The mineralogy of tropical weathering illustrated by some west Malaysian soils, *Q.J. Engng Geol.* 3, 25–40

Weyman, D. R. 1970: Throughflow on hillslopes and its relation to the stream hydrograph, *Publs int. Ass. scient. Hydrol.* 15, 25–32

Weyman, D. 1973: Measurement of downslope flow of water in a soil, *J. Hydrol.* 20, 267–88

Whalley, W. B. 1976: *Properties of materials and geomorphological explanation*, Oxford University Press, 64 pp.

Whipkey, R. Z. and Kirkby, M. J. 1978: Flow within the soil, in *Hillslope hydrology* (ed. M. J. Kirkby), Wiley, Chichester, 121–44

White, S. E. 1976: Is frost action really only hydration shattering? *Arct. alp. Res.* 8, 1–6

Witkind, J. J., Myers, W. B., Hadley, J. B., Hamilton, W. and Fraser, G. D. 1962: Geologic features of the earthquake at Hegben Lake, Montana, August 17 1959, *Bull. seism. Soc. Am.* 52, 163–80

Wilhelmy, H. 1958: *Klimamorphologie der Massengesteine*, Westerman, Braunschweig, 238 pp.

Williams, A.T. 1971: An analysis of some factors involved in the depth of disturbance of beach sand by waves, *Mar. Geol.* 11, 145–58

Williams, G. 1964: Some aspects of the eolian saltation load, *Sedimentology* 3, 257–87

Williams, H. 1942: The geology of Crater Lake National Park, Oregon, with a reconnaissance of the Cascade Range southward to Mount Shasta, *Publs Carnegie Instn* 540, 162 pp.

Williams, J. E. 1949: Chemical weathering at low temperatures, *Georgl Rev.* 39, 129–35

Williams, M. A. J. 1968: Termites and soil development near Brocks Creek, Northern Territory, *Aust. J. Sci.* 31, 153–4

Williams, M. A. J. 1974: Surface rock-creep on sandstone slopes in Northern Territory, Australia, *Aust. Geogr.* 12, 419–24

Williams, P. J. 1961: Climatic factors controlling the distribution of certain frozen ground phenomena, *Geogr. Annlr* 43, 339–47

Williams, P. J. 1967: Properties and behaviour of freezing soils, *Norw. geotech. Inst. Publs* (Oslo) 72, 119 pp.

Willman, H. B. 1944: Resistance of Chicago area dolomites to freezing and thawing, *Bull. Ill. St. geol. Surv.* 68, 249–62

Wilson, G. 1952: Influence of rock structures on coastal cliff development, *Proc. Geol. Ass.* 63, 20–48

Wilson, I. G. 1973: Equilibrium cross-sections of meandering and braided rivers, *Nature, Lond.* 241, 393–4

Wilson, L. 1968: Morphogenetic classification, in *Encyclopedia of geomorphology* (ed. R. W. Fairbridge), Rheinhold, New York, 3, 717–29

Wiman, S. 1963: A preliminary study of experimental frost weathering, *Geogr. Annlr* 45, 113–21

Winkler, E. M. 1970: The importance of air pollution in the corrosion of stone and metals, *Engng Geol.* 4, 327–34

Winkler, E. M. and Wilhelm, E. J. 1970: Salt burst by hydration pressures in architectural stone in urban atmosphere, *Bull. geol. Soc. Am.* 81, 567–72

Wippermann, F. 1969: The orientation of vortices due to instability of the Ekman-boundary layer, *Beitr. Phys. Atmos.* 42, 255–44

Wolman, M. G. 1955: The natural channel of Brandywine Creek, Pennsylvania, *U.S. geol. Surv. Prof. Pap.* 271, 56 pp.

Wolman, M. G. and Gerson, R. 1978: Relative scales of time and effectiveness of climate in watershed geomorphology, *Earth Surface Processes* 3, 189–208

Wolman, M. G. and Miller, J. P. 1960: Magnitude and frequency of forces in geomorphic processes, *J. Geol.* 68, 54–74

Woo, D. C. and Brater, E. F. 1962: Spatially varied flow from controlled rainfall, *J. Hydraul. Div. Am. Soc. civ. Engrs* 88 (HY6), 31–56

Wood, A. 1942: The development of hillside slopes, *Proc. Geol. Ass.* 53, 128–40

Wood, A. M. M. 1955: Folkestone Warren landslips: investigations, 1948–50, *Proc. Instn civ. Engrs*, Railway Paper 56, 410–28

Wood, W. L. 1976: Three-dimensional conditions of surf, *Proc. 15th Conf. Coastal Processes*, Honolulu, 525–38

Woodcock, A. H. 1940: Convection and soaring over the open ocean, *J. mar. Res.* 3, 248–53

Wooding, R. A. 1965: A hydraulic model for the catchment-stream problem. I. Kinematic wave theory, *J. Hydrol.* 3, 254–67

Woolhiser, D. A. 1975: Simulation of unsteady flow, in *Unsteady flow in open channels* (ed. K. Mahmood and V. Yevjevich), Water Resources Publs, Fort Collins, 2, 485–508

Woolhiser, D. A. and Liggett, J. A. 1967: Unsteady one-dimensional flow over a plain—the rising hydrograph, *Wat. Resour. Res.* 3, 753–71

Wolman, M. G. and Miller, J. P. 1960: Magnitude and frequency of forces in geomorphic processes, *J. Geol.* 68, 54–74

Yaalon, D. H. 1960: Some implications of fundamental concepts of pedology in soil classification, *Trans. 7th int. Conf. Soil Sci.* 4, 119–23

Yaalon, D. H. 1974: Accumulation and distribution of loess-derived deposits in the semi-desert fringe areas of Israel, *Z. Geomorph., Suppl. Bd* 20, 91–105

Yaalon, D. H. and Ganor, E. 1974: Dust in the environment: the source and modes of transport from deserts, *Pap. 5th Congr. Israel ecol. Soc.*

Yair, A. 1973: Theoretical considerations on the evolution of convex hillslopes, *Z. Geomorph., Suppl. Bd* 18, 1–9

Yair, A. and Klein, M. 1973: The influence of surface properties on flow and erosion processes on debris-covered slopes in an arid area, *Catena* 1, 1–18

Yalin, M. S. 1971: On the formation of dunes and meanders, *Proc. 14th int. Congr. hydraul. Res. Ass.*, Paris, 3, C13, 1–8

Yalin, M. S. 1977: *Mechanics of sediment transport*, Pergamon Press, Oxford, 298 pp.

Yang, C. T. 1971: Formation of riffles and pools, *Wat. Resour. Res.* 7, 1567–74

Yang, C. T. 1972: Unit stream power and sediment transport, *J. Hydraul. Div. Am. Soc. civ. Engrs.* 98, HY 10, 1805–26

Yatsu, E. 1967: Some problems of mass movement, *Geogr. Annlr* A, 49, 396–401

Yen, B. C. 1969: Stability of slopes undergoing creep deformation, *J. Soil Mech. Fdns Div. Am. Soc. civ. Engrs* 95, 1075–93

Yermolayev, M. M. 1932: Geologitscheskii i geomorfologitscheskii otscherk ostrowa B. L. Ljachowskogo. Sbornik Poljarnoi geofizitscheskoi stancii na ostrowe Bolschom Ljachowskom Nr. 1, *Izd. Wseso. Arkt. Inst.* (Leningrad)

Young, A. 1960: Soil movement by denudation processes on slopes, *Nature, Lond.* 188, 120–2

Young, A. 1963: Deductive models of slope evolution, *Rep. Int. geogr. Un., Slopes Commission* 3, 45–66

Young, A. 1974: The rate of slope retreat, in *Progress in geomorphology* (ed. E. H. Brown and R. S. Waters), *Spec. Publs Inst. Br. Geogr.* 7, 65–78

Young, R. A. and Mutchler, C. K. 1969: Soil and water movement in small tillage channels, *Trans. Am. Soc. agric. Engrs* 12, 543–5

Zaruba, Q. and Mencl, V. 1969: *Landslides and their control*, Academia and Elsevier, Prague, 205 pp.

Zaslavsky, D. and Rogowski, A. 1969: Hydrologic and morphologic implications of anisotropy and infiltration in soil profile development, *Proc. Soil Sci. Soc. Am.* 33, 594–9

Zhigarev, L. A. and Kaplina, T. N. 1960: Solifluktsionnye formy reliefa na severovostoke SSSR, *Trudy Inst. Merzlotov.* 16, 46–59

Zumberge, J. H. 1955: Glacial erosion in tilted rock layers, *J. Geol.* 63, 149–58

Index